Cellulose-Based Biosensing Platforms

Cellulose-Based Biosensing Platforms

Editors

Eden Morales-Narváez
Alfredo de la Escosura-Muñiz
Hamed Golmohammadi
Erhan Zor

MDPI • Basel • Beijing • Wuhan • Barcelona • Belgrade • Manchester • Tokyo • Cluj • Tianjin

Editors

Eden Morales-Narváez
Centro de Investigaciones en Óptica A.C.
Mexico

Alfredo de la Escosura-Muñiz
University of Oviedo
Spain

Hamed Golmohammadi
Chemistry and Chemical Engineering
Research Center of Iran
Iran

Erhan Zor
Necmettin Erbakan University
Turkey

Editorial Office
MDPI
St. Alban-Anlage 66
4052 Basel, Switzerland

This is a reprint of articles from the Special Issue published online in the open access journal *Biosensors* (ISSN 2079-6374) (available at: https://www.mdpi.com/journal/biosensors/special_issues/cellulose_bio).

For citation purposes, cite each article independently as indicated on the article page online and as indicated below:

LastName, A.A.; LastName, B.B.; LastName, C.C. Article Title. *Journal Name* **Year**, *Volume Number*, Page Range.

ISBN 978-3-0365-2400-9 (Hbk)
ISBN 978-3-0365-2401-6 (PDF)

Contents

About the Editors

Eden Morales-Narváez is a Young Professor in the Department of Photonics and Head of the Biophotonic Nanosensors Laboratory at Centro de Investigaciones en Óptica (CIO), Mexico. He received a degree in Bionics Engineering from the National Polytechnic Institute of Mexico in 2006 and a Ph.D. degree in Biomedical Engineering from the Polytechnic University of Catalonia, Spain, in 2013. He was a Postdoctoral Researcher at the Catalan Institute of Nanoscience and Nanotechnology and started his independent career in late 2016. His research is focused on novel (bio)sensing platforms, particularly using nanophotonics at the cutting-edge. He has published in leading scientific journals on Advanced Materials, Nanotechnology, Multidisciplinary Chemistry, Miniaturization, Biosensing, and Wearable Devices. He is a Lecturer in Biophotonics and Photonic Materials at the CIO. He also serves as an Editor to the *Journal of Physics: Photonics*, *Biosensors* (MDPI), *Green Analytical Chemistry* (Elsevier), and *Frontiers in Bioengineering and Biotechnology*. He has also been internationally recognized as an Emerging Leader 2020 by IOP Publishing.

Alfredo de la Escosura-Muñiz holds a PhD in Chemistry (2006) from the University of Oviedo (Spain). His post-doctoral period (2007–2010) was granted by a Juan de la Cierva Research Fellowship at the Institute of Nanoscience of Aragon (Spain) in collaboration with the Catalan Institute of Nanotechnology (Spain), including a stay at the Chalmers University of Technology (Sweden). He then joined Prof. Merkoçi's group at the Catalan Institute of Nanoscience and Nanotechnology (ICN2, Spain) as Senior Researcher (2011–2017) followed by a stay at the Polytechnic University of Catalonia (Spain) (2017–2018).

As of June 2018, he holds a Ramon y Cajal Research Fellowship in the Nanobioanalysis Group, Department of Physical and Analytical Chemistry, University of Oviedo (Spain), where he is also a Lecturer in Chemistry.

Dr. De la Escosura-Muñiz has broad experience in biosensing systems based on nanoparticles and nanochannels for point-of-care diagnostic applications. He has been involved in +25 national and international projects and is coauthor of over 80 scientific publications (+2600 citations; h-index: 30, according to Isi Web of Science—October 2021) and 4 patents. He has also supervised 6 PhD and 14 master/end degree works has and presented communications in +80 congresses, among other merits.

Researcher ID: M-2221-2014
Scopus ID: 6505947368
ORCID ID: 0000-0002-9600-0253

Hamed Golmohammadi is currently an Assistant Professor at the Chemistry and Chemical Engineering Research Center of Iran (CCERCI), where he leads the Nanosensor Bioplatforms Laboratory/Research Group. His research activities focus on the development of novel (bio)sensors, particularly using plasmonic, colorimetric, photoluminescent nanoparticles, and nature-based materials on (nano)paper-based bioplatforms/microfluidics, in combination with digital technologies including smartphones, the Internet of Things (IoT), and so on, as smart chemical/biological kits and wearables for applications in (bio)sensing technology, whose results aim at displaying analytical information related to various fields such as clinical/medical diagnostics, environmental monitoring, food safety, etc., at the point-of-need. He is also a member of the American Chemical Society and the Scientific Elite Federation of Iran.

Erhan Zor has been working as a research scientist and lecturer at Necmettin Erbakan University, Turkey. He received his Ph.D in chemistry from Selçuk University in the topic of electrochemical chiral discrimination based on graphene-related materials. During his Masters and Ph.D., he had worked as a visiting researcher at European research centers in the field of nano(bio)sensors at Nanotechnas-Nanotechnology and Material Research Center at Vilnius University in Vilnius, Lithuania, and at the Catalan Institute of Nanoscience and Nanotechnology (ICN2) in Barcelona, Spain. His current research interest is focused on graphene and related materials, nanocellulose, chiral sensors, (nano)biosensors, optical sensors, electrochemical sensors, paper/plastic-based diagnostics, biomedical diagnostics, as well as nanoscience and nanotechnology education.

Preface to "Cellulose-Based Biosensing Platforms"

As the most abundant renewable biopolymer in nature, cellulose is a convenient family of materials to design low-cost devices. In addition, cellulose-based materials are flexible, biocompatible, biodegradable, and amenable to straightforward functionalization, as well as mass production. These unrivaled features of cellulosic substrates and their fascinating simplicity of fabrication and coupling with ubiquitous technologies such as smartphones make them tailor-made biosensing platforms. Furthermore, cellulose-based biosensing approaches can meet the World Health Organization's REASSURED criteria (real-time connectivity, ease of specimen collection, affordable, sensitive, specific, user-friendly, rapid and robust, equipment-free, and deliverable to end-users) for ideal diagnostic assays/devices. Hence, cellulose endows the biosensing community with exquisite materials to envisage innovative analytical devices.

This Special Issue is focused on cutting-edge approaches dealing with the design, fabrication, and advantageous analytical performance of cellulose-based biosensing platforms. The included review articles will help the reader to understand and build an overview of the state of the art in cellulose-based (bio)sensing. Additionally, original research on nano-enabled signal enhancement, clinical diagnostics, health care, point-of-care-testing, and environmental monitoring was carefully peer-reviewed to be part of this article collection.

Eden Morales-Narváez, Alfredo de la Escosura-Muñiz, Hamed Golmohammadi, Erhan Zor

Editors

biosensors

Review

Nanobioengineered Sensing Technologies Based on Cellulose Matrices for Detection of Small Molecules, Macromolecules, and Cells

Divya, Supratim Mahapatra, Vinish Ranjan Srivastava and Pranjal Chandra *

Laboratory of Bio-Physio Sensors and Nanobioengineering, School of Biochemical Engineering, Indian Institute of Technology (BHU) Varanasi, Varanasi 221005, Uttar Pradesh, India; divya.rs.bce20@itbhu.ac.in (D.); supratimmahapatra.rs.bce20@itbhu.ac.in (S.M.); vinishrsrivastava.bce20@iitbhu.ac.in (V.R.S.)
* Correspondence: pranjal.bce@iitbhu.ac.in

Abstract: Recent advancement has been accomplished in the field of biosensors through the modification of cellulose as a nano-engineered matrix material. To date, various techniques have been reported to develop cellulose-based matrices for fabricating different types of biosensors. Trends of involving cellulosic materials in paper-based multiplexing devices and microfluidic analytical technologies have increased because of their disposable, portable, biodegradable properties and cost-effectiveness. Cellulose also has potential in the development of cytosensors because of its various unique properties including biocompatibility. Such cellulose-based sensing devices are also being commercialized for various biomedical diagnostics in recent years and have also been considered as a method of choice in clinical laboratories and personalized diagnosis. In this paper, we have discussed the engineering aspects of cellulose-based sensors that have been reported where such matrices have been used to develop various analytical modules for the detection of small molecules, metal ions, macromolecules, and cells present in a diverse range of samples. Additionally, the developed cellulose-based biosensors and related analytical devices have been comprehensively described in tables with details of the sensing molecule, readout system, sensor configuration, response time, real sample, and their analytical performances.

Keywords: nanobioengineering; cellulose; matrix design; biosensors; cytosensing; human health

Citation: Divya; Mahapatra, S.; Srivastava, V.R.; Chandra, P. Nanobioengineered Sensing Technologies Based on Cellulose Matrices for Detection of Small Molecules, Macromolecules, and Cells. *Biosensors* **2021**, *11*, 168. https://doi.org/10.3390/bios11060168

Received: 11 April 2021
Accepted: 17 May 2021
Published: 24 May 2021

Publisher's Note: MDPI stays neutral with regard to jurisdictional claims in published maps and institutional affiliations.

1. Introduction

Cellulose is a biopolymer of β-1,4 D-glucose units, a widely used biomaterial that exhibits unique properties and is used in various industries including textiles, electronics, biomedical, etc. [1]. Cellulose is the most commonly occurring biomaterial on earth [2], obtained mainly from different plant species [3] and some form of bacteria [4], i.e., *Acetobacterxylinum* [5]. Glucose molecules are bonded with van der Waal forces and hydrogen bonding to form a parallel stacking of cellulose microfibrils into crystalline cellulose [6]. Due to its structural form, cellulose exhibits unique properties such as high young's modulus, biocompatibility, biodegradability, high mechanical strength, transparency, and thermal stability, which allows high chemical modifications [7–9]. As well as appealing piezoelectricity, mechanical performance, dialectricity, and convertibility [10] have also been recorded. Cellulose has its application in a broad range of fields, such as in the pharmaceutical industry for coating of tablets, pellets, beads, granules [11], and in various other industries [12]. In addition to this, cellulose has also been widely used in the immobilization of different molecules, in wound healing, tissue engineering [13–15], and also has tremendous applications in flexible printed bioelectronics [16]. In recent times, further modification of cellulose with the nanostructures leads to its application as a matrix material for biosensor fabrication, which is generally called "nanocellulose". Nanocellulose is a sustainable, renewable, and eco-friendly nanomaterial with remarkable properties such as high strength, low density, high specific surface area, high aspect ratio, etc. [17]. It is also worth mentioning that these properties are tunable, and they majorly rely on the chemical modification steps and type of nanomaterials integrated with cellulose matrix. These nano-cellulosic systems can be constructed using various methods such as chemical activation, grafting, coating, impregnation, covalent binding, cryocrushing, micro-fluidization, high-intensity ultrasonication and various other wet chemistry procedures [18]. They can be cellulose nanofibers (CNF), cellulose nanocrystal (CNC), or bacterial nanocellulose [19]. The remarkable properties of nanocellulose make the polymer of cellulose one of the most fascinating and advanced materials with implementation in various fields [20]. Cellulosic material possesses the complete degradation and biocompatible property due to which it is also called "green cellulose" [21] and is used for the fabrication of disposable and degradable sensors. The hydrogen bonding present in the core structure of cellulose provides both mechanical strength and flexibility at the same time; this unique property helps in the fabrication of a flexible matrix for sensors [22–24]. Due to such multidimensional properties and cost-effectiveness, cellulose has been considered as one of the most fascinating and widely used materials for the development of biosensors. For improving the sensitivity of the cellulose surface, novel modifications can be performed using conducting materials such as gold nanoparticles (AuNPs), carbon nanotubes (CNTs), gold-nanorods, graphene oxide (GO), and other nanocomposites [9,25].

In this review, we have critically evaluated the cellulose-based sensors that have been fabricated through distinct modification techniques and their varying applications for analyzing diverse sample types including small molecules, ions, macromolecules, and cells. While the studies on cellulose-based biosensors continue, major advances have been achieved in this field. Thus, we have performed a scientific survey for the research documentation concerning cellulose-based sensors through the online database "Scopus", the result of which is quite a convincing indicator that cellulose-based biosensors have gained a major focus in recent times considering their multipotent abilities. According to this survey, it was found that more than 7000 research articles have been published in the last decade, following an exponential increase (Figure 1). In the following sections, we have listed the introductory concepts of cellulose biosensors followed by their nanobioengineering design aspects for the detection of diverse molecules in various real sample matrices.

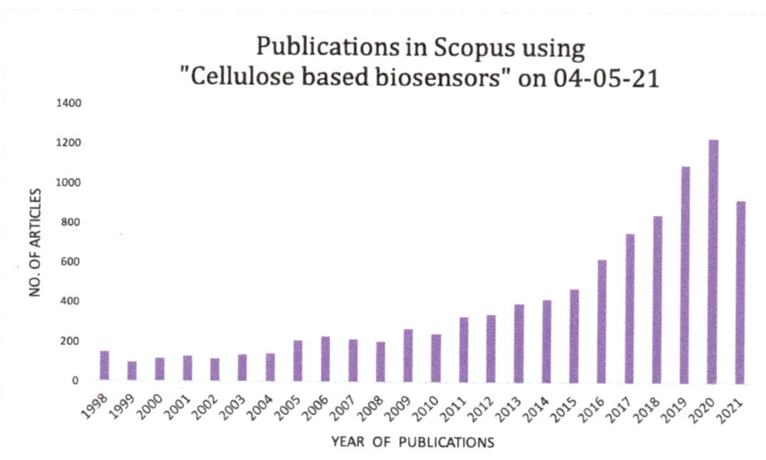

Figure 1. The number of articles published in consequent years in the online database "Scopus" using the keyword "cellulose-based biosensors".

2. Cellulose: Structure and Biosensing Design Aspects

A biosensor is defined as an analytical device that detects the biological molecule through electrical, thermal, or optical signals [26,27]. It consists of four components —the bio-recognition element, transducer, amplifier, and detector. Biorecognition elements are composed of enzymes, cells, DNA, proteins, tissue, organelles, antibodies, aptamers, etc. [26,28]. The analyte, when starting to sinter, acts with a biorecognition element to produce a signal that is converted to readable form by the transducer [29]. Depending upon the types of transducer, biosensors could be categorized commonly into different kinds such as optical, electrochemical, electromechanical, calorimetric, acoustic biosensors, etc. Optical biosensors detect the change in optical properties such as phase change, polarization, and change in frequency of light. Mechanical biosensors detect the variation in mechanical properties such as mass, force, motion, etc. Electrochemical biosensors analyze the change in chemical stimuli to readable electrical signals [26,30,31]. Colorimetric biosensors detect the change in color through naked eyes or by using simple optical detector [32]. To date, several cellulose-based sensors have been developed based on the source of cellulose and various modifications were performed according to the required function. Interestingly, the cellulose and its composite have been successfully used as a component of various transducing surfaces for the development of these sensors. Cellulose and its derivatives are designed to introduce the desired characteristics through chemical functionalization and also by changing the inherited inter and intra hydrogen-bonding pattern. Various nanomaterials are also explored for enzyme-like properties to enhance the electrocatalytical activity in the designing of biosensors.

In the structure of cellulose, each glucose unit consists of three hydroxyl groups at which chemical modifications are performed. These modifications can be majorly performed on the surface without disturbing the main chain, which results in the formation of the required cellulosic matrix [33]. The immobilization of cellulose with AuNPs [34], CNTs [35], reduced GO [36], quantum dots (QDs) [37], and other nanomaterials help in the fabrication of biosensors with enhanced properties in a cost-effective manner [38,39]. CNTs and rGO provide enhanced electrical conductivity and high active surface area that helps in large number of bio-receptor immobilization. AuNPs helps to establish a plasmonic field, color change, fluorescence quenching, and also assist in enhancing the surface area as well as electron transfer of the sensing surface [40–42]; QDs have unique luminescence and electronic properties [43], so they are also used in cellulose-based sensing platforms. Bacterial cellulose (BC) is an eco-friendly, natural three-dimensional nanostructure, and a low-cost molecule derived from bacteria, which is also used as a matrix material in biosensor development [44]. Biosensors based on BC show various advantages over conventional cellulose-based sensors because of their immense surface area, higher crystallinity, good biocompatibility, and great mechanical strength [45]. Cellulose-based biosensors are widely accepted and used for the detection of various sorts of molecules such as small molecules, macromolecules, metals, and cells [9], etc. CNFs are used for microfluidic channel preparation and enzyme derivatization in the microfluidic-sensor matrices for effective biomedical diagnosis [46,47]. These microfluidic sensors are inexpensive, disposable, lightweight, and a rapid method of diagnostics that supports rheological modification and exhibit mechanical strength and advantageous dimensional stability [48,49]. In addition to these, cellulose in another form, such as paper, has also been explored, for example, Whatman filter paper [50,51], glossy papers [52], paper towels [53], etc. Glossy paper can be used in a paper-based flexible sensing device, as it is composed of cellulose fibers blended with inorganic compounds. The main advantage of using glossy paper is that it is easy to engineer its surface properties.

3. Biosensors for the Detection of Small Molecules and Metals

Small molecules are the low molecular weight organic molecules, normally less than 900 Daltons. They perform various types of biological functions such as a signaling molecule, drugs, effector, and even sometimes changes the function of the target. Due to their size, they readily diffuse across the membranes. Due to their wide range of functions, it is required to monitor the small molecules involved in maintaining the biological functions. In this section, cellulose-based biosensors that are developed for the determination of numerous small molecules have been discussed. One of the most important small molecules is glucose, which has a tremendous role in clinical laboratories and other industries. Glucose is a monosaccharide ubiquitously present in humans and acts as a fuel of human body, and when it is present in the blood, is called blood sugar [54]. To detect glucose, various cellulose-based sensors have been developed in recent years. A blood glucose biosensor for simple and affordable monitoring has been developed recently for its direct detection in sweat and saliva samples [55]. In this case, sulfated (S-CNC) and non-sulfated (N-CNC) cellulose nanocrystal/magnetite film was used to determine the instant color change in the presence of glucose and 2,2′-azino-bis (3-ethylbenzothiazoline-6-sulphonic acid) (ABTS). The authors used the peroxidase-like characteristics of (N-CNC)-Fe_3O_4 and (S-CNC)-Fe_3O_4 nanoparticles to determine glucose, using glucose oxidase (GOx) and H_2O_2 (Figure 2B). Here, ABTS was used as a substrate for the peroxidase. Both sulfated and non-sulfated types of biosensors detect the concentration of glucose as low as 5 mM showing high sensitivity and concentration compared to the glucose level present in biological fluids. Even sulfated nanoparticles showed 1.5 to 2 times more reactivity than that of non-sulfated systems. An ampero metric glucose biosensor has been developed by Lawrence et al., 2014, through immobilization of GOx, extracted from *Aspergillus niger*, in a paper disk matrix (diameter 1 cm) positioned on the surface of a screen-printed carbon electrode (SPCE) [21]. For glucose oxidation, ferrocene monocarboxylic acid was used as an intermediate molecule. The developed paper-based biosensor was used for detecting the glucose at an extremely low operating volume of 5 μL and concentration range between 1 to 5 mM. The limit of detection (LOD) of the developed sensor was 180 μM (n = 5 at 90% confidence level). In addition to these, other glucose biosensors are mentioned in Table 1 with information related to the sensor's design, readout system, and other analytical details.

In addition to glucose, other molecules are also detected using a cellulose-based sensor such as phenol, which is a major component of various industries and industrial waste that leads to its deteriorating effects on the environment. Phenol is one such toxic pollutant present in wastewater, released from industrial products, and causes harmful effects to aquatic life, plants, and also humans, sometimes leading to organ damage [56,57]. An electrochemical cellulose-based biosensor has been developed for the sensing of phenol by Manan et al., 2019 [58]. The biosensor was fabricated through hybridization of cetyltrimethylammonium bromide (CTAB) on nano-crystalline cellulose (NCC) and further hybridization of CTAB-NCC to QDs capped with 3-mercaptopropionic acid (MPA). The developed nanocomposite material was immobilized with tyrosinase enzyme (Tyr) to form CTAB-NCC/MPA-QDs/Tyr for the determination of phenol (Figure 2D). The developed biosensor shows the dynamic range from 5 to 40 μM (0.47 to 3.76 mg/L) and the observed LOD of the sensor was 82 nM (7.7 μg/L). It shows high sensitivity up to 0.078 μA/μM and can be a potential biosensor for detection of phenol in surrounding samples. Apart from these, cellulosic matrix is also being used in the sensing of amines, which are important materials employed in the food industry, biological processes, and are harmful to the atmosphere and public health [59].

A cellulose-based fluorescent biosensor has been developed for the sensing of various amines by Nawaz et al., 2020. The biosensor was manufactured by using the phenanthroline (Phen) as a color-imparting molecule and 4,4′-methylene diphenyl diisocyanate (MDI) immobilized on cellulose acetate (CA) to form a Phen-MDI-CA sensor for the visualization of amines [60]. Different colors of the fluorescent molecule were observed for different amines on Phen-MDI-CA paper by using UV-visible light. For example, the blue color fluorescence of Phen-MDI-CA modified to light blue when triethylamine (TEA) was used as a substrate, to green for diethyl amine (DEA), and to cyan for methylamine. The observed LOD of the manufactured biosensor was found to be 900 nmol, 990 nmol, 1700 nmol for TEA, EDA, methylamine, respectively. Not only these molecules but cellulosic matrices has been also applied for the sensing of pesticides such as fluazinam, which is a low-toxic fungicide that poses a threat to the environment and food safety [61]. Wang et al., 2020 prepared a paper-based biosensor for sensing and digital analysis of fluazinam. The biosensor was fabricated by crosslinking the disulfide MoS_2 QDs into cellulose membranes. The developed biosensor shows the detection range between 10 to 800 μM. The LOD observed for the fluazinam sensor was 2.26 μM [62]. Apart from fluazinam, cellulose-based sensors were explored for the sensing of other toxins. A label-free electrochemical biosensor was developed for the quantification of aflatoxin B1 (AFB1) in wheat samples by Huang et al., 2020. The author used CNFs derived from BC and coupled them with AuNPs. It produced a unique 3D porous structure that helps in the speedy diffusion of electrolytes, providing a greater electrochemical working are a that helps in obtaining higher current magnitude in a typical differential pulse voltammogram (DPV). The developed sensor shows the detection of AFB1 in a broad concentration range from 50 to 25×10^3 pg mL^{-1} having a determination coefficient of 0.995. The estimated LOD for the immunosensor was observed to be 27 pg/mL [63]. The author also checked the reproducibility, stability, and selectivity of the biosensor and it was found to be suitable. As mentioned earlier, cellulose in its other forms, such as paper, which can be derived from wood, grasses, rags, and other sources, has also been widely used in sensing devices.

The cellulosic paper provides unique geometry and hydrophilic nature which are explored to develop paper-based analytical devices [64,65]. Phenylalanine (Phe) is an essential amino acid and failure of its metabolism causes phenylketonuria, which is due to the deficiency of phenylalanine hydroxylase enzyme, which converts Phe into tyrosine, due to which Phe is deposited in the human body and causes brain injury and neurocognitive dysfunction [66]. Sun et al., 2021, manufactured a paper-based biosensor for the identification of Phe using phenylalanine ammonia lyase hybrid nanoflowers [67]. The developed biosensor was capable of sensing Phe concentration in urine samples with detection range of 60 to 2400 μM within 10 min. In addition to these examples, sensors have also been developed for the detection of other small molecules, which has been comprehensively described in Table 1.

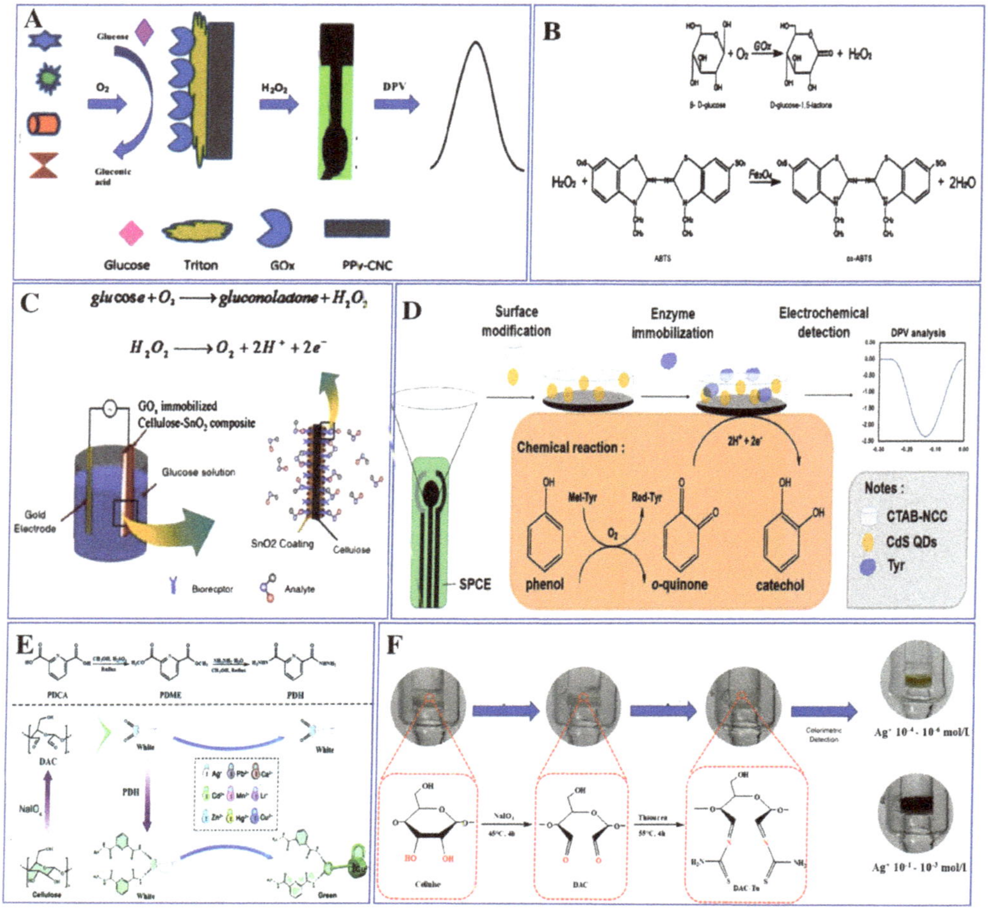

Figure 2. Biosensors for detection of small-molecule and metal ion. (**A**) Pictorial representation of poly pyrrole-cellulose nanocrystal (PPy-CNC)-based biosensor for the analysis of glucose and response recorded with a DPV (reproduced with permission from [68]). (**B**) Representation of the reaction mechanism of cellulose nanocrystal/magnetite glucose biosensor (Reproduced with permission from [55]). (**C**) Representation of cellulose–SnO_2 nanocomposite-based biosensor for the detection of glucose molecule (reproduced with permission from [54]). (**D**) Pictorial representation of the development of CTAB-NCC/MPA-QDs/Tyr sensor through immobilization of CTAB-NCC/QDs on SPCE (screen-printed carbon electrode) and further with Tyr enzyme for the detection of phenol and response evaluated using DPV (reproduced with permission from [58]). (**E**) Fabrication scheme of DAC-PDH-based colorimetric sensor through chemical alteration and discriminatory identification of Cu^{2+} by DAC-PDH within 30 s (reproduced with permission from [69]). (**F**) Preparation of DAC-Tu selective colorimetric sensor and the identification of Ag^{+} from aqueous solutions according to color change based on the varying concentrations of Ag+ (reproduced with permission from [70]).

Cellulose-based biosensors are not simply used for the sensing of small molecules as discussed above; they are also used for the screening and detection of heavy metals. Heavy metals are detrimental to living beings beyond a particular concentration as they are not biodegradable and can accumulate in the body [71]. Cellulose-based biosensors have been designed for the sensing of various metal ions as well. A few examples that are specific and sensitive to a particular ion even in the presence of interfering molecules are discussed here. Copper ion (Cu^{2+}) plays a significant role in different physiological processes, excess of which causes liver and kidney damage, and which is one of the frequently occurring ions causing the contamination of drinking water. A cellulose-based biosensor has been developed for the unaided detection of Cu^{2+} ions present in biological fluids by Wang et al., 2019 [69]. The biosensor was fabricated without the requirement of the probe using natural cellulose through a facile one-pot process, and structural characterization was carried out by Fourier transform infrared spectroscopy (FTIR) The sensitivity of the biosensor was detected in serum, urine, and water through the UV–vis absorption intensity and it shows the color change within 10 s. The LOD was concluded to be 1.9309, 1.9154 and 1.185 ppm in urine, serum, and tap water, respectively. The developed biosensor was established to be very effective in sensing Cu^{2+} ions in the biological fluids and showed rapid response time, high sensitivity, and selectivity (Figure 1E). Apart from Cu^{2+}, other cellulose-based sensing systems have been developed for the other metal ion such as silver (Ag^{+}), which is present in water and its high concentration causes adverse effects in the environment and on public health [72]. A cellulose-based colorimetric sensor has been fabricated for the detection of Ag^{+} in an aqueous solution by Wang et al., 2020 [70]. The biosensor was fabricated by embed ding thiourea (Tu) onto the surface of Eucalyptus cellulose, which is modified to dialdehyde cellulose (DAC) through a chemical grafting method. The visual detection method depends upon the chelation of Ag^{+} with N and S atoms of DAC-Tu to form N-Ag, S-Ag, and Ag_2S and it shows different colors based on different concentrations of Ag^{+} (Figure 2F). The coloring mechanism between Ag^{+} and DAC-Tu in aqueous solution was characterized by nuclear magnetic resonance, transmission electron microscopy (TEM), and X-ray photoelectron spectroscopy (XPS) analysis. The signal response of the sensor reduces with an increase in the concentration of Ag^{+} and the LOD was found to be 10^{-6} mol/L within 10 min. Apart from these ions, cellulose-based sensing devices are also fabricated for the identification of zinc ion (Zn^{2+}), which is found abundantly in the environment due to industrial processes and causes harm to human health. Such a cellulose-based optical biosensor was designed for the sensing of Zn^{2+} by Daniyal et al., 2019 [73]. The surface plasmon resonance (SPR)-based optical sensor was fabricated through enhancement of thin gold film with the help of NCC. The SPR signal was examined both with the unmodified and modified thin film of gold for determining the sensing potential. The Zn^{2+} interacts with the negative charge present on the modified-gold film and such an interaction has been investigated by XPS. The XPS scans into the region of 0 to 1400 eV and the recorded data were equipped with the Gaussian–Lorentzian curve program. The developed biosensor possessed a wide dynamic range with the LOD of 0.01 ppm and illustrates the sensitivity of about 1.892 ppm^{-1}.

Table 1. List of developed biosensors using cellulose matrix material for the detection of small molecules and metal ions. (NR: not reported).

Sr. No	Analyte	Detection	Sensor Configuration	Response Time	Real Sample	Detection Range	Limit of Detection	Reference
1	Glucose	Amperometric, CV	GOx immobilized on cellulose paper	NR	Soda beverages	1 to 5 mM	0.18 mM	[21]
2	Glucose	Conductometric	GOx immobilized on nanocomposite of cellulose–tin oxide	NR	NR	0.5 to 12 mM	NR	[54]
3	Glucose	Colorimetric	Sulphated and non-sulfated cellulose nanocrystal/magnetite film	NR	Sweat and saliva	low as 5 mM	NR	[55]
4	Glucose	Voltammetric	Polypyrrole-cellulose nanocrystal-based composites with GOx	NR	NR	1.0 to 20 mM	(50 ± 10) µM	[68]
5	Glucose	Cyclic Voltammetry (CV)	Ag@ SiO_2-PEG metalloid polymer nanoparticle functionalized with graphene oxide	NR	Urine, Serum	0.1 to 20 mM	NR	[75]
6	Phenol	Voltammetric	Nanocrystalline cellulose CdS QDs tyrosine-based biosensor	NR	Water	5 to 40 µM	0.082 µM	[58]

Table 1. *Cont.*

Sr. No	Analyte	Detection	Sensor Configuration	Response Time	Real Sample	Detection Range	Limit of Detection	Reference
7	Triethylamine	Fluorescence	Cellulose-based Phen-MDI-CA sensor	NR	NR	NR	0.90 μM	[60]
8	Ethylenediamine	Fluorescence	Cellulose-based Phen-MDI-CA sensor	NR	NR	NR	0.99 μM	[60]
9	Fluazinam	Fluorescence	Disulfide quantum dots (MoS_2 QDs) cross-linked into cellulose membrane	NR	Food	10 to 800 μM	2.26 μM	[62]
10	Aflatoxin B1	Amperometric	Carbon nanofibers derived from BC and coupled with AuNPs	NR	Wheat	0.05 to 25 ng mL^{-1}	0.027 $ngmL^{-1}$	[63]
11	Phenylalanine (Phe)	Colorimetric	Paper-based detection of Phe using phenylalanine ammonia-lyase hybrid nanoflowers	10 min	Urine	60 to 2400 μM	NR	[67]
12	H_2O_2	Amperometric	Gold nanoparticles–bacterial cellulose nanofibers (Au–BC)-based sensor	1 s	Disinfector	0.3 μM to 1.00 mM	0.1 μM	[76]
13	Ethanol	Amperometric	Paper-based sensor modified with CB/PBNPs nanocomposite	NR	Beer	up to 10 mM	0.52 mM	[77]
14	Atrazine	Optical	Paper-based algal sensor for nanoencapsulated atrazine	NR	Water	0.5 to 200 nM	4 pM	[78]
15	Cu^{2+}	Colorimetric	Liquid cellulose biosensor using a facile one-pot process	10 s	Urine and serum	NR	1.9309 and 1.9154 ppm	[69]
16	Ag^{+}	Colorimetric	Cellulose modified to DAC-Tu biosensor	10 min	Water	NR	10^{-6} mol/L	[70]
17	Zn^{2+}	SPR optical	Gold thin film modified with a nanocrystalline cellulose	NR	NR	low as 0.01 ppm	NR	[73]
18	Hg^{2+}	Fluorescent	Cells-alginate hydrogel paper-based sensor	5 min	Wastewater	NR	NR	[74]
19	Mn^{7+}	Fluorescence	Nitrogen, aluminium co-doped cellulose-based carbon dots (N/Al-CDs)	NR	Water	0 to 100 μM	46.8 nM	[79]
20	Fe^{3+}	Fluorescence	Nitrogen-doped carbon dots anchored on BC	10 min	NR	0.5 to 600 μM	84 nM	[80]
21	Carbonfuran	Colorimetric	Whatman paper used in ULOC device	3 min	Apple	0.01–5.00 mg L^{-1}	0.05 mg Kg^{-1}	[81]
22	Ochratoxin A	CV, EIS	Cellulose nanofibrous matrix labelled with aptamer probe	NR	Coffee	0.002–2 ng mL^{-1}	0.81 pg mL^{-1}	[82]

Interestingly, not only these ions but cellulose-based sensors have also been fabricated for the identification of mercury (Hg^{2+}), which is a major threat to human health due to its toxicity. Timely detection is necessary to decrease the level of Hg^{2+} in industrial waste. Zheng et al., 2021, have prepared a paper-based fluorescent biosensor for the determination of Hg^{2+} [74] by immobilizing the cells in the alginate hydrogel, forming a cells-alginate hydrogel encapsulated system which was attached to the paper strip. The developed biosensor was able to detect Hg^{2+} at micro molar concentration in presence of other molecules and ions within 5 min. In addition to these examples, other developed biosensors for the determination of metal ions have been discussed in Table 1 with various information such as the readout system, sensor configuration, response time, detection range, and LOD.

4. Biosensors for Detection of Macromolecules

Macromolecules are relatively high molecular mass polymers composed of thousands of molecules. Macromolecules also show unusual properties that usually do not occur in small molecules. In living organisms, mainly three types of macromolecules are found which help in biological functions such as DNA, RNA, and proteins which are composed of monomer units, and other non-polymer types of macromolecule are also present such as lipid moiety and macrocycles. The developed cellulose-based sensors for the detection of various macromolecules are discussed in this section. A cellulose paper-based sensor has been fabricated for the visual detection of DNA by Jirakittiwut et al., 2015. The biosensor was developed by immobilization of acpc PNA (D-prolyl-2-aminocyclopentane-carboxylic acid PNA) on cellulose paper by divinyl sulfone-mediated conjugation [83]. PNA is an artificial mimic of the DNA that acts as a probe for the sensing of DNA. The interaction between PNA and DNA is based on the difference of charges on these molecules. PNA is a neutral molecule while DNA is a negatively charged molecule and thus provides a unique chance of interaction. The developed biosensor is highly specific and was even able to differentiate between the genes having the single base-pair mutation. The activity of the biosensor is checked using human leukocyte antigen and 26th mutations of thalassemia. The signal detection of the biosensor has been coupled with the cationic dye, Azure A, which lowers the detection limit (Figure 3A). Another paper-based biosensor was designed by Mohanraj et al., 2020, for the investigation of double-stranded DNA (dsDNA). The sensor was fabricated using graphene nanosheets through electrochemical exfoliation of the biomass, derived from corncob [84]. This paper-based graphene sensor can directly detect electrolytes without the requirement of sample preparation. The sensing of dsDNA was based on the oxidation of adenine and guanine within the detection range from 2×10^{-4} ng mL^{-1} to 50×10^{-4} ng mL^{-1} and the LOD of 0.68 pg mL^{-1}. More information about other developed biosensors for nucleotide detection has been tabulated in Table 2.

Cellulose-based sensors are not only for the nucleotides but are also developed for the sensing of different types of proteins. In this direction, several proteins have been detected using a novel sensor system. In one such example, the transcription factor (TF) has been detected, which is a DNA binding protein essential for gene regulation. Lin et al., 2019, have developed a paper-based biosensor for the visual detection of TF, which was fabricated through dopamine coating onto the cellulose paper and was used to obtain the analytical signals [85]. Characterization of the developed sensor was performed by FTIR and other techniques. The developed biosensor was analyzed for the target NF-κB p50, which is based on the Exo III-mediated cycling amplification reaction and the response was investigated by a color change. It was also hypothesized that the proposed biosensor is generic and can be extended towards different biomolecules by changing the recognition system, and thus may provide a low-cost, disposable, portable biosensing device. Another type of protein molecule is alkaline phosphatase (ALP), a metalloprotein that is inherently present in milk and acts as a biomarker for the investigation of pasteurized milk. In this direction, Mahato et al., 2019, developed a paper-based biosensor for the visual identification of ALP integrated with a smartphone system [50]. The sensor was fabricated through immobilization of the ALP antibody on the top of the paper. The ALP was detected through an immune complexation reaction between probe and ALP that forms a blue-green precipitate in the presence of 5-bromo-4-chloro 3-indolyl phosphate (Figure 3E). The quantification was performed using the digital image colorimetry technique and the detection range was observed between 10 to 1000 U/mL; LOD was found to be 0.87 (±0.07) U/mL. This study helps in developing an affordable biosensor for checking the quality of milk in miniaturized and personalized settings.

Apart from these, cellulose-based devices have also been explored for the sensing of glycoproteins, which plays a critical role in cell division, cell signaling, cell migration, and also as a biomarker for various disease diagnoses [86]. Another paper-based electrochemical biosensor has been developed for the ultra-sensitive identification of glycoprotein ovalbumin (OVA) by Sun et al., 2019. The biosensor was fabricated through the introduction of Au nanorods on the cellulose paper, which acted as a matrix for the preparation of boron ate-based molecularly imprinted polymers (MIPs) [87]. For biosensor fabrication, AuNPs were embedded on the top of SiO_2 nanoparticles and the formed SiO_2@Au were anchored with the dsDNA to enhance the signal. CeO_2 nanoparticles were used as an indicator that binds with the dsDNA, which leads to the formation of the SiO_2@Au/dsDNA/CeO_2 signal tag. The boron ate affinity-based MIPs were immobilized on a paper matrix to recognize the target glycoprotein OVA, through the covalent bonding formation between the boronic acid and glycoprotein. The detection range of the developed biosensor was measured to be 0.001 ng/mL to 1µg/mL and the LOD was 0.87 pg/mL (S/N = 3). Cellulose-based sensors have also been fabricated for the identification of another protein such as Bilirubin, which is related to jaundice and other clinical conditions. Tabatabaee et al., 2019, have prepared a photo lumines cent nano paper-based biosensor for early diagnosis of jaundice through the detection of Bilirubin in infants' blood samples. This provides an easy, effective, non-toxic, disposable, and inexpensive biosensor with a smartphone readout [88]. Smartphone readout systems provide great potential for point-of-care and point-of-need platforms [89]. It was fabricated by embedding the photoluminescent carbon dot sensing probes in BC nano paper substrate. Quenching of the photoluminescence was observed in the presence of Bilirubin, which acts as a quencher and was selectively recovered upon blue light (λ = 470 nm) exposure. The resulting intensity of the biosensor was found to be linearly proportional to the amount of Bilirubin present in the sample with a detection range of 2 to 20 mg dL^{-1}.

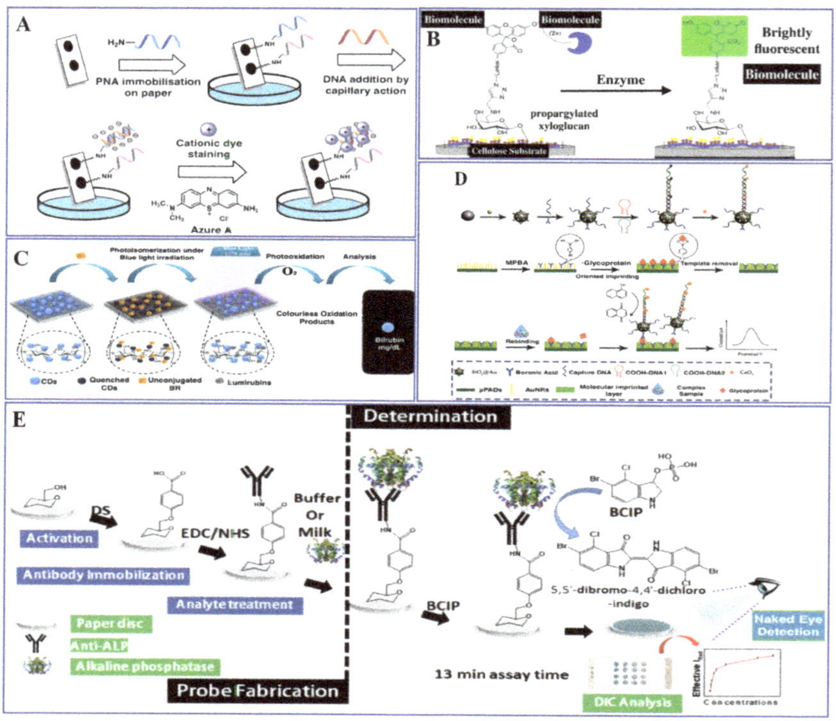

Figure 3. Biosensors for detection of macromolecules. (**A**) Fabrication of DNA sensor by immobilization of acpc PNA (D-prolyl-2-aminocyclopentanecarboxylic acid—peptide nucleic acid) through covalent bonding on the cellulose paper and cationic dye Azure A is used for signal detection (reproduced with permission from [83]). (**B**) Surface tethering of the cellulose-based sensor for the identification of Esterase enzyme by using the flu orogen (reproduced with permission from [90]). (**C**) Illustration of bilirubin sensor using photoluminescent carbon dot sensing probes (reproduced with permission from [88]). (**D**) Fabrication pattern of the paper-based biosensor for determination of glycoprotein (reproduced with permission from [87]). (**E**) Schematic representation of a fabrication and detection method of ALP biosensor (reproduced with permission from [50]).

Furthermore, the cellulose-based matrix is explored for fabricating the biosensor for the detection of numerous types of enzymes. A cellulose-based calorimetric biosensor was manufactured by Ling et al., 2019, for the identification of the enzyme human neutrophil elastase (HNE) [91]. HNE is a serine protease secreted from neutrophil at the time of chronic wounds and leads to the breakdown of proteins responsible for healing. The biosensor was developed through immobilization of HNE peptide to the cotton and wood nanocellulose. Cotton CNCs show a higher degree of sensing than wood cellulose nanofibrils. The sensitivity of cotton CNCs colorimetric sensor was observed to be less than 0.005 U/mL. Apart from the HNE, cellulose-based sensors also developed for other enzymes such as acetylcholinesterase (AChE), which plays a significant role in the hydrolysis of the neurotransmitter acetylcholine into choline and acetic acid. AChE is present at the neuromuscular junction and in the chemical synapse, which terminates the synaptic transmission [92]. A cellulose nanofiber-based biosensor has been manufactured for the identification of AChE by Wang et al., 2021, by grafting the DNA aptamer onto CNF. For the detection of AChE activity, CNF-DNA was combined with silver to develop CNF-DNA-AgNCs and chemical characterization was performed by FTIR, XPS, scanning electron microscopy (SEM), and TEM analysis [93]. Acetylthiocholine (ATCh) has been used as a substrate for AChE, causing hydrolysis of ATCh, converting it to thiocholine that reacted with the CNF-DNA-AgNCs. This reaction overall generated the analytical signals that were calibrated to detect the target compound. The sensitivity of the developed biosensor for AChE concentration was observed to be 0.053 mU/mL. Both these and the cellulosic matrix were also designed for the sensing of various macromolecules, as described in Table 2 with their design parameters and analytical performances.

Table 2. List of developed biosensors using cellulose matrix material for the detection of macromolecules. (NR: not reported).

Sr. No	Analyte	Detection	Sensor Configuration	Response Time	Real Sample	Detection Range	Limit of Detection	Reference
1	DNA	Colorimetric	Acpc PNA on cellulose paper by DVS conjugation	NR	Human leukocyte antigen alleles	low as 200 nm	NR	[83]
2	dsDNA	CV, EIS	Paper-based modified electrode sensor	NR	NR	0.2 pg/mL to 5 pg/mL	680 fg mL^{-1}	[84]
3	miRNA	Voltammetric	PNA-based paper biosensor	<1 h	Serum	up to 100 nm	6 nm	[94]
4	miR-21	Voltammetric	Cerium dioxide—Au@ glucose oxidase paper-based sensor	NR	Serum	0.001 pm to 1 pm	0.434 fm	[95]
5	ALP	Colorimetric	Paper-based naked eye detection	NR	Milk	10 to 1000 U/mL	0.87 (±0.07) U/mL	[50]
6	Esterase	Fluorescence	Chemoenzymatic method used for modification of cellulose matrix	NR	NR	NR	NR	[90]
7	Transcription factor	Colorimetric	Dopamine coated on the surface of cellulose paper	20 s	NF-κB p50 in biological fluids	NR	NR	[85]
8	Glycoprotein	Voltammetric	Paper-based biosensor for glycoprotein based on boronate affinity tag	NR	Ovalbumin	0.001 ng/mL to 1 µg/mL	870 fg/mL	[87]
9	Bilirubin	Photoluminescence	BC nanopaper-based biosensor through embedding of carbon dot sensing probes	NR	Infant's blood	2 to 20 mg dL^{-1}	NR	[88]
10	Human neutrophil elastase (HNE)	Colorimetric	Immobilizing HNE peptide to the cotton and wood nanocellulose	NR	Chronic wound fluid	Less than 0.005 U/mL	NR	[91]
11	Acetylcholinesterase (AChE)	Fluorescence	DNA aptamer immobilized on the surface of cellulose nanofiber	NR	NR	NR	NR	[93]

Table 2. *Cont.*

Sr. No	Analyte	Detection	Sensor Configuration	Response Time	Real Sample	Detection Range	Limit of Detection	Reference
12	Interleukin-6	Colorimetric	Paper sensor for IL-6 detection in COVID-19 patients	10 min	Respiratory	up to 10^{-1} ng mL^{-1}	1 fg mL^{-1}	[96]
13	Suppression of Tumorigenicity 2	CV, EIS	Graphite paper-based disposable sensor through modification of fullerene C_{60}	NR	Serum	as low as 414 ag mL^{-1}	124 ag mL^{-1}	[97]
14.	Bovine haptoglobin	Colorimetric	AuNP/MWCNT-anti-Hpnanobioconjugate paper-based sensor	NR	Serum	0.01 to 0.9 mg/mL	28 µg/mL	[98]

5. Biosensors for Detection of Cells

In addition to small molecules, macromolecules, and metal ions detection cellulose-based biosensors are also being developed for the visualization of cells. These biosensors are effective in determining both prokaryotic and eukaryotic cells. Here, we have discussed the previously developed sensors, focusing upon the identification of various kinds of bacteria, viruses, fungi, and also some cancer cells. In recent years, cellulose-based biosensors have been developed for bacteria detection where some interesting work has been carried out towards *Staphylococcus aureus* which produces enterotoxins and is the main reason for food poisoning and skin infections [99,100]. For the safety of food and to prevent humans from bacterial infection, it is required to focus on quick responsive detection of *S. aureus*. An electrochemical cellulose-based biosensor was manufactured for the early determination of living *S. aureus cells* from a sample of food containing both live and dead cells by Farooq et al., 2020. The biosensor was fabricated by immobilizing the bacteriophage onto a modified surface of the BC matrix. BC has a poriferous and fibrous structure that provides a larger effective surface for the impregnation of carboxy late multiwalled carbon nanotubes (c-MWCNTs), which enables greater density and phage immobilization. BC/c-MWCNTs nanocomposites modified through surface polymerization with polyethylene eimine provide a positive charge, which assists in the correct orientation of the phage on the matrix. Detection of anti-staphylococcus activity of the immobilized phage was measured through optical density and the density was determined through confocal microscopy. The developed biosensor effectively detects *S. aureus* up to 3 and 5 CFU$\cdot$mL^{-1} in the phosphate buffer saline and milk sample, respectively, using DPV in 30 min [101]. The biosensor shows high sensitivity, specificity, accuracy, and stability up to six weeks at 4 °C (Figure 4A). The investigation of bacterial pathogens is a main problem for the food industry and also for public health. *Escherichia coli*, *Staphylococcus aureus*, and *Listeria monocytogenes* are the three most crucial microorganisms which disperse through food and cause food borne diseases [102,103]. Another paper-based biosensor was fabricated for the identification of pathogenic bacteria by Liu et al., 2015. The biosensor relied on multiplex asymmetric PCR and a paper-based matrix. The probe was conjugated with AuNPs for the visual sensing of genomic DNA of the pathogenic microorganism. The detection limit of the developed method was found to be 1 pg/µL genomic DNA in standard conditions [103]. Additionally, another paper-based sensor was developed for the detection of a highly pathogenic microorganism: *Escherichia coli* O157:H7 present in food samples through naked eye detection by You et al., 2021. The biosensor was fabricated by using Poly-L-lysine-coated starch magnetic particles (PLL@SMPs) for magnetic separation. The signal amplification is performed by using an antibody conjugated with Horseradish peroxidase and 3, 3′, 5, 5′-tetramethylbenzidine. The developed PLL@SMPs show an efficiency of >90% in a large volume for the target bacteria. The LOD of the biosensor was observed to be as low as 30.8 CFU/mL with a probability of 95% [104]. This technique may offer an easy, sensitive, and specific process for the evaluation of the food and surrounding samples. Apart from the bacteria, interestingly, cellulose-based sensing devices have also been used for the detection of viruses. Pseudorabies virus (PRV) causes porcine pseudorabies, which is an acute virulent that is infectious to livestock and wildlife. Due to the swine infection with PRV, there is a great financial loss for the pig industry which requires the urgent diagnosis of PRV. Recently, Huang et al., 2021, have manufactured a paper-based biosensor integrated with a smartphone for the diagnosis of wild-type PRV infection [105]. For detection, latex beads have been used for labeling PRV and the coated test line with PRV gE-m Abs. The signals were recorded and processed using the smartphone's light detector. The developed biosensor indicates excellent sensitivity, selectivity, and rapid detection of PRV within 15 min.

Nowadays, cytosensing is also extensively used for the detection of multiple types of cancer cells [106,107]. There are various surface biomarkers such as MUC-1 (Mucin 1), carcino embryonic antigen (CEA), BRCA1, BRCA2, HERS, folic acid receptors, sialic-acid associated glycoprotein, and many other protein/glycoprotein biomarkers [108] that are recognized by aptamers [109], antibodies, and nano-enzymes immobilized on cellulose matrix [110]. More emphasis has been given on whole cell-system-based biosensing and a microfluidic system for targeting different antigens expressed by cancer cells at the onset of the disease. Cellulose-based multiplexed biosensor systems and microfluidic paper-based biosensing devices are utilized nowadays for detecting multiple surface antigen-specific to cancer synchronously. Kumar et al. (2016) portrayed electrochemical biosensors for the investigation of CEA cancer biomarker [111]. The sensor was developed using a novel strategy comprising systematic layering of the components. The performance of the sensor was comparable and shows the CEA in clinical levels. Liang et al. (2016) detected the cancer cells by multiplexed sensing panel, which is based on a microfluidic paper-based biosensor [112]. The author uses mesoporous silica nanoparticles that were encapsulated with cadmium telluride QDs and DNA aptamers were immobilized onto it, offer ing great sensitivity, specificity, and efficiency against MCF-7, HL-60, K562 cancerous cells using the quenching ability of GO [112]. When tumor cells were injected on the microfluidic paper analytical device, the fluorescence intensity of QDs increases [113]. This kind of microfluidics paper analytical device could be utilized for the identification of metastatic phases of cancerous cells. In another study, Shi, Ma et al., 2019, used an SPCE for the sensing of cancer antigen 125, which was fabricated on a pure cellulosic paper [114]. Dickert et al. created an ABO blood grouping-based biosensor system. In this device, they fabricated the polyurethane imprints on which different surface antigens were identified based on the type of blood group [115]. In an interesting study by Feng, Liu, et al., 2014, a paper-based electrochemiluminescence (ECL) biosensor has been developed for the detection of HL-60 cancer cells. The biosensor was fabricated by coating the AuNPs and graphene on the surface of porous filter paper in a systematic manner. The aptamer used was flagged with silica nanoparticles conjugated with $Ru(bpy)_3^{2+}$ and showed a high affinity towards the target cell (Figure 4B). ECL intensity directly shows the quantity of the HL-60 cancer cells with the detection range of 56 to 56 $\times$ 10^5 cells/mL [116].

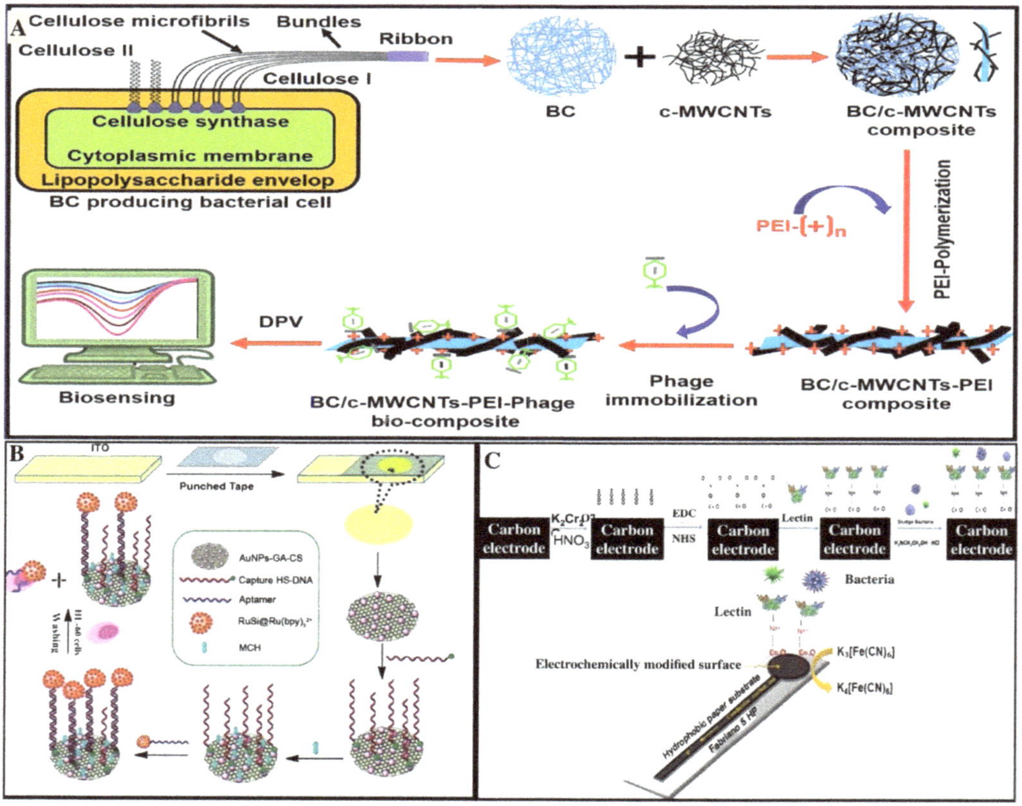

Figure 4. Biosensors for detection of cells. (**A**) Fabrication scheme of the sensor for identification of *S. aureus* cells using (**B**,**C**) matrix (reproduced with permission from [101]). (**B**) Pictorial depiction of paper-based ECL biosensor for the sensing of HL-60 cancer cells (reproduced with permission from [116]). (**C**) Representation of steps involved in electrode surface modification and the bacterial detection by the SPCE (Reproduced with permission from [117]).

Severe acute respiratory syndrome coronavirus 2 (SARS-CoV-2) outbreaks were announced as a global pandemic by WHO. Rapid and sensitive detection of the virus has become the need of the hour and various methods have been explored by scientists across the globe [118–122]. In recent research, a paper-based electrochemical biosensor has been fabricated for the investigation of SARS-CoV-2 [123]. A label-free paper-based biosensor was manufactured by Yakoh et al., 2021, targeting the SARS-CoV-2 antibodies even excluding the need for a specific antibody. The appearance of SARS-CoV-2 antibodies would obstruct the redox reaction of the redox indicator, and the response was measured as a decrease in the current. The recorded result was proven effective when compared with clinical reports, showing satisfactory results [123]. This example clearly suggests the importance of cellulose matrix even in the design and development of biosensors for pandemics. Further examples of cellulosic sensors for the sensing of cells are mentioned in Table 3 with their analytical details.

Table 3. List of developed biosensors using cellulose matrix material for the detection of cells. (NR: not reported).

Sr. No	Analyte	Detection	Sensor Configuration	Response Time	Real Sample	Detection Range	Limit of Detection	Reference
1	*Staphylococcus aureus*	Voltammetric	Immobilization of bacteriophage onto EC	30 min	Milk, PBS	5 CFU mL^{-1}, 3 CFU mL^{-1}	NR	[101]
2	*Staphylococcus aureus*	Optical	Paper-based biosensor using a primer-based asymmetric PCR	NR	*nuc* gene	low as 1 pg/μL	NR	[103]
3	*Listeria monocytogenes*	Optical	Paper-based biosensor using a primer-based asymmetric PCR	NR	*Hlya* gene	low as 1 pg/μL	NR	[103]
4	*Escherichia coli*	Optical	Paper-based biosensor using a primer-based asymmetric PCR	NR	*rbfE* gene	low as 1 pg/μL	NR	[103]
5	*Escherichia coli*	Colorimetric	PLL@SMPs-based paper sensor	NR	Food	NR	30.8 CFU/mL	[104]
6	Pseudorabies virus	Colorimetric	Latex beads paper-based sensor using PRV gE-mAb	15 min	Pig serum	NR	NR	[105]
7	HL-60 cancer cell	Electrochemiluminescence	$Ru(bpy)^2$ 3^+-conjugated silica nanoparticle-based	NR	NR	56–5.6 × 10^6 cells/mL	NR	[116]
8	SARS-CoV-2	Voltammetric	Label-free paper-based biosensor	30 min	Serum	1 ng/mL	NR	[123]
9	*Listeria monocytogenes*	Chemiluminescence	Paper-based sensing device with an immobilized DNA probe	NR	*hlyA*gene	0.194 pmol/L to 19.4 × 10^3 pmol/L	6.3 × 10^{-2} pmol/L	[124]
10	Papillomavirus	Voltammetric	acpcPNA and graphene-polyaniline modified paper-based biosensor	NR	SiHa cells	10 to 200 nM	2.3 nM	[125]

6. Conclusions and Future Prospects

Cellulose is a widely used material showing various advantages over other materials. Cellulose exhibits unique properties such as biocompatibility, flexibility, mechanical strength, biodegradability, electrical properties, and cost-effectiveness. Apart from the cellulose derived from plants, bacterial cellulose, which is obtained from bacteria, shows more advantageous properties such as mechanical strength, water-holding capacity, and biocompatibility. These advantages of cellulose have made it a significant matrix substance in recent times for the fabrication of biosensors. Many applications have been reported concerning cellulose-based sensing of biomarkers for various disease conditions such as cancer, diabetes, liver and kidney disorders, etc., and also in the detection and monitoring of food quality and various pollutants and heavy metals present in surrounding samples, such as in wastewater. Owing to the significant advances, various methods have been examined to modify cellulose to nanocomposite material, which shows unique structures and desirable properties. In this review, a detailed discussion has been carried out on various modifications required for the formation of cellulose matrix and engineering aspects of the cellulose-based sensors for their applications in different areas.

These sensors are used for sensing the numerous types of small molecules, macromolecules, metal ions, and cells, which are discussed through the numerous examples of sensors. We have tried to include the fabricated sensors to date for various categories of molecules possible, and the techniques/methods used for their modifications. Despite the great progress achieved, there are still some issues that need to be addressed. Recent advances in cellulose-based biosensors typically exhibiting a single function type only and lack the multi-functionality, which leads to its limitation in function and also affects the performances and customer satisfaction. In the future, numerous innovative technologies could be introduced in developing cost-effective multifunctional cellulose-based biosensors.

Author Contributions: Conceptualization, P.C.; writing—original draft preparation, D.; P.C.; writing—review and editing, D., S.M., V.R.S., P.C.; visualization, P.C.; supervision, P.C. All authors have read and agreed to the published version of the manuscript.

Funding: This research was funded by IIT (BHU) Varanasi seed grant, grant number (IIT (BHU)/Budget/19-(14)/2020-21/18792).

Institutional Review Board Statement: Not applicable.

Informed Consent Statement: Not applicable.

Data Availability Statement: Not applicable.

Acknowledgments: Pranjal Chandra acknowledge Pramod Kumar Jain, Director of the Indian Institute of Technology (BHU) Varanasi, India for encouragement and providing the necessary facility for completion of this work. Author D. acknowledge her PhD research fellowship by IIT (BHU).

Conflicts of Interest: The authors declare no conflict of interest.

References

1. Hallac, B.B.; Ragauskas, A.J. Analyzing cellulose degree of polymerization and its relevancy to cellulosic ethanol. *Biofuels Bioprod. Biorefining* **2011**, *5*, 215–225. [CrossRef]
2. Li, S.; Bashline, L.; Lei, L.; Gu, Y. Cellulose synthesis and its regulation. *Arab. Book* **2014**, *12*, e0169. [CrossRef] [PubMed]
3. Maleki, S.S.; Mohammadi, K.; Ji, K.S. Characterization of cellulose synthesis in plant cells. *Sci. World J.* **2016**, *2016*, 8641373. [CrossRef] [PubMed]
4. Rangaswamy, B.E.; Vanitha, K.P.; Hungund, B.S. Microbial cellulose production from bacteria isolated from rotten fruit. *Int. J. Polym. Sci.* **2015**, *2015*, 280784. [CrossRef]
5. Ghozali, M.; Meliana, Y.; Chalid, M. Synthesis and characterization of bacterial cellulose by Acetobacter xylinum using liquid tapioca waste. *Mater. Today* **2021**, *44*, 2131–2134. [CrossRef]
6. Song, D.; Shen, J.; Li, L. Characterization of cellulose synthase complexes in Populus xylem differentiation. *New Phytol.* **2010**, *187*, 777–790. [CrossRef] [PubMed]
7. Morales-Narváez, E.; Golmohammadi, H.; Naghdi, T.; Yousefi, H.; Kostiv, U.; Horak, D.; Pourreza, N.; Merkoçi, A. Nanopaper as an optical sensing platform. *ACS Nano* **2015**, *9*, 7296–7305. [CrossRef]
8. Kamel, S.; Khattab, A.T. Recent advances in cellulose-based biosensors for medical diagnosis. *Biosensors* **2020**, *10*, 67. [CrossRef]
9. Ratajczak, K.; Stobiecka, M. High-performance modified cellulose paper-based biosensors for medical diagnostics and early cancer screening: A concise review. *Carbohydr. Polym.* **2020**, *229*, 115463. [CrossRef]
10. Wang, Y.; Hou, S.; Li, T.; Jin, S.; Shao, Y.; Yang, H.; Wu, D.; Dai, S.; Lu, Y.; Chen, S.; et al. Flexible capacitive humidity sensors based on ionic conductive wood-derived cellulose nanopapers. *ACS Appl. Mater. Interfaces* **2020**, *12*, 41896–41904. [CrossRef]
11. Kamel, S.; Ali, N.; Jahangir, K.; Shah, S.M.; El-Gendy, A.A. Pharmaceutical significance of cellulose: A review. *Express Polym. Lett.* **2008**, *2*, 758–778. [CrossRef]
12. Lavanya, D.; Kulkarni, P.; Dixit, M.; Raavi, P.K.; Krishna, L.N.V. Sources of cellulose and their applications—A review. *Int. J. Drug Formul. Res.* **2011**, *2*, 19–38.
13. Park, H.-J.; Yu, S.J.; Yang, K.; Jin, Y.; Cho, A.-N.; Kim, J.; Lee, B.; Yang, H.S.; Im, S.G.; Cho, S.-W. Based bioactive scaffolds for stem cell-mediated bone tissue engineering. *Biomaterials* **2014**, *35*, 9811–9823. [CrossRef] [PubMed]
14. Lim, W.Y.; Goh, B.T.; Khor, S.M. Microfluidic paper-based analytical devices for potential use in quantitative and direct detection of disease biomarkers in clinical analysis. *J. Chromatogr. B* **2017**, *1060*, 424–442. [CrossRef] [PubMed]
15. Prathapan, R.; McLiesh, H.; Garnier, G.; Tabor, R.F. Surface engineering of transparent cellulose nanocrystal coatings for biomedical applications. *ACS Appl. Bio Mater.* **2018**, *1*, 728–737. [CrossRef]
16. Zhu, G.; Yin, X.; Jin, D.; Zhang, B.; Gu, Y.; An, Y. Paper-based immunosensors: Current trends in the types and applied detection techniques. *TrAC Trends Anal. Chem.* **2019**, *111*, 100–117. [CrossRef]
17. Gopakumar, D.A.; Thomas, S.; Owolabi, F.; Thomas, S.; Nzihou, A.; Rizal, S.; Khalil, H.A. Nanocellulose based aerogels for varying engineering applications. *Encycl. Renew. Sustain. Mater.* **2020**, *2*, 155–165.
18. Tortorella, S.; Buratti, V.V.; Maturi, M.; Sambri, L.; Franchini, M.C.; Locatelli, E. Surface-modified nanocellulose for application in biomedical engineering and nanomedicine: A review. *Int. J. Nanomed.* **2020**, *15*, 9909. [CrossRef]
19. Klemm, D.; Kramer, F.; Moritz, S.; Lindstrom, T.; Ankerfors, M.; Gray, D.; Dorris, A. Nanocelluloses: A new family of nature-based materials. *Angew. Chem. Int. Ed. Engl.* **2011**, *50*, 5438–5466. [CrossRef]
20. Moon, R.J.; Martini, A.; Nairn, J.; Simonsen, J.; Youngblood, J. Cellulose nanomaterials review: Structure, properties and nanocomposites. *Chem. Soc. Rev.* **2011**, *40*, 3941–3994. [CrossRef] [PubMed]
21. Kuek Lawrence, C.S.; Tan, S.N.; Floresca, C.Z. A "green" cellulose paper based glucose amperometric biosensor. *Sens. Actuators B* **2014**, *193*, 536–541. [CrossRef]
22. Zhao, D.; Zhu, Y.; Cheng, W.; Chen, W.; Wu, Y.; Yu, H. Cellulose-based flexible functional materials for emerging intelligent electronics. *Adv. Mater.* **2020**, 2000619. [CrossRef]

23. Oishi, Y.; Nakaya, M.; Matsui, E.; Hotta, A. Structural and mechanical properties of cellulose composites made of isolated cellulose nanofibers and poly (vinyl alcohol). *Compos. Part A* **2015**, *73*, 72–79. [CrossRef]
24. Ramezani, M.G.; Golchinfar, B. Mechanical properties of cellulose nanocrystal (cnc) bundles: Coarse-grained molecular dynamic simulation. *J. Compos. Sci.* **2019**, *3*, 57. [CrossRef]
25. Baranwal, A.; Mahato, K.; Srivastava, A.; Maurya, P.K.; Chandra, P. Phytofabricated metallic nanoparticles and their clinical applications. *RSC Adv.* **2016**, *6*, 105996–106010. [CrossRef]
26. Purohit, B.; Vernekar, P.R.; Shetti, N.P.; Chandra, P. Biosensor nanoengineering: Design, operation, and implementation for biomolecular analysis. *Sens. Int.* **2020**, *1*, 100040. [CrossRef]
27. Mahato, K.; Prasad, A.; Maurya, P.; Chandra, P. Nanobiosensors: Next generation point-of-care biomedical devices for personalized diagnosis. *J. Anal. Bioanal. Tech.* **2016**, *7*, e125.
28. Chandra, P. Advances in clinical diagnosis through electrochemical aptamer sensors. *J. Bioanal. Biomed.* **2013**, *5*, e119. [CrossRef]
29. Mahato, K.; Maurya, P.K.; Chandra, P. Fundamentals and commercial aspects of nanobiosensors in point-of-care clinical diagnostics. *3 Biotech* **2018**, *8*, 149. [CrossRef]
30. Purohit, B.; Kumar, A.; Mahato, K.; Chandra, P. Smartphone-assisted personalized diagnostic devices and wearable sensors. *Curr. Opin. Biomed. Eng.* **2020**, *13*, 42–50. [CrossRef]
31. Chandra, P. Miniaturized multiplex electrochemical biosensor in clinical bioanalysis. *J. Bioanal. Biomed.* **2013**, *5*, e122.
32. Zhao, V.X.T.; Wong, T.I.; Zheng, X.T.; Tan, Y.N.; Zhou, X. Colorimetric biosensors for point-of-care virus detections. *Mater. Sci. Energy Technol.* **2020**, *3*, 237–249. [CrossRef]
33. Khan, A.; Abas, Z.; Kim, H.S.; Kim, J. Recent progress on cellulose-based electro-active paper, its hybrid nanocomposites and applications. *Sensors* **2016**, *16*, 1172. [CrossRef]
34. Saeed, A.A.; Abbas, M.N.; Singh, B.; Abou-Zeid, R.E.; Kamel, S. Cellulose nanocrystals decorated with gold nanoparticles immobilizing GOx enzyme for non-invasive biosensing of human salivary glucose. *Anal. Methods* **2019**, *11*, 6073–6083. [CrossRef]
35. Ahmad, R.; Khare, S.K. Immobilization of Aspergillus Niger cellulase on multiwall carbon nanotubes for cellulose hydrolysis. *Bioresour. Technol.* **2018**, *252*, 72–75. [CrossRef]
36. Lee, J.; Kim, J.; Kim, S.; Min, D.H. Biosensors based on graphene oxide and its biomedical application. *Adv. Drug Deliv. Rev.* **2016**, *105*, 275–287. [CrossRef] [PubMed]
37. Ma, F.; Li, C.C.; Zhang, C.Y. Development of quantum dot-based biosensors: Principles and applications. *J. Mater. Chem. B* **2018**, *6*, 6173–6190. [CrossRef] [PubMed]
38. Chandra, P.; Das, D.; Abdelwahab, A.A. Gold nanoparticles in molecular diagnostics and therapeutics. *Digest J. Nanomater. Biostructures* **2010**, *5*, 363–367.
39. Koh, W.C.A.; Chandra, P.; Kim, D.-M.; Shim, Y.-B. Electropolymerized self-assembled layer on gold nanoparticles: Detection of inducible nitric oxide synthase in neuronal cell culture. *Anal. Chem.* **2011**, *83*, 6177–6183. [CrossRef]
40. Stobiecka, M.; Chalupa, A. Modulation of plasmon-enhanced resonance energy transfer to gold nanoparticles by protein survivin channeled-shell gating. *J. Phys. Chem. B* **2015**, *119*, 13227–13235. [CrossRef] [PubMed]
41. Purohit, B.; Kumar, A.; Mahato, K.; Chandra, P. Electrodeposition of metallic nanostructures for biosensing applications in health care. *J. Sci. Res.* **2020**, *64*, 68–73. [CrossRef]
42. Mahato, K.; Nagpal, S.; Shah, M.A.; Srivastava, A.; Maurya, P.K.; Roy, S.; Jaiswal, A.; Singh, R.; Chandra, P. Gold nanoparticle surface engineering strategies and their applications in biomedicine and diagnostics. *3 Biotech* **2019**, *9*, 57. [CrossRef] [PubMed]
43. Valizadeh, A.; Mikaeili, H.; Samiei, M.; Farkhani, S.M.; Zarghami, N.; Akbarzadeh, A.; Davaran, S. Quantum dots: Synthesis, bioapplications, and toxicity. *Nanoscale Res. Lett.* **2012**, *7*, 1–14. [CrossRef]
44. Ullah, H.; Wahid, F.; Santos, H.A.; Khan, T. Advances in biomedical and pharmaceutical applications of functional bacterial cellulose-based nanocomposites. *Carbohydr. Polym.* **2016**, *150*, 330–352. [CrossRef] [PubMed]
45. Gomes, N.O.; Carrilho, E.; Machado, S.A.S.; Sgobbi, L.F. Bacterial cellulose-based electrochemical sensing platform: A smart material for miniaturized biosensors. *Electrochim. Acta* **2020**, *349*, 136341. [CrossRef]
46. Prasad, A.; Mahato, K.; Chandra, P.; Srivastava, A.; Joshi, S.N.; Maurya, P.K. Bioinspired composite materials: Applications in diagnostics and therapeutics. *J. Mol. Eng. Mater.* **2016**, *4*, 1640004. [CrossRef]
47. Mahato, K.; Kumar, A.; Purohit, B.; Mahapatra, S.; Srivastava, A.; Chandra, P. Nanomaterial functionalization strategies in bio-interface development for modern diagnostic devices. In *Biointerface Engineering: Prospects in Medical Diagnostics and Drug Delivery*; Springer: Singapore, 2020; pp. 195–214.
48. Chinnadayyala, S.R.; Park, J.; Le, H.T.N.; Santhosh, M.; Kadam, A.N.; Cho, S. Recent advances in microfluidic paper-based electrochemiluminescence analytical devices for point-of-care testing applications. *Biosens. Bioelectron.* **2019**, *126*, 68–81. [CrossRef] [PubMed]
49. Akyazi, T.; Basabe-Desmonts, L.; Benito-Lopez, F. Review on microfluidic paper-based analytical devices towards commercialisation. *Anal. Chim. Acta* **2018**, *1001*, 1–17. [CrossRef] [PubMed]
50. Mahato, K.; Chandra, P. Paper-based miniaturized immunosensor for naked eye ALP detection based on digital image colorimetry integrated with smartphone. *Biosens. Bioelectron.* **2019**, *128*, 9–16. [CrossRef]
51. Kumar, A.; Hens, A.; Arun, R.K.; Chatterjee, M.; Mahato, K.; Layek, K.; Chanda, N. A paper based microfluidic device for easy detection of uric acid using positively charged gold nanoparticles. *Analyst* **2015**, *140*, 1817–1821. [CrossRef]

52. Arena, A.; Donato, N.; Saitta, G.; Bonavita, A.; Rizzo, G.; Neri, G. Flexible ethanol sensors on glossy paper substrates operating at room temperature. *Sens. Actuators B* **2010**, *145*, 488–494. [CrossRef]
53. Cinti, S.; Mazzaracchio, V.; Cacciotti, I.; Moscone, D.; Arduini, F. Carbon black-modified electrodes screen-printed onto paper towel, waxed paper and parafilm M((R)). *Sensors* **2017**, *17*, 2267. [CrossRef] [PubMed]
54. Mahadeva, S.K.; Kim, J. Conductometric glucose biosensor made with cellulose and tin oxide hybrid nanocomposite. *Sens. Actuators B* **2011**, *157*, 177–182. [CrossRef]
55. Tracey, C.T.; Torlopov, M.A.; Martakov, I.S.; Vdovichenko, E.A.; Zhukov, M.; Krivoshapkin, P.V.; Mikhaylov, V.I.; Krivoshapkina, E.F. Hybrid cellulose nanocrystal/magnetite glucose biosensors. *Carbohydr. Polym.* **2020**, *247*, 116704. [CrossRef]
56. Guan, H.; Liu, X.; Wang, W. Encapsulation of tyrosinase within liposome bioreactors for developing an amperometric phenolic compounds biosensor. *J. Solid State Electrochem.* **2013**, *17*, 2887–2893. [CrossRef]
57. Zhang, J.; Lei, J.; Liu, Y.; Zhao, J.; Ju, H. Highly sensitive amperometric biosensors for phenols based on polyaniline-ionic liquid-carbon nanofiber composite. *Biosens. Bioelectron.* **2009**, *24*, 1858–1863. [CrossRef]
58. Manan, F.A.A.; Hong, W.W.; Abdullah, J.; Yusof, N.A.; Ahmad, I. Nanocrystalline cellulose decorated quantum dots based tyrosinase biosensor for phenol determination. *Mater. Sci. Eng. C Mater. Biol. Appl.* **2019**, *99*, 37–46. [CrossRef]
59. Cao, X.; Ding, Q.; Zhao, N.; Gao, A.; Jing, Q. Supramolecular self-assembly system based on naphthalimide boric acid ester derivative for detection of organic amine. *Sens. Actuators B* **2018**, *256*, 711–720. [CrossRef]
60. Nawaz, H.; Zhang, J.; Tian, W.; Jin, K.; Jia, R.; Yang, T.; Zhang, J. Cellulose-based fluorescent sensor for visual and versatile detection of amines and anions. *J. Hazard. Mater.* **2020**, *387*, 121719. [CrossRef]
61. Younes, N.A.; Dawood, M.F.A.; Wardany, A.A. The phyto-impact of fluazinam fungicide on cellular structure, agro-physiological, and yield traits of pepper and eggplant crops. *Environ. Sci. Pollut. Res. Int.* **2020**, *27*, 18064–18078. [CrossRef]
62. Wang, X.; Hou, J.; Lan, S.; Shen, C.; Huo, D.; Ji, Z.; Ma, Y.; Luo, H.; Zhang, S.; He, Q.; et al. MoS2 QDs-based sensor for measurement of fluazinam with triple signal output. *Anal. Chim. Acta* **2020**, *1108*, 152–159. [CrossRef]
63. Huang, Y.; Zhu, F.; Guan, J.; Wei, W.; Zou, L. Label-free amperometric immunosensor based on versatile carbon nanofibers network coupled with au nanoparticles for aflatoxin B1 detection. *Biosensors* **2020**, *11*, 5. [CrossRef]
64. Ahmed, S.; Bui, M.P.; Abbas, A. Paper-based chemical and biological sensors: Engineering aspects. *Biosens. Bioelectron.* **2016**, *77*, 249–263. [CrossRef] [PubMed]
65. Mahato, K.; Srivastava, A.; Chandra, P. Paper based diagnostics for personalized health care: Emerging technologies and commercial aspects. *Biosens. Bioelectron.* **2017**, *96*, 246–259. [CrossRef]
66. Levy, H.L.; Sarkissian, C.N.; Scriver, C.R. Phenylalanine ammonia lyase (PAL): From discovery to enzyme substitution therapy for phenylketonuria. *Mol. Genet. Metab.* **2018**, *124*, 223–229. [CrossRef]
67. Sun, B.; Wang, Z.; Wang, X.; Qiu, M.; Zhang, Z.; Wang, Z.; Cui, J.; Jia, S. Paper-based biosensor based on phenylalnine ammonia lyase hybrid nanoflowers for urinary phenylalanine measurement. *Int. J. Biol. Macromol.* **2021**, *166*, 601–610. [CrossRef] [PubMed]
68. Esmaeili, C.; Abdi, M.M.; Mathew, A.P.; Jonoobi, M.; Oksman, K.; Rezayi, M. Synergy effect of nanocrystalline cellulose for the biosensing detection of glucose. *Sensors* **2015**, *15*, 24681–24697. [CrossRef] [PubMed]
69. Wang, R.J.; Zhang, L.W.; Liu, R.; Liu, L.; Yao, J.M. Ultra-fast and probe-free cellulose biosensor for visual detection of Cu(2+) ions in biological samples. *Carbohydr. Polym.* **2019**, *223*, 115117. [CrossRef] [PubMed]
70. Wang, L.; Zhang, C.; He, H.; Zhu, H.; Guo, W.; Zhou, S.; Wang, S.; Zhao, J.R.; Zhang, J. Cellulose-based colorimetric sensor with N, S sites for Ag(+) detection. *Int. J. Biol. Macromol.* **2020**, *163*, 593–602. [CrossRef] [PubMed]
71. Jaishankar, M.; Tseten, T.; Anbalagan, N.; Mathew, B.B.; Beeregowda, K.N. Toxicity, mechanism and health effects of some heavy metals. *Interdiscip. Toxicol.* **2014**, *7*, 60–72. [CrossRef] [PubMed]
72. Jin, J.-C.; Wang, B.-B.; Xu, Z.-Q.; He, X.-H.; Zou, H.-F.; Yang, Q.-Q.; Jiang, F.-L.; Liu, Y. A novel method for the detection of silver ions with carbon dots: Excellent selectivity, fast response, low detection limit and good applicability. *Sens. Actuators B* **2018**, *267*, 627–635. [CrossRef]
73. Daniyal, W.M.E.M.M.; Fen, Y.W.; Anas, N.A.A.; Omar, N.A.S.; Ramdzan, N.S.M.; Nakajima, H.; Mahdi, M.A. Enhancing the sensitivity of a surface plasmon resonance-based optical sensor for zinc ion detection by the modification of a gold thin film. *RSC Adv.* **2019**, *9*, 41729–41736. [CrossRef]
74. Zheng, Y.; Wei, L.; Duan, L.; Yang, F.; Huang, G.; Xiao, T.; Wei, M.; Liang, Y.; Yang, H.; Li, Z.; et al. Rapid field testing of mercury pollution by designed fluorescent biosensor and its cells-alginate hydrogel-based paper assay. *J. Environ. Sci.* **2021**, *106*, 161–170. [CrossRef]
75. Veerapandian, M.; Seo, Y.T.; Shin, H.; Yun, K.; Lee, M.H. Functionalized graphene oxide for clinical glucose biosensing in urine and serum samples. *Int. J. Nanomed.* **2012**, *7*, 6123–6136. [CrossRef]
76. Wang, W.; Zhang, T.J.; Zhang, D.W.; Li, H.Y.; Ma, Y.R.; Qi, L.M.; Zhou, Y.L.; Zhang, X.X. Amperometric hydrogen peroxide biosensor based on the immobilization of heme proteins on gold nanoparticles-bacteria cellulose nanofibers nanocomposite. *Talanta* **2011**, *84*, 71–77. [CrossRef]
77. Cinti, S.; Basso, M.; Moscone, D.; Arduini, F. A paper-based nanomodified electrochemical biosensor for ethanol detection in beers. *Anal. Chim. Acta* **2017**, *960*, 123–130. [CrossRef] [PubMed]
78. Scognamiglio, V.; Antonacci, A.; Arduini, F.; Moscone, D.; Campos, E.V.R.; Fraceto, L.F.; Palleschi, G. An eco-designed paper-based algal biosensor for nanoformulated herbicide optical detection. *J. Hazard. Mater.* **2019**, *373*, 483–492. [CrossRef]

79. Jayaweera, S.; Yin, K.; Hu, X.; Ng, W.J. Fluorescent N/Al Co-doped carbon dots from cellulose biomass for sensitive detection of manganese (VII). *J. Fluoresc.* **2019**, *29*, 1291–1300. [CrossRef] [PubMed]
80. Lv, P.; Yao, Y.; Li, D.; Zhou, H.; Naeem, M.A.; Feng, Q.; Huang, J.; Cai, Y.; Wei, Q. Self-assembly of nitrogen-doped carbon dots anchored on bacterial cellulose and their application in iron ion detection. *Carbohydr. Polym.* **2017**, *172*, 93–101. [CrossRef]
81. Tsagkaris, A.S.; Pulkrabova, J.; Hajslova, J.; Filippini, D. A hybrid lab-on-a-chip injector system for autonomous carbofuran screening. *Sensors* **2019**, *19*, 5579. [CrossRef] [PubMed]
82. El-Moghazy, A.Y.; Amaly, N.; Istamboulie, G.; Nitin, N.; Sun, G. A signal-on electrochemical aptasensor based on silanized cellulose nanofibers for rapid point-of-use detection of ochratoxin A. *Microchim. Acta* **2020**, *187*, 1–11. [CrossRef]
83. Jirakittiwut, N.; Panyain, N.; Nuanyai, T.; Vilaivan, T.; Praneenararat, T. Pyrrolidinyl peptide nucleic acids immobilised on cellulose paper as a DNA sensor. *RSC Adv.* **2015**, *5*, 24110–24114. [CrossRef]
84. Mohanraj, J.; Durgalakshmi, D.; Rakkesh, R.A.; Balakumar, S.; Rajendran, S.; Karimi-Maleh, H. Facile synthesis of paper based graphene electrodes for point of care devices: A double stranded DNA (dsDNA) biosensor. *J. Colloid. Interface Sci.* **2020**, *566*, 463–472. [CrossRef] [PubMed]
85. Lin, H.; Wang, X.; Wu, J.; Li, H.; Li, F. Equipment-free and visualized biosensor for transcription factor rapid assay based on dopamine-functionalized cellulose paper. *J. Mater. Chem. B* **2019**, *7*, 5461–5464. [CrossRef]
86. Parker, R.B.; Kohler, J.J. Regulation of intracellular signaling by extracellular glycan remodeling. *ACS Chem. Biol.* **2010**, *5*, 35–46. [CrossRef] [PubMed]
87. Sun, X.; Jian, Y.; Wang, H.; Ge, S.; Yan, M.; Yu, J. Ultrasensitive microfluidic paper-based electrochemical biosensor based on molecularly imprinted film and boronate affinity sandwich assay for glycoprotein detection. *ACS Appl. Mater. Interfaces* **2019**, *11*, 16198–16206. [CrossRef] [PubMed]
88. Tabatabaee, R.S.; Golmohammadi, H.; Ahmadi, S.H. Easy diagnosis of jaundice: A smartphone-based nanosensor bioplatform using photoluminescent bacterial nanopaper for point-of-care diagnosis of hyperbilirubinemia. *ACS Sens.* **2019**, *4*, 1063–1071. [CrossRef]
89. Roda, A.; Michelini, E.; Zangheri, M.; Di Fusco, M.; Calabria, D.; Simoni, P. Smartphone-based biosensors: A critical review and perspectives. *TrAC Trends Anal. Chem.* **2016**, *79*, 317–325. [CrossRef]
90. Derikvand, F.; Yin, D.T.; Barrett, R.; Brumer, H. Cellulose-based biosensors for esterase detection. *Anal. Chem.* **2016**, *88*, 2989–2993. [CrossRef] [PubMed]
91. Ling, Z.; Xu, F.; Edwards, J.V.; Prevost, N.T.; Nam, S.; Condon, B.D.; French, A.D. Nanocellulose as a colorimetric biosensor for effective and facile detection of human neutrophil elastase. *Carbohydr. Polym.* **2019**, *216*, 360–368. [CrossRef] [PubMed]
92. Adler, M.; Sweeney, R.E.; Hamilton, T.A.; Lockridge, O.; Duysen, E.G.; Purcell, A.L.; Deshpande, S.S. Role of acetylcholinesterase on the structure and function of cholinergic synapses: Insights gained from studies on knockout mice. *Cell. Mol. Neurobiol.* **2011**, *31*, 909–920. [CrossRef] [PubMed]
93. Wang, L.; Guo, W.; Zhu, H.; He, H.; Wang, S. Preparation and properties of a dual-function cellulose nanofiber-based bionic biosensor for detecting silver ions and acetylcholinesterase. *J. Hazard. Mater.* **2021**, *403*, 123921. [CrossRef] [PubMed]
94. Moccia, M.; Caratelli, V.; Cinti, S.; Pede, B.; Avitabile, C.; Saviano, M.; Imbriani, A.L.; Moscone, D.; Arduini, F. Paper-based electrochemical peptide nucleic acid (PNA) biosensor for detection of miRNA-492: A pancreatic ductal adenocarcinoma biomarker. *Biosens. Bioelectron.* **2020**, *165*, 112371. [CrossRef] [PubMed]
95. Sun, X.; Wang, H.; Jian, Y.; Lan, F.; Zhang, L.; Liu, H.; Ge, S.; Yu, J. Ultrasensitive microfluidic paper-based electrochemical/visual biosensor based on spherical-like cerium dioxide catalyst for miR-21 detection. *Biosens. Bioelectron.* **2018**, *105*, 218–225. [CrossRef]
96. Adrover-Jaume, C.; Alba-Patino, A.; Clemente, A.; Santopolo, G.; Vaquer, A.; Russell, S.M.; Baron, E.; Gonzalez Del Campo, M.D.M.; Ferrer, J.M.; Berman-Riu, M.; et al. Paper biosensors for detecting elevated IL-6 levels in blood and respiratory samples from COVID-19 patients. *Sens. Actuators B* **2021**, *330*, 129333. [CrossRef]
97. Demirbakan, B.; Kemal Sezginturk, M. An impedimetric biosensor system based on disposable graphite paper electrodes: Detection of ST2 as a potential biomarker for cardiovascular disease in human serum. *Anal. Chim. Acta* **2021**, *1144*, 43–52. [CrossRef]
98. Weng, X.; Ahmed, S.R.; Neethirajan, S. A nanocomposite-based biosensor for bovine haptoglobin on a 3D paper-based analytical device. *Sens. Actuators B* **2018**, *265*, 242–248. [CrossRef]
99. Hennekinne, J.A.; De Buyser, M.L.; Dragacci, S. Staphylococcus aureus and its food poisoning toxins: Characterization and outbreak investigation. *FEMS Microbiol. Rev.* **2012**, *36*, 815–836. [CrossRef]
100. Akhtar, S.; Sarker, M.R.; Hossain, A. Microbiological food safety: A dilemma of developing societies. *Crit. Rev. Microbiol.* **2012**, *40*, 348–359. [CrossRef]
101. Farooq, U.; Ullah, M.W.; Yang, Q.; Aziz, A.; Xu, J.; Zhou, L.; Wang, S. High-density phage particles immobilization in surface-modified bacterial cellulose for ultra-sensitive and selective electrochemical detection of Staphylococcus aureus. *Biosens. Bioelectron.* **2020**, *157*, 112163. [CrossRef]
102. Bintsis, T. Foodborne pathogens. *AIMS Microbiol.* **2017**, *3*, 529–563. [CrossRef] [PubMed]
103. Liu, F.; Liu, H.; Liao, Y.; Wei, J.; Zhou, X.; Xing, D. Multiplex detection and genotyping of pathogenic bacteria on paper-based biosensor with a novel universal primer mediated asymmetric PCR. *Biosens. Bioelectron.* **2015**, *74*, 778–785. [CrossRef]
104. You, S.-M.; Jeong, K.-B.; Luo, K.; Park, J.-S.; Park, J.-W.; Kim, Y.-R. Paper-based colorimetric detection of pathogenic bacteria in food through magnetic separation and enzyme-mediated signal amplification on paper disc. *Anal. Chim. Acta* **2021**, *1151*, 338252. [CrossRef]

105. Huang, L.; Xiao, W.; Xu, T.; Chen, H.; Jin, Z.; Zhang, Z.; Song, Q.; Tang, Y. Miniaturized paper-based smartphone biosensor for differential diagnosis of wild-type pseudorabies virus infection versus vaccination immunization. *Sens. Actuators B* **2021**, *327*, 128893. [CrossRef]
106. Chandra, P. Electrochemical nanobiosensors for cancer diagnosis. *J. Anal. Bioanal. Tech.* **2015**, *6*, e119. [CrossRef]
107. Chandra, P.; Noh, H.-B.; Shim, Y.-B. Cancer cell detection based on the interaction between an anticancer drug and cell membrane components. *Chem. Commun.* **2013**, *49*, 1900–1902. [CrossRef] [PubMed]
108. Chandra, P.; Suman, P.; Mukherjee, M.; Kumar, P. HER2 protein biomarker based sensor systems for breast cancer diagnosis. *J. Mol. Biomark. Diagn.* **2013**, *5*, e119. [CrossRef]
109. Zhu, Y.; Chandra, P.; Shim, Y.-B. Ultrasensitive and selective electrochemical diagnosis of breast cancer based on hydrazine-AuNP-aptamer bioconjugate. *Anal. Chem.* **2013**, *85*, 1058–1064. [CrossRef] [PubMed]
110. Purohit, B.; Kumar, A.; Mahato, K.; Roy, S.; Chandra, P. Cancer cytosensing approaches in miniaturized settings based on advanced nanomaterials and biosensors. In *Nanotechnology in Modern Animal Biotechnology*; Elsevier: Amsterdam, The Netherlands, 2019; pp. 133–147.
111. Kumar, S.; Umar, M.; Saifi, A.; Kumar, S.; Augustine, S.; Srivastava, S.; Malhotra, B.D. Electrochemical paper based cancer biosensor using iron oxide nanoparticles decorated PEDOT:PSS. *Anal. Chim. Acta* **2019**, *1056*, 135–145. [CrossRef]
112. Liang, L.; Su, M.; Li, L.; Lan, F.; Yang, G.; Ge, S.; Yu, J.; Song, X. Aptamer-based fluorescent and visual biosensor for multiplexed monitoring of cancer cells in microfluidic paper-based analytical devices. *Sens. Actuators B* **2016**, *229*, 347–354. [CrossRef]
113. Chandra, P.; Zaidi, S.A.; Noh, H.-B.; Shim, Y.-B. Separation and simultaneous detection of anticancer drugs in a microfluidic device with an amperometric biosensor. *Biosens. Bioelectron.* **2011**, *28*, 326–332. [CrossRef] [PubMed]
114. Fan, Y.; Shi, S.; Ma, J.; Guo, Y. A paper-based electrochemical immunosensor with reduced graphene oxide/thionine/gold nanoparticles nanocomposites modification for the detection of cancer antigen 125. *Biosens. Bioelectron.* **2019**, *135*, 1–7. [CrossRef] [PubMed]
115. Mujahid, A.; Dickert, F.L. Blood group typing: From classical strategies to the application of synthetic antibodies generated by molecular imprinting. *Sensors* **2015**, *16*, 51. [CrossRef] [PubMed]
116. Feng, Q.-M.; Liu, Z.; Chen, H.-Y.; Xu, J.-J. Paper-based electrochemiluminescence biosensor for cancer cell detection. *Electrochem. Commun.* **2014**, *49*, 88–92. [CrossRef]
117. Rengaraj, S.; Cruz-Izquierdo, Á.; Scott, J.L.; Di Lorenzo, M. Impedimetric paper-based biosensor for the detection of bacterial contamination in water. *Sens. Actuators B* **2018**, *265*, 50–58. [CrossRef]
118. Mahapatra, S.; Chandra, P. Clinically practiced and commercially viable nanobio engineered analytical methods for COVID-19 diagnosis. *Biosens. Bioelectron.* **2020**, *165*, 112361. [CrossRef] [PubMed]
119. Chandra, P. Miniaturized label-free smartphone assisted electrochemical sensing approach for personalized COVID-19 diagnosis. *Sens. Int.* **2020**, *1*, 100019. [CrossRef]
120. Mahapatra, S.; Baranwal, A.; Purohit, B.; Roy, S.; Mahto, S.K.; Chandra, P. Advanced biosensing methodologies for ultrasensitive detection of human coronaviruses. In *Diagnostic Strategies for COVID-19 and Other Coronaviruses*; Springer: Singapore, 2020; pp. 19–36.
121. Baranwal, A.; Mahapatra, S.; Purohit, B.; Roy, S.; Chandra, P. Insights into novel coronavirus and COVID-19 outbreak. In *Diagnostic Strategies for COVID-19 and Other Coronaviruses*; Springer: Singapore, 2020; pp. 1–17.
122. Chandra, P.; Roy, S. *Diagnostic Strategies for COVID-19 and Other Coronaviruses*; Springer: Singapore, 2020.
123. Yakoh, A.; Pimpitak, U.; Rengpipat, S.; Hirankarn, N.; Chailapakul, O.; Chaiyo, S. Paper-based electrochemical biosensor for diagnosing COVID-19: Detection of SARS-CoV-2 antibodies and antigen. *Biosens. Bioelectron.* **2021**, *176*, 112912. [CrossRef]
124. Liu, F.; Zhang, C. A novel paper-based microfluidic enhanced chemiluminescence biosensor for facile, reliable and highly-sensitive gene detection of Listeria monocytogenes. *Sens. Actuators B* **2015**, *209*, 399–406. [CrossRef]
125. Teengam, P.; Siangproh, W.; Tuantranont, A.; Henry, C.S.; Vilaivan, T.; Chailapakul, O. Electrochemical paper-based peptide nucleic acid biosensor for detecting human papillomavirus. *Anal. Chim. Acta* **2017**, *952*, 32–40. [CrossRef]

biosensors

MDPI

Review

Disposable Paper-Based Biosensors for the Point-of-Care Detection of Hazardous Contaminations—A Review

Mohammad Mahdi Bordbar [1], Azarmidokht Sheini [2], Pegah Hashemi [3], Ali Hajian [4] and Hasan Bagheri [1,*]

1 Chemical Injuries Research Center, Systems Biology and Poisonings Institute, Baqiyatallah University of Medical Sciences, Tehran 19945, Iran; mohammadmahdibordbar@gmail.com
2 Department of Mechanical Engineering, Shohadaye Hoveizeh Campus of Technology, Shahid Chamran University of Ahvaz, Dashte Azadegan 78986, Iran; azar.sheini@gmail.com
3 Research and Development Department, Farin Behbood Tashkhis Ltd., Tehran 16471, Iran; pegi.hashemi@gmail.com
4 Institute of Sensor and Actuator Systems, TU Wien, Gusshausstrasse 27-29, 1040 Vienna, Austria; ali.hajian@tuwien.ac.at
* Correspondence: h.bagheri@bmsu.ac.ir

Abstract: The fast detection of trace amounts of hazardous contaminations can prevent serious damage to the environment. Paper-based sensors offer a new perspective on the world of analytical methods, overcoming previous limitations by fabricating a simple device with valuable benefits such as flexibility, biocompatibility, disposability, biodegradability, easy operation, large surface-to-volume ratio, and cost-effectiveness. Depending on the performance type, the device can be used to analyze the analyte in the liquid or vapor phase. For liquid samples, various structures (including a dipstick, as well as microfluidic and lateral flow) have been constructed. Paper-based 3D sensors are prepared by gluing and folding different layers of a piece of paper, being more user-friendly, due to the combination of several preparation methods, the integration of different sensor elements, and the connection between two methods of detection in a small set. Paper sensors can be used in chromatographic, electrochemical, and colorimetric processes, depending on the type of transducer. Additionally, in recent years, the applicability of these sensors has been investigated in various applications, such as food and water quality, environmental monitoring, disease diagnosis, and medical sciences. Here, we review the development (from 2010 to 2021) of paper methods in the field of the detection and determination of toxic substances.

Keywords: paper sensors; toxic substances; biological receptors; optical detection; electrochemical methods; rapid tests

Citation: Bordbar, M.M.; Sheini, A.; Hashemi, P.; Hajian, A.; Bagheri, H. Disposable Paper-Based Biosensors for the Point-of-Care Detection of Hazardous Contaminations—A Review. *Biosensors* **2021**, *11*, 316. https://doi.org/10.3390/bios11090316

Received: 5 August 2021
Accepted: 1 September 2021
Published: 4 September 2021

1. Introduction

One of the major challenges in developed countries is the uncontrolled spread of hazardous contaminations, due to the activity of industrial centers or microorganisms. They can be classified by source (plant, animal, mineral, or chemical agents), nature (metal, toxin, microorganism, or organic compound), and their uses (insecticides, food additives, or fungicides). The contaminations are classified in the chemical, environmental, agricultural, medical, and radioactive categories [1]. Among them, the environmental and agricultural contaminations were considered in this review. These contaminants are different in nature: chemical (metal and organic compounds), biological (pathogen bacteria and virus), and physical (energy) [1]. Focusing on the chemical and biological compounds, these contaminants can affect the ecosystem of an area by penetrating water, soil, and air remaining in the environment and entering the human life cycle through inhalation, skin absorption, and swallowing [2]. Depending on the toxicity degree and contamination exposure duration, the toxic substances can result in different influences on human health. These influences may be temporary, leading to headaches, nausea, lung failure (due to the inhalation of volatile gases), blood poisoning, liver and kidney failure, and even cardiovascular failure [3]. Moreover, high doses of toxins may weaken the immune system, as observed for compounds such as carbon disulfide, mercury, manganese, arsenic, lead, and cadmium [4–6]. On the other hand, some compounds (e.g., aflatoxins and organophosphates or some metals, such as lead and cadmium) may remain in the body for a long time and their excretion process may be prolonged, leading to nervous system dysfunction [7].

Since toxins with extremely low concentrations can also pose serious hazards, it is important to detect them using a sensitive method. In fact, the detection is carried out in a real sample consisting of thousands of chemical species. Therefore, the detection method needs to be highly selective, identifying the species in the presence of similar compounds. The detection analysis is performed with large and small analytical devices. The former is based on chromatographic methods (e.g., high-performance liquid chromatography and gas chromatography) or spectroscopic methods (e.g., infrared, ultraviolet (UV)-visible, mass, and nuclear magnetic resonance). These methods provide a unique response for each real sample, thereby detecting the presence of the desired analyte in the sample and quantifying its concentration. Moreover, the detection methods are able to measure extremely small amounts of toxic species. Nevertheless, their measurement process is time-consuming and expensive. Meanwhile, analyzing and interpreting the results requires a skilled operator and sufficient knowledge [8].

Alternatively, the small devices makes it possible to perform analytical experiments using the lowest volume (up to picomolar levels) of indicators and analytes [9]. In turn, this makes the analysis process rapid and cost-effective, as the small devices are not complex and do not require special laboratory conditions [10,11]. Therefore, it is possible to use them in the sampling sites and by individuals who need the analysis. Mostly, the detection methods based on these devices are called point-of-care test (POCT) [12], being widely used in diagnosing diseases, examining food control, monitoring environmental pollution, etc. [13]. The commercial types of POCTs are available in the market such as diagnosing prostate cancer, intestinal cancer, infectious diseases, pregnancy diagnosis, drinking water quality control, detection of food spoilage, food adulterations, etc. [14,15]. Profits from the production of POCTs are expected to reach 39.96 billion dollars by the end of 2021 [16]. Basically, POCTs consist of different parts: the sensor substrate, the sample inlet, the receptor, and the response transducer. Receptors can be chemical compounds (e.g., inorganic complexes, organic markers, polymers, and nanoparticles) or biological species (e.g., antibodies, antigens, aptamers, enzymes, or part of a plant or animal tissue).

In terms of selectivity, bio-POCTs outperform chemical POCTs [17]. Bioreceptors respond mainly to a specific analyte, thus increasing the sensor's ability to determine a species in the presence of additives and other contaminants [17]. The responses generated by bio-POCTs are accurate and reliable, determining extremely low concentrations of the analyte [18]. Nevertheless, compared to chemical POCTs, they need special storage conditions, in a narrow range of parameters, or they suffer from the complexity of the storage process [19]. The activity of the bioreceptors are reduced by mechanical and environmental changes, thus leading to the inefficiency of the resultant sensor [19]. Furthermore, the cost of fabricating biosensors is much higher than that of a chemical sensor [19].

However, in recent years, the development of bio-POCTs has increased in both the laboratory and commercial fields [16]. In this study, we will review different types of bio-POCTs, while also investigating their application in detecting and determining hazardous contaminations, such as mycotoxin, organophosphate, bacteria, and heavy metal ions. This category has not been reported in previous studies. Figure 1 shows an overview of the bio-POCTs classification in this study.

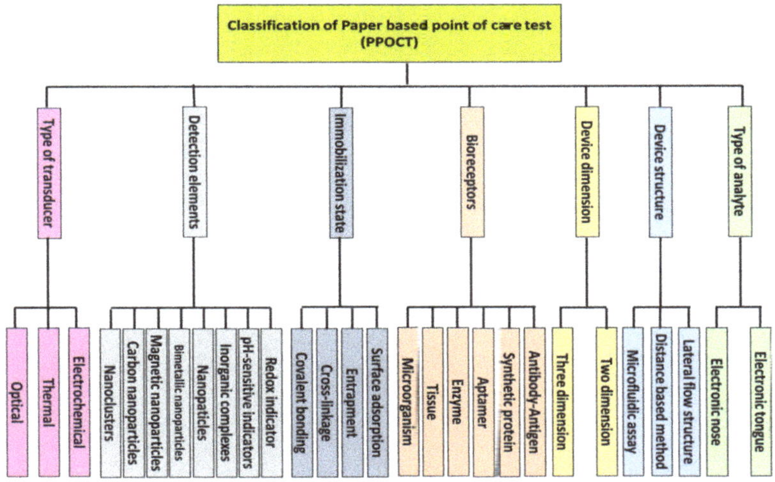

Figure 1. Schematic diagram for PPOCT classification, which is described in this review.

2. Bio-POCT

To design a bio-POCT, a sensing element (being primarily a biological compound) is initially coupled to a detection element. A transducer is connected to the detection element, converting its changes into an intelligible signal. This structure can be installed on a well-plate or substrate. It is clear that performing a well-plate test requires a time-consuming preparation process, as well as a skilled operator for the laboratory conditions and tools, for a typical test. Additionally, moving the designed device for on-site analysis is cumbersome. The components of a bio-POCT can be immobilized on a substrate, providing a portable structure, which requires low amounts of materials for performing a test.

2.1. Substrates Used in the Bio-POCT

To select the substrate, different features, including flexibility, biocompatibility, biodegradability, availability, cost-effectiveness, surface modification, permeability, and portability of the sample are considered [20]. To this end, substrates such as glass, polydimethylsiloxane (PDMS), silicone, and paper have been used [21]. Among them, the use of paper as a substrate is very popular, due to its fibrous structure, enabling us to easily modify it [22]. Moreover, paper has a capillary nature, making the liquid sample flow easily on the substrate, while also providing the possibility of the penetration of gaseous samples into its textures [23]. The paper selected as the substrate must be so flexible that it does not break or tear when fabricating sensors with a three-dimensional (3D) design [24]. This substrate should have a thickness of 10–100 μm, consuming less volume of the sample (in the microliter level) [25].

The paper substrate needs to have a soft texture that can be easily attached to solids, collecting small amounts of the sample [26]. Additionally, it needs to be a strong absorber, storing an exact volume of a sample for the subsequent displacement of a chemical [27]. The paper substrate should also be permeable to air and gas, with a network structure, in order to separate the analyte from the contaminated matrix, by filtering disturbing components [28]. By having a high surface-to-volume ratio, the paper substrate is capable of immobilizing a large number of sensing elements on its surface [29]. Moreover, it should be compatible with biological samples. In some cases, applying heat treatments may be required to immobilize the enzyme or antibody coated on the nanoparticles [30]. Thus, the paper substrate must be heat-resistant [31]. Overall, it should be inert against physical and chemical changes [32].

2.2. Types of Paper

To select paper as a substrate in the preparation of a paper-based POCT (PPOCT), the following three factors must be considered: the purpose of the determination, the specificity of the analyte, and the characteristics of the measurement method. Paper substrates should be selected in such a way as to play a positive role in the sensitivity, selectivity, and reduction of the interfering effect on the method response and the stability of the sensor. Depending on the purpose of the study, the paper substrate can vary, in terms of thickness, pore size, permeability, and capillary nature, as well as the flow rate of the sample on the substrate, smoothness, and softness [33].

In recent years, most paper-based sensors have been made of cellulose substrates [34]. The most popular cellulosic substrates are filter, chromatography, and blotting papers, which have different grades [35]. For example, Whatman grade 1 filter paper is made of 98% cellulose, with a uniform and smooth surface and a thickness of 0.18 mm. Liquids flow on these papers at moderate speeds. They also have pores with a size of 0.11 μm. As an adsorbent substrate, the retention rate for this paper is fine. Higher grades of Whatman paper have larger pore sizes, causing the increase in the sample retention rate. For example, Watman grade 4 paper has a pore size of 25 μm. The weight of Whatman filter paper changes from 85 to 100 $g{\cdot}m^{-2}$, based on the grade of the paper [36].

Cellulose papers have a specific surface area of 1.4 $m^2{\cdot}g^{-1}$. The porosity of the papers is high. It has good hydrophilic properties, mechanical strength, and is easily degradable [37]. One way to create cellulose substrates is to use bacteria. Compared to other cellulose substrates, bacterial cellulose has specific advantages, such as renewability and biocompatibility. The paper porosity increases up to 92% [37].

The properties of cellulose paper can be changed by adding some compounds, such as surfactants, polymers, aldehydes, and epoxy groups [38,39]. Notably, the hydrophilicity, porosity, retention of the sample, and the flow rate of liquid on the paper varies, depending on the type and amount of modifiers [40]. Cellulose surface modification can improve the physical and chemical properties of the paper substrate. For example, surface area and adsorption capacity can be increased to 172.49 $m^2{\cdot}g^{-1}$ and 158.98 $mg.{\cdot}g^{-1}$, respectively [37].

It is possible to produce nitrocellulose by nitrating cellulose partially [37]. The nitration process increases the porosity and hydrophobicity of cellulose, forming membranes for suitably immobilizing biological species, such as enzymes, proteins, and antibodies [37]. Nevertheless, they have a more fragile structure than cellulose substrates [37]. Here, an electrostatic interaction occurs, in which positively charged biological species are adsorbed to the negatively charged surface of nitrocellulose. The pore size of these papers varies from 9 to 35 μm. Additionally, the porosity rate is in the range of 75 to 81%, depending on the type of paper. Using nitrocellulose papers, the absorption ability improves by up to 2038 $mg.{\cdot}g^{-1}$. Of course, this claim applies to substrates most commonly used in biosensor structures [37].

The glossy paper is another substrate used to prepare paper sensors. This type of paper is made of cellulosic fibers bonded to an inorganic material, giving rise to flexible, non-degradable, and relatively smooth substrates, whose surface can be easily modified with other compounds, such as nanomaterials [41]. In turn, the hydrophobicity of the resulting paper increases, which can be used mainly for colorimetric experiments [40]. Recently, nanocellulose has been used to produce transparent papers with very high aspect ratios, including cellulose nanofibers, crystalline nanocellulose, and bacterial nanocellulose [42]. The density and fiber resistance of the transparent papers increase, making them more resistant to moisture and heat [43]. In paper devices, the substrates should be as insensitive to ambient humidity as possible. Since the humidity changes during the day (or on different days), the use of hydrophobic polymer substrates is preferred, being inert in the relative humidity range of 10–100% [44]. Notably, polyethylene terephthalate, polyvinylidene difluoride, and polypropylene substrates show high chemical resistance to gases, acids, and bases [44]. These substrates have been used extensively to detect volatile species in the vapor phase [45].

3. Classification of Bio-PPOCT, Based on the Analyte Type

The analysis of samples can be performed in gas and liquid phases. In the gas phase, the goal is to detect volatile analytes existing naturally in the environment (e.g., air pollutants) or caused by the decomposition of a material. Volatile compounds are mainly analyzed as metabolites in exhaled breath [46], sweat [47], and saliva [48] secretions, as well as blood [49] and urine vapors [50], diagnosing a disorder in the body. These compounds can also be formed from the breakdown of proteins, fats, and carbohydrates in food, so that they can be used to control the quality of food products [51–54]. It is even possible to determine the amount of impurities present in petroleum products, supplements, and medicines qualitatively and quantitatively [44]. Of late, bacteria and fungi have been identified based on their volatile compounds [55,56]. In this regard, the paper sensor is exposed to the sample vapors, having a high chemical and mechanical resistance to moisture, along with good permeability to penetrate the analyte in its paper texture, in order to interact with the indicators [44]. The pore size, thickness, and surface to volume ratio of the paper are the considered factors to have a practical substrate for storing the vapors [57]. The resulting devices can contain one or more sensors, being capable of detecting one or more gas samples. Since their performance is similar to the olfactory system, the sensor devices are called electronic noses [58].

In the liquid phase, the analyte is either a pure liquid or a component dissolved in a solution. In this respect, the purpose of analysis is the diagnosis of a disease, detection of an environmental pollutant, evaluation of a food product, and so on [34]. To analyze the samples in the liquid phase, the paper is either immersed in the liquid sample or part of the sample is injected into the surface of the paper [59]. In the latter case, the sample is transferred to the detection zones through embedded channels or moves along the paper strip (arising from the capillary nature of the paper) to react with the indicator [59]. Accordingly, the paper with high hydrophilicity and low permeability should be chosen [23]. Moreover, the flow rate of the sample should be appropriate for transporting the liquid samples from injection zones to detection ones [23], also allowing for the interaction between the analyte and the indicator [23]. These sensors mostly use Whatman grade 1 paper [60]. In this case, the analysis can be single species or multispecies, having a structure similar to the taste system, which is the so-called electronic tongue [61]. One important point that must be considered in the fabrication of these sensors is the lack of displacement of the sensing element. To this end, the surface of the detection zones is modified with polymeric compounds (e.g., chitosan and polyvinyl alcohol) or protein compounds (e.g., BSA), maintaining the indicator stationary on the surface. It is also possible to mix the indicator with hydrophobic or plasticizing compounds, in order to fix it on the paper surface without having a negative effect on the sensing ability and sensitivity of sensor [62,63].

4. Classification of Bio-PPOCT, Based on the Device Structure

The simplest configuration for a paper device is the dipstick, in which the sensor is immersed in the solution, in order to detect analytes [64]. This device is mostly used for qualitative detection and employed as paper strips sensitive to medium pH, urinary infections, metabolites, urinary proteins, and water contaminants [65]. While the design of these sensors is apparently simple, the reagents must be placed on paper and do not leak into the solution during immersion. Moreover, the species suspended in the solution should not be adsorbed on the texture of the sensor, thus obtaining the sensor response efficiently. However, the use of dipstick devices is limited, partly due to the high adhesion and viscosity of the liquid. Sometimes the goal of the study is the determination of several analytes simultaneously; therefore, several detection reagents should immobilize on the surface of paper without merging together. In some other cases, appropriate reagents may not be available (thus making it necessary to convert the analyte to another species) or it is not possible to carry out the analysis, due to the presence of foreign species. Therefore, a series of preparatory processes are required to be performed on the sample prior to the measurement. The dipstick design does not address these limitations.

4.1. Lateral Flow Structure

The lateral flow structures complies with the principles of enzyme-linked immunosorbent assays (ELISA), made of rectangular paper strips that are typically a width of 6 mm and length of 7 cm [66]. The sample moves along the paper on a series of consecutive pads [66]. As illustrated in Figure 2, the four main components of these sensors are the sample pad, conjugate pad, detection pad, and absorbent pad (embedded along the paper strip) [66]. The sample is injected into the sensor through the sample pad, storing a large part of the liquid, while also directing it to the conjugate pad [67].

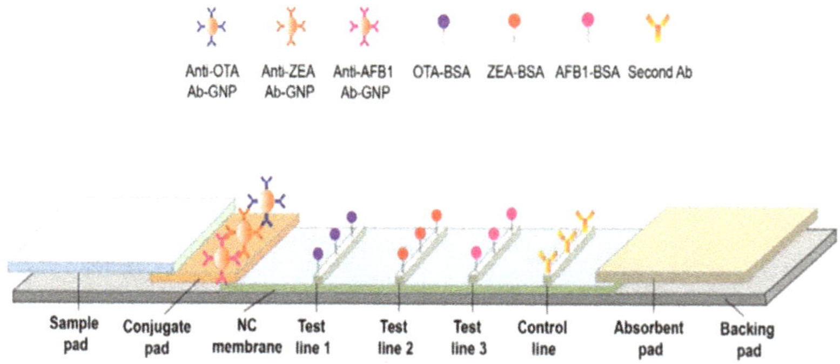

Figure 2. Different components of a lateral flow structure. This structure consists of sample pad for receiving the sample, conjugated pad for interacting analyte, and labeled detection element; detection pad containing test lines for forming complexes between capture bioreceptor and labeled analyte and absorbent pad for terminating the sample flow. The gray rectangular shows the back pad. This schematic was proposed by Chen et al., for the simultaneous determination of aflatoxin B1, zearalenone, and ochratoxin A in agriculture products (reprinted with permission from [68]; copyright (2016) Elsevier).

In order to improve the performance of the sensor, the sample pad is pre-treated using a buffer solution, at a certain pH and ionic strength, before adding the sample. The interaction between the analyte and the receptor can be influenced by the pH and ionic strength [69–71]. To prevent non-specific interactions between the sample and the paper, while also facilitating the sample transfer, the sample pad is modified by detergents, such as SDS, Tween 20, and Triton [67]. Blocking agents such as BSA or Casein can even be used to eliminate non-specific bonds [67]. In order to remove microbial contamination, the sample pad is impregnated with sodium azide [67]. Sometimes, a filter is placed on the sample pad to remove the interfering species containing analytes, including proteins and blood cells [72]. The thickness of the paper should be taken into consideration when choosing a sample pad. In other words, the thicker the paper, the slower and more stable the flow [72]. Since the pad may affect the measurement target, it needs to be free of chemical impurities. The sample pad can be made of cellulose fibers or glass fibers [67]. While the cellulose fibers are thicker and cheaper than the glass ones, they are difficult to handle. In contrast, glass fibers with good tensile strength are capable of uniformly distributing the sample on their surface, thereby acting as a filter. Nevertheless, glass fibers with lower cutting ability are more expensive than paper fibers and can be contaminated with environmental chemicals during the fabrication process [73].

On the other hand, the conjugate pad is made of nitrocellulose, immobilizing the bioreceptor on its surface [74]. This pad can formed by the other membranes such as nylon, polyvinylidene fluoride, or polyvinylidene difluoride membranes, but nitrocellulose is more attractive because of some of its advantages, such as having a high capacity for immobilizing the biological compounds and having low costs [75]. In this case, the target analyte (e.g., antigen) interacts with the bioreceptor (e.g., antibody). The solution is directed to the conjugated pad, based on the capillary nature of the paper. The bioreceptor is conjugated to color compounds with unique optical and electrical properties [74]. These compounds mainly comprise of carbon dots, as well as metal (e.g., Au), upconversion, and magnetic nanoparticles [67]. Among them, Au nanoparticles (AuNPs) produce a stable red color (being observable with the naked eye), show good physicochemical stability that can be easily functionalized, and have low toxicity [76]. Thus, AuNPs are employed as a label on the conjugate pad. Since the nanoparticles used are colloidal suspensions and their stability is affected by the ionic strength of the solution, the conjugated pad is modified by a buffering agent [67]. For this purpose, some sugars, such as sucrose and trehalose, are mostly used [67].

In the conjugate pad, the target analyte interacting with the labeled bioreceptor creates a complex, moving toward the detection pad. Two lines are created on the detection pad: one is the test line and the other is the control line [77]. In the former, the capture bioreceptor interacts with the labeled analyte, indicating the existence of the analyte in the environment. In the latter, the correct performance of the designed system is evidenced [77]. The analyte detection is carried out on the detection pad, involving the two following principles: competitive or inhibitory methods and sandwich methods [67]. In the competitive method, the target in the sample competes with the one labeled in the conjugated pad to interact with the capture bioreceptor, mostly used for analytes with a small size and high concentration [78]. In contrast, the sandwich method is very popular for detecting medium- and large-sized analytes, including proteins, antibodies, cells, and bacteria [79]. In the sandwich method, the analyte is sandwiched between the detection (primary) and capture (secondary) bioreceptors in the test line. These bioreceptors can be both monoclonal. In some cases, the detection bioreceptor is monoclonal and the capture bioreceptor is polyclonal [67]. A high concentration of the capture bioreceptor in the test line is recommended for the sandwich method [67]. Finally, the absorbent pad is the last part of a lateral flow system with a sufficient bed volume, thereby terminating the sample flow [80]. The lateral flow components are pasted to a polymer substrate, via the help of a pressure-sensitive adhesive. This substrate, known as the backing pad, is mainly made of polystyrene or plastic materials. The strength and flexibility of the created strips depends on the material of this pad [75].

The flow velocity and pore size of the membrane affects the assay sensitivity in the lateral flow system. The high sensitivity is achieved via the strong interaction between the labeled analyte and test line antibody. For this purpose, the membranes should have a small pore size with a slow sample flow rate [81]. This prolongs the experiment time, which is between 10 and 30 min for a simple test. In addition, the lateral flow system of sandwich format suffered from the hook effect. The hook effect is a phenomenon in which free analytes in the media compete with the labeled analyte for binding to the test line antibody. This has a negative effect on the color intensities and, consequently, the sensor responses [81]. Flow-through immunoassay (immunofiltration assay) can be used to reduce these limitations. In the alternative assay, a larger sample volume is consumed, thus improving the kinetics and sensitivity. Hooke phenomenon is not observed in these methods. An immunofiltration assay can be performed by passive and alternative approaches. In the former, the lateral flow pads are layered by stacking method so that the detection pad is located on the top of the conjugate and absorbent pad. In the later, the membrane is embedded into the syringe filter holder after modification with a bioreceptor. Reagents and samples flow vertically over the membrane through the syringe [81].

4.2. Distance-Based Method

The method of stain length measurement uses another 2D strip structure, in which a strip of Whatman grade 1 paper (with dimensions of 0.5 cm × 3–5 cm) is used and impregnated with detection reagents [82]. The sample is added to the sensor from the injection site, moving along the sensor to react with the reagent, while also changing its color [49]. The distance moved is measured by a ruler and depends on the concentration of the analyte [83]. In this way, unlike the lateral flow methods, the entire analysis process is performed on a single pad with a simpler design, detecting small molecules. In addition to qualitative diagnosis, the stain length measurement can also be used for quantitative analysis [83].

4.3. Microfluidic Assay

Another widely used 2D configuration is microfluidic structures, in which an extremely small volume of the sample (in the range of 10^{-6}–10^{-18} L) is consumed [84]. In these structures, the sample flows in channels with a width of 10 μm [84]. Certainly, the behavior of liquids at the micro-scale is different from that at the macro-scale and can be influenced by several factors, such as the surface tension and fluid resistance [84]. In this respect, the effect of surface forces is found to be greater than that of volumetric forces [85]. In microfluidic structures, all the components needed for the liquid entrance, pumps, valves, and mixers, along with the detection devices (i.e., transducers and indicators), are mounted on a very small substrate. Depending on the driving force of the liquid transfer, the substrate used, and the system configuration, microfluidic devices can be categorized as open microfluidic [86], continuous flow microfluidic [87], drop microfluidic [88], digital microfluidic [89], paper microfluidic [90], and microfluidic particle detection systems [91]. Among these devices, the microfluidic paper system, designed by Whitesides, enables the flow of the liquid on the surface of a porous substrate, based on the capillary nature [90]. With this system, no external driving force is required to transfer the liquid. In these sensors, the channels, the injection, and the detection zones can be created between hydrophobic barriers [90]. This matter can be one of the limitations of the paper-based methods because the width of the channels may be blocked by the hydrophobic barriers, so that only 50% of the actual sample volume may reach the detection zone. Despite all the limitations, paper microfluidic sensors are one of the most popular methods for fabricating the point-of-care instruments.

5. Classification of Bio-PPOCT Based on the Device Dimension

In fact, the configuration design is based on the direction of the sample flow on the paper [23]. This flow can be along the direction of the paper (i.e., the horizontal direction, forming a 2D configuration) [92] or along the depth of the paper (i.e., the vertical direction, creating 3D structures) [93,94]. To design each sensor structure, the desired pattern is drawn using design software, such as Photoshop, Illustrator, CorelDraw, InDesign, AutoCAD, etc. Accordingly, it is possible to shape and resize channels and detection zones with the above software.

5.1. Two-Dimensional Configuration

To have a 2D configuration sensor, the designed patterns should be executed on paper using physical or chemical methods [95].

5.1.1. Physical Methods

Paper sensors are physically fabricated by the following four methods: knife plotter, craft cutting, embossing, and laser cutting [96].

In the knife plotter method, the cutting process was carried out by an automatic cutter and controlled by a computer program [97]. To prevent the paper from tearing, the cutting process was continuous, thus being performed in several consecutive steps [97]. The above-mentioned method is the simplest and least expensive one for preparing paper sensors. Additionally, the fabrication process was not time consuming, which enabled us to adjust the computer program to cut paper with different sizes and thicknesses [98]. This method was used only for the fabrication of hydrophilic areas. So, the complementary treatment using hydrophobic materials was necessary. Additionally, a large portion of the paper used was wasted [98].

In the craft cutting method, the paper is initially glued to a thin sheet, and the paper strips are then cut into different dimensions and sizes using a craft cutter [99]. The advantage of this method is the production of flexible, portable, and disposable substrates [100,101]. Sometimes the strips formed are modified by fluoroalkyl trichlorosilan, thereby increasing the hydrophobicity of the paper [96]. In turn, this reduces the capillary nature of the paper, while also requiring an external pump to transfer the liquid [100]. Additionally, it is not possible to create variable channels, based on the craft cutting method [96].

In the embossing method, the designed pattern is engraved on paper [102]. The paper is also moistened with ethanol and placed between two molds made of plastic. The surface of the paper can be modified by silane and sealed between two adhesive tapes, forming open channels with porous walls. This process creates flexible and foldable substrates, although they are prone to the absorption of contaminants, such as ambient gases, due to their porous structure [102].

Laser cutting is the last physical method in the fabrication of paper sensors, employing CO_2 laser cutting to create the selected designs on paper [103]. This method is only able to cut a part of the paper thickness. The advantage of the laser cutting is the reproducible production of paper substrates in the shortest possible time, by using inexpensive tools. Similar to other physical methods, it is also prone to the absorption of environmental pollutants, due to the porous structure of the paper [104].

5.1.2. Chemical Methods

In chemical methods, hydrophobic barriers are created by blocking holes in the paper. The most widely used chemical methods can be categorized as follows: photolithography, wax printing, inkjet printing, laser printing, flexographic printing, stamping, chemical vapor deposition, screen printing, and spraying [105].

The general chemical method is photolithography, in which the paper is initially impregnated with a photoresist (e.g., SU-8) and then exposed to UV light to selectively engrave the pattern on it [106]. The photoresist used in the engraved pattern is removed using organic solvents, such as propylene glycol monomethyl ether acetate and propanol [96]. The rest of the photoresist is removed with the help of oxygen plasma [90]. In this way, channels (with a width in the range of 80–200 μm) and hydrophilic zones are created between hydrophobic barriers, facilitating the flow of the liquid in them. One can also use TiO_2 nanoparticles and light-sensitive polymers, mixed with silane instead of SU-8 [107,108]. Despite all of its advantages, the photolithography method requires expensive equipment and reagents, while also having a complex testing process. Additionally, the photoresists used have low mechanical resistance and can be cracked and broken [109]. In order to overcome these limitations, a simple UV lamp and a heating plate can be used, along with flexible UV-sensitive materials [33].

Another alternative approach is to use wax instead of photoresist compounds. Wax can be immobilized on paper using a pen, printer, or a metal mold [110,111]. Note that the placement of the wax is based on the designed pattern. By heating the paper, the wax penetrates the texture of the paper, closing the holes and creating a hydrophobic barrier [110]. On the other hand, one can use metal molds, in which the desired pattern is engraved. The paper is placed between two metal molds or between a metal mold and thin glass, followed by immersing it in molten paraffin for a few seconds [112,113]. Accordingly, the metal mold pattern is engraved on the paper, leading to the formation of a hydrophilic substrate. The other parts of the paper are impregnated with paraffin, making the hydrophobic barriers [114]. Although the above-mentioned methods introduce simple and inexpensive processes to fabricate paper sensors, it is not possible to mass-produce them efficiently. One approach to overcoming this limitation is to employ an inkjet printer, engraving the pattern on paper by dropping ink droplets [115]. In this regard, the type of ink, cartridges, and nozzles inside the printer are selected according to the usage type of the device [115]. Polystyrene and alkyl ketene dimers have been used as reagents to create hydrophobic barriers. These reagents are mixed with solvents, such as toluene and heptane [22]. The process of creating a pattern on paper is completed by heating it at high temperatures [116]. One drawback of this process is the use of toxic and environmentally hazardous organic compounds, which can also damage the structure of the printer [22]. In this respect, alternative compounds including acrylate, surfactant, and gel have been used [117]. However, most reagents must be dissolved in volatile solvents, likely clogging the nozzle or causing a non-uniform distribution of the reagent on paper over time.

In addition to the inkjet printing, flexographic printing is also used to mass-produce printing sensors. Unlike the inkjet printing, multiple layers of printing are required in the flexographic printing to create hydrophobic barriers on paper, printing the design continuously on successive rolls [118]. In this case, polystyrene is used as a reagent, which is dissolved in a toluene solvent or xylene group [119]. As well, there is no need to heat the paper to make it hydrophobic [119]. However, flexographic printers are not cost-effective, and their cleaning and preparation processes are complex. The uniformity of the paper surface also affects the print quality [96].

In recent years, commercial laser printers have been used to print designed patterns [120]. The printed paper is placed on a hot plate, at a temperature range of 150–200 °C, in order to perform the hydrophobic process [121]. While the printing limitations are reduced by the laser printers, an additional heating step is required to create hydrophobic barriers [121]. The inks used in these printers are also water soluble. If the hydrophobicity process and penetration into the texture of the paper are not carried out properly, the inks can be dissolved in the injected solution, destroying the pattern created on the surface of the paper. Therefore, the advantages of using laser printers are limited by choosing the appropriate ink [121].

Screen printing is one of the printing-based methods most used to produce electrodes in electrochemical systems [122]. The printing process is carried out with the help of a stencil [122]. The hydrophobic barriers are made by wax, UV-sensitive polymer compounds, polystyrene, TiO_2 nanoparticles, and conductive paints using different stencils [122]. The working and reference electrodes are positioned in hydrophilic areas. While the fabrication process of the resulting sensors is simple, it is not possible to produce hydrophobic barriers uniformly [123]. Additionally, different patterns need to be designed to perform different electrochemical processes. To create different patterns, different economically unviable stencils must be designed [124]. Screen printing is mainly used to fabricate ion-selective and glucose-sensitive electrodes [124].

Additionally, 3D printers are used to create microfluidic patterns [125]. A layer of ink is scanned on the paper using a computer program. The gaps created are covered with the help of PDMS. UV-sensitive polymers and waxes are employed as inks, allowing for the mass production of paper sensors. Anhydrous alcohol is used to remove the non-hydrophobic substrate [126]. The hydrophilicity of channels and other detection zones increases by modifying the paper surface with compounds such as cellulose [126]. The paper also needs to be heated at a certain temperature [126]. Since the price of a 3D printer is high, it may not be suitable for users.

One of the limitations of the printing methods was their need for space-consuming, expensive printing tools and cartridges, hindering the rapid production of paper sensors ubiquitously. This problem was solved by introducing the stamping method, in which a stamp made of PDMS was immersed in a stable, indelible ink and stamped on paper within less than a few seconds [127]. Unlike previous methods, no modifications were made to remove hydrophobic agents in the hydrophilic channels [127]. In addition to PDMS, stamps can be made of metal and foam materials [128]. Although the resulting stamps are capable of storing inks and easily fabricated under laboratory experimental conditions, their repeated fabrication is not desirable. This leads to the fabrication of different paper-based devices that affect the performance of the sensor.

Spraying is another method that uses no printing process [129]. To carry out the spraying method, the paper is initially covered with a stencil, and the hydrophobic material is then sprayed on its surface [130]. It should be noted that Whatman grade 1 paper is not suitable for this purpose, as it prevents the penetration of the hydrophobic material into the paper texture, thus causing it to leak [130]. Increasing the paper grade from 1 to 4 enhances the possibility of the material penetration into the substrates, thereby creating more hydrophobic barriers [130]. Similar to the stamping method, it is not possible to create microfluidic patterns with high reproducibility, due to the non-uniform spraying.

In the chemical vapor deposition method, a monomeric compound is first vaporized under vacuum conditions to produce radical particles. By polymerizing the particles on the paper surface, together with the help of a mask, hydrophobic barriers are created [131]. The type of the polymers can be poly (chloro-p-xylene), poly (perfluorodecyl acrylate), poly (fluorocarbon), poly (octafluorocyclo butane), and poly (hydroxybutyrate) [132]. There is no need to wash the polymer compound excessively with solvent, unless factors such as metal salts prevent the performance of polymerization [96,133]. One approach to dealing with this issue is to immerse the polymer-impregnated paper in an ethanol bath, thus allowing for the accumulation of the polymer on its surface. The paper is then covered with a mask and exposed to UV light to create channels and hydrophilic zones [134]. However, these methods require a special laboratory equipped with expensive devices [131].

Sensors with 2D structure have been used for different applications because they can be designed in a variety of formats. Some experimental processes require several pre-preparation steps such as separation, preconcentration, filtration, and mixing. The test may also involve the production of gas that must be stored and measured (impossible to perform in a planner condition). A portion of detection reagent can be washed by the sample flow and reduce a part of the sensor response. To solve this problem, the sensor structure needs to be changed from 2D to 3D.

5.2. Three-Dimensional Configuration

In 3D structures, the sample is perpendicular to the surface of the paper, passing through various overlapping layers [135]. Each layer is responsible for performing a chemical reaction and transferring the corresponding product to the next layer. The detection element is embedded in the last layer, in order to indicate changes that occurred in the analyte amount [136]. To fabricate a 3D paper device, different methods, including stacking, origami, and double-sided 3D printing, have been employed [23].

In the stacking method (Figure 3a), the patterns plotted are implemented on paper using one of the above-mentioned methods of fabricating 2D structures. Different layers of paper are then glued to each other via a double-sided adhesive [93]. The most important limitation of the stacking method is the adjustment of the layers, so that the upper layer patterns match those in the lower layers. The misalignment effects can cause the sample, solvent, and reagent to be wasted, partially resulting in a negative error in the sensor response. Moreover, the adhesives used may even cover some of the hydrophobic areas, preventing the liquid moving [137].

In origami structures (Figure 3b), the flexibility of paper is used to fold different layers created on top of each other. In this respect, hydrophilic zones and hydrophobic barriers are formed using wax printing, ink printing, and screen printing on the paper, separating each layer by a line. The layers are folded over these lines, solving the problems of layer misalignment and hydrophobic area blockage [138].

Finally, double-sided 3D printing (Figure 3c) is a new way of creating 3D structures that has been introduced recently [139]. In this method, the filter paper is exposed to the printer after being modified by resin and PDMS or a light-sensitive polymer, thereby creating hydrophilic areas. This process is repeated for both sides of the paper. The engraved paper is then immersed in an organic solvent (e.g., ethanol), in order to eliminate excess polymeric material from the paper surface, providing a 3D pattern [139]. Accordingly, it is possible to create hydrophilic channels or zones with different widths, lengths, and depths using the double-sided 3D printing method [139].

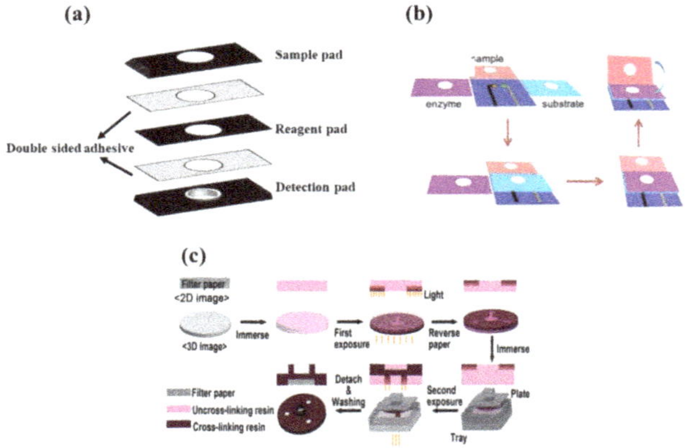

Figure 3. Fabricating methods for three-dimensional configuration. (**a**) Stacking method, (**b**) origami method (reprinted with permission from [140], copyright (2016) John Wiley and Sons), and (**c**) double-sided 3D printing method (reprinted with permission from [139], copyright (2018) The Royal Society of Chemistry).

6. Classification of Bio-PPOCT, Based on Bioreceptors

The biosensors use bioreceptors that have the ability to interact specifically with the target analyte, affecting the intrinsic properties of the detection elements [141]. Thus, the ability of a sensor depends on the selectivity of the bioreceptors. A suitable bioreceptor can detect a specific species in a complex matrix, in the presence of interfering species, without needing isolation and purification processes. Nevertheless, the bioreceptor selection depends on the type of analyte, the purpose of the analysis, and the type of transducer used [142–144].

6.1. Antibody-Based Bioreceptors

The development of immunoassays are based on the interaction of an antibody with a specific compound, such as an antigen [145]. The mechanism of the interaction is based on the lock–key principle, meaning that a particular antigen has an affinity for a corresponding antibody [145]. Hence, these bioreceptors are called affinity bioreceptors and are used to fabricate rapid and serological tests for diagnosing diseases, while also qualitatively detecting contaminants in food and the environment [146]. The interaction between the bioreceptors and the analyte does not lead to an intelligible signal. Therefore, the bioreceptors need to be labeled with a detection element, including nanoparticles, as well as fluorescent, electroactive, and radioactive compounds with the capability of generating the signal [145].

6.2. Synthetic Protein-Based Bioreceptors

The affinity constant of the antibody-antigen complex is 10^8 L per mole, resulting from an irreversible interaction [147]. However, some limitations, such as high molecular weight, low stability of the antibodies at ambient temperature, and the need for special storage conditions, make them difficult to employ in the biosensors. Additionally, their production cost is not affordable [148]. An alternative is to use small synthetic protein structures that can interact with antigens. The affinity constant of the resulting complex is in the range of about 10^2–10^4 L per mole, giving rise to a reversible interaction. In other words, one can recover the receptor and reuse it. Unlike antibodies, the protein structures are low in molecular weight, stable in environmental conditions, and inexpensive to produce. Nevertheless, they are mainly used for in vitro techniques [149].

6.3. Aptamer-Based Bioreceptors

Another class of bioreceptors is aptamers, consisting of a double-helix structure of nucleotide bases, including adenine, thymine, cytosine, and guanine, that can interact with their complementary bases. Depending on the type of the analyte base, a complementary bioreceptor is selected to induce the hydrogen interaction between adenine/thymine and cytosine/guanine [150]. These structures are called hybrid bioreceptors [151], accurately detecting sensitive and selective microorganisms, such as viruses and bacteria, in the field of spectral detection and food spoilage [152].

6.4. Enzyme-Based Bioreceptors

Enzymes are bioreceptors that act as catalysts in a specified reaction; consequently, they are not consumed in the analysis process and can be used for a long time [153]. This feature, along with other advantages, such as the ability of the bioreceptors to pair with different optical and electrochemical transducers, high-detection efficiency of a single or group of analytes, and good compatibility (as a participant in both activation and inhibitory reactions), has made enzymes the most popular receptor in the fabrication of biosensors [154]. The following events occur in enzyme-based reactions: (i) the analyte activates the activity of the enzyme or, conversely, inhibits its performance, and (ii) the analyte is converted to a compound during enzymatic reaction that can be detected by the transducer [155]. Among enzymes, oxidative structures are more commonly used because they are stable compounds that do not require coenzymes. Notwithstanding, a serious problem for these bioreceptors is their reduced activity over time, considerably influencing the sensitivity of the sensor [156].

6.5. Microorganisms-Based Bioreceptors

Along with microorganisms, cell organs, and tissues, enzymes fall into the category of catalytic bioreceptors [151]. Organs such as lysosomes, chloroplasts, and mitochondria have been used as receptors [157]. Each of these organs has a different metabolic activity that can be altered by the interaction with the analyte [157]. Microorganisms, such as bacteria, fungi, algae, and yeasts, are receptors that can be easily immobilized on the surface and are resistant to environmental changes. Thus, they can be stable for a long time [158]. Since the microorganisms are so sensitive to changes in the environment, they can be used to detect a wide range of toxic analytes (e.g., herbicides), food spoilage, biological oxygen demand, and even the effectiveness of drugs in treating diseases [159]. The sensor response is obtained based on the amount of analyte absorbed on the receptor or the determination of changes in the receptor's respiratory performance in a metabolic process [159]. Although the microorganisms are less selective than enzymes, they can be easily recovered by immersion in a nutrient solution.

6.6. Tissue-Based Bioreceptors

Tissues are also used to fabricate biosensors. Unlike other receptors, tissues with high environmental stability are more cost-effective and available, making it easier to stabilize them on a solid substrate [160]. Moreover, the tissues do not need to extract or purify enzymes, due to their enzyme-rich environment [161]. However, the tissues do not have good selectivity because they can respond to a wide range of analytes, while also requiring more time to receive the sensor response than other sensors [161].

7. Classification of Bio-PPOCT, Based on Immobilization State of Bioreceptors

Since the bioreceptors located on the paper surface are in the path of the sample, part of these materials may dissolve in the sample solution if they are not firmly immobilized on the substrate. Accordingly, the analyte detection process may be disrupted, causing an error in the sensor response. In other words, the sensor practically loses its efficiency. Depending on the type of receptor and analyte, the interaction between them, the detection method, and the sample matrix, the immobilization process of the receptor is carried out by physical (e.g., surface adsorption, entrapment, and cross-linkage) and chemical (e.g., covalent bonding) methods [162–164].

The surface adsorption is the simplest method of immobilization, in which the paper substrate is first coated with protein, polymer, and paste layers, followed by placing the receptor on them via van der Waals adsorption forces [164]. BSA, chitosan, polyvinyl alcohol, and polyvinyl pyrrolidone are the most common adsorbents used for this purpose [163]. By the surface adsorption method, the chemical modification of reagents is not required, and recovering the surface is obtained by washing with a suitable solvent [165]. Unfortunately, the resulting sensors do not show good stability against environmental changes such as pH, temperature, and ionic strength of the sample matrix [30].

In the entrapment method, the receptor is trapped inside a microscopic hollow space made of protein, gel, ink, or synthetic polymer, preventing the components from entering or leaving it [166]. In the cross-linkage method, the receptors are interconnected using a double-functionalized reagent, thus forming a carrier-free macroparticle [165]. The resultant macroparticle can have a crystalline or aggregated structure. A high purity receptor is required to form crystalline macroparticles, whereas aggregated species are formed by adding salts, organic solvents, or nonionic polymers to the solution, resulting in the precipitation of bioreceptors [167]. The precipitated species are covalently bonded together. The cross-linkage should not result in blocking the active sites in the receptor structure, avoiding a decrease in the receptor activity and, consequently, in the efficiency of the sensor [167].

Chemical adsorption is performed by establishing a covalent bond between the functional groups present in the receptor structure (i.e., those that do not interact with the analyte) and the active sites at the substrate surface [168]. The surface of the paper is modified by organic polar solvents, such as ethanol [169]. The hydroxyl groups in the substrate bind to the amino, carboxylic, thiol, or hydroxyl groups present in the receptor structure [169]. The chemical adsorption method has better stability than physical methods, leading to the highest immobilization and lowest receptor leakage.

8. Classification of Bio-PPOCT, Based on Detection Elements

The sensor must be able to provide an intelligible signal after the interaction of analyte and bioreceptor. In this regard, detection elements with unique optical, electrical, thermal, and acoustic properties need to be immobilized separately on a hydrophilic zone of the paper reacting with the analyte-bioreceptor reaction product. Sometimes, the bioreceptor is physically or chemically coated by the detection element; thus, changes in the reaction caused variations in the structural, spatial, and environmental characteristics of the receptor. Consequently, the physiochemical properties of the detection element change as well. Detection elements fall into different groups, such as organic dyes, inorganic dyes, and nanoparticles [59]. Organic dyes include redox indicators, pH-sensitive detectors, and chemo-responsive dyes [44].

8.1. Redox Indicator

The redox indicators cause significant electrical or color changes, due to the conversion of the reduced form to the oxidized one [170]. These indicators must be able to generate a rapid and reversible response, establishing a rapid chemical equilibrium between the reduced and oxidized forms [171]. Additionally, electro-optical properties of the redox indicators should not be affected by those of other sensor components, such as the receptors, amplifiers, inhibitors, etc. [59]. Given the above-mentioned conditions, only a limited number of organic compounds (e.g., potassium iodide (KI) [172], 2,2′-azino-bis (3-ethylbenzothiazoline-6-sulphonic acid) (ABTS) [173], 3,3′-5,5′tetramethylbenzidine (TMB) [174], 3,3′diaminobenzidine (DAB) [175], o-phenylenediamine (OPD) [176], thionine [177], methylene blue [178], and indigo carmine [179]) can be used for this purpose.

8.2. The pH-Sensitive Indicators

In some cases, the interaction between analyte and bioreceptors leads to a change in the acidic and alkaline conditions of the reaction medium, which can be monitored using pH-sensitive indicators [180]. They are weak organic acids and bases whose color changes in the test solution depend on the proton (H_3O^+) concentration [181]. The most popular paper sensor (containing pH indicators) is litmus paper, which is made of 7-hydroxyphenoxazone [182]. Sometimes the reaction leads to a change in the physical or chemical nature of the reaction medium, including temperature, solvent polarity, and electrophilic or nucleophilic interactions. Thereby, chemo-responsive dyes can be used depending on the type of the reaction [183]. These dyes are in the categories of Lewis acids and bases, as well as solvatochromics or thermochromics [184].

8.3. Inorganic Complexes

Inorganic complexes that are a combination of a transition metal and a ligand can also be used as a detection element [185]. In this case, the product of the analyte-receptor reaction with the complex gives an alternative or a combinatory reaction. In the substitution reaction, the product replaces the ligand of the complex composition, which is known as the indicator displacement method (IDM) [185]. To have this mechanism, the metal-ligand complex formation constant should be less than that of the metal product [186]. In a combinatory reaction, the product interacts with the complex to form a ternary structure [135]. The ternary structure is induced by d–d interactions between the product and the metal, or by electrostatic, hydrogen, covalent, and charge transfer interactions between the product and the ligand [135]. Both alternative and combinatory reactions change the electro-optical properties of the initial complex. It is worth noting that the selectivity and sensitivity of the sensor in the presence of inorganic complexes are higher than organic indicators [185].

8.4. Nanopaticles

The other category is nanoparticles, having different physical, optical, electrical, catalytic, and biological properties than their bulk counterparts [187]. This arises from the small size of nanoparticles (ranged between 10 and 100 nm) and the dominance of surface forces over volumetric ones [187]. The properties of nanoparticles depend on the type of central metal, reducing and coating agents, size, shape, surface electrical charge, and their distribution state [188]. So far, a wide range of metallic and non-metallic nanoparticles have been used in analytical studies, among which gold, silver, copper, palladium, platinum, carbon, cobalt, and metal oxide nanoparticles have been employed in the fabrication of biosensors [189]. Gold, silver, and copper nanoparticles have shown better optical properties than other nanoparticles, arising from their surface plasmon resonance effect [8]. In fact, a high-intensity absorption band has been observed in the UV-visible region for the nanoparticles, due to the surface plasmon resonance, giving rise to a higher (up to 10^4 times) molar absorption coefficient than organic indicators [8]. Meanwhile, gold, platinum, and palladium nanoparticles have been employed in the fabrication of electrochemical biosensors, due to their high electrochemical properties and conductivity [190].

Various chemical compounds are used as reducing, coating, and preserving agents in the synthesis of nanoparticles. Some of these compounds are toxic and carcinogenic, limiting their biological and medical applications. Nowadays, the use of green methods is strongly preferred, giving rise to the production of environmentally friendly compounds [191]. In this regard, actinomycetes, bacteria, fungi, yeasts, tissues, and plant extracts have been employed as green reducing agents [191]. The synthesis of chemical compounds using plant extracts is more cost-effective and faster than other reducing agents, avoiding the involvement of isolation, purification, preparation, and maintenance of the culture medium [191]. As a result, large volumes of nanoparticles can be synthesized based on the green methods [191].

8.5. Bimetallic Nanoparticles

Nanoparticles can be combined to form a bimetallic nanoparticle, possessing new physicochemical properties, in addition to those found in single metal nanoparticles [192]. Therefore, the figures of merit of the resulting sensor (such as sensitivity, selectivity, and linear amplitude) are enhanced [192]. Most bimetallic nanoparticles fall into the following two categories: core-shell and alloy structures [193]. In the former, a metal is initially reduced to form a core, followed by the precipitation of the second metal on the primary core. This forms a thin layer called the shell [193]. In the latter, both metals are reduced simultaneously using a reducing agent, inducing an intertwined structure. In this way, the properties of alloy nanoparticles change, by varying percentages of metal cores [193].

8.6. Magnetic Nanoparticles

Magnetic nanoparticles use a magnetic substance as the core and a chemical compound as a reducing or coating agent [194]. Generally, the metal cores are composed of iron, nickel, and cobalt [195]. The magnetism of the particles is activated in the presence of an external field and it is lost by removing the magnet [194]. Sensors fabricated based on the magnetic nanoparticles are used in a variety of biological applications, as well as in the detection of environmental pollutants [196]. Depending on whether the biological study is carried out inside or outside the body, the use of magnetic nanoparticles can be variable [196]. Notably, the nanoparticles are used to treat (e.g., tracking a drug) or diagnose a disease in the in vivo conditions, whereas they isolate a specific species from the sample matrix or catalyze a chemical reaction in the in vitro conditions [196].

8.7. Carbon Nanoparticles

Alternatively, the use of carbon nanoparticles in the fabrication of biosensors, drug tracking, cancer diagnosis and treatment, and imaging the inside of the body have been reported in numerous articles [197]. The carbon-nanoparticle-based biosensors are often coupled to electrochemical transducers [197]. The nanoscale carbon is synthesized in the form of nanotubes, graphene oxide, graphene quantum dots, and fullerene [198]. One of the carbon structures is graphene, consisting of a network of sp^2 hybrid carbon, in the form of a flat plate [199]. The length of the carbon–carbon bond in the graphene structure is 1.42 Å [199]. Furthermore, the high surface-to-volume ratio of graphene structures, along with the ability to modify their surface with different functional groups, makes it possible to detect a wide range of compounds in extremely low concentrations with high selectivity [199]. Graphene monolayers can be placed on top of each other with the help of van der Waals forces, forming a 3D graphite structure. The gap between the layers is 3.42 Å [200]. By rotating the graphene layer around its axis, hollow cylinders (so-called carbon nanotubes) are created [201]. If a cylinder is made of a high-grade, single-layer graphite sheet, it is called a single-walled nanotube. On the other hand, a multi-walled nanotube is made of several sheets of graphite rolled together [202]. In this case, the distance between the graphite sheets, relative to each other, is 3.4 Å [202]. The multi-walled nanotubes outperform the single-walled ones, in terms of electrical conductivity and mechanical stability. Additionally, the multi-walled nanotubes can have various electrical and structural behaviors for different applications [202]. Fullerene is another form of the carbon structure created by the spherical rotation of graphene layers [197]. Depending on the number of layers, graphene can be divided into thin and thick categories, possessing different electrical, optical, and mechanical properties. To form graphene quantum dots, the layers with lateral dimensions of less than 100 nm are placed side by side, leading to good physical and chemical stability, low toxicity, and high photoluminescence emission of graphene, which can be used in a wide range of wavelengths from infrared to UV [203]. As a detection element, carbon structures have good conductivity and electron transfer rate, allowing for the suitable immobilization of bioreceptors on their surface. In turn, this increases the sensor performance [197].

8.8. Nanoclusters

Nanoclusters are particles with sizes less than 5 nm, having different physicochemical properties, compared to nanoparticles [204]. Moreover, nanoclusters do not show plasmonic properties, behaving like a molecule (in the range between atoms and nanoparticles), in which electron structures are discrete, due to the proximity of nanoclusters to the Fermi wavelength of metals. Thus, the enhanced magnetic properties, conductivity, and luminescence of nanoclusters are achieved [205]. The emission intensity of the nanoclusters can also be changed by controlling their size or by selecting an appropriate coating agent, giving rise to good optical stability and biocompatibility [205]. The advantage of nanoclusters, over organic fluorophores, is that they do not suffer from photobleaching. Additionally, the toxicity and physical size of nanoclusters are less than those of quantum dots, allowing for their use in in vivo studies [204].

9. Classification of Bio-PPOCT, Based on the Detection Method

The interaction between the analyte and bioreceptor in the sensor structure leads to changes in the optical, electrical, and thermal properties of the detection element, which can be converted into an intelligible signal by a transducer.

9.1. Electrochemical Transducer

If the analyte detection is associated with the production or consumption of electrons, an electrochemical transducer will be used, involving working, reference, and auxiliary electrodes [206]. The analyte is detected on the surface of the working electrode, enabling the rapid detection of low-risk species. It is possible to increase the surface and conductivity of the electrode by modifying it, thus improving and amplifying the signal [206]. In some cases, the reaction of an electroactive species is carried out at a certain potential, while also measuring the current produced (i.e., the amperometric method). In fact, the current changes are proportional to the concentration of the species in the sample [206]. Amperometric sensors are classified into three categories: (i) sensors that monitor the amount of oxygen consumed in a reaction using a Clark oxygen electrode; (ii) sensors that use a redox intermediate to transfer electrons between the bioreceptor and the electrode; and (iii) sensors that directly transfer electrons (acting as a catalyst) between the bioreceptor and the electrode [207]. An example of a sensor of the first category is one to determine glucose using an enzyme system (glucose oxidase) [207]. In order to determine the current, cyclic voltammetry, normal pulse voltammetry, and differential pulse voltammetry, methods have been used. In potentiometric sensors, performing a chemical reaction leads to a change in the potential, involving the electrode surface with high sensitivity to the desired species or environmental conditions, such as pH [208]. The changes in the electrode potential are measured against the potential of a reference electrode, pertaining to the logarithm of the concentration [208]. It is possible to place an amplifier in the sensor structure to amplify the resulting signal [208]. The sensitivity, accuracy, and speed of measurement of the potentiometric methods are less than amperometric techniques [206]. The measurements can be carried out by observing the resistance vs. current behavior in the circuit, which changes over time [209]. To this end, a bridge circuit with three electrical resistors (having two known resistors and one variable resistor) is used to calculate the unknown resistance. The variable resistance value is adjusted so that the potential difference in the circuit becomes zero, eliminating nonspecific changes in the unknown resistance, due to variations in the temperature, concentration of dissolved gases, and evaporation. Impedance-based biosensors are mostly used to detect pathogens, such as bacteria, in food or biological samples [209]. Chemical reactions can give rise to the production of a charged species, thus changing the conductivity of the solution, measured by a conductometer. Since these sensors do not have good selectivity and sensitivity, they are used less, compared to other electrochemical sensors [210]. Another applicable method that has attracted the attention of researchers nowadays is the electroluminescence method, exciting the target species by an electrochemical stimulus [211]. Unlike the photoluminescence method, the interference caused by light scattering and luminescence background in the electroluminescence method is minimized, thereby increasing the signal-to-noise ratio. In turn, this enhances the sensitivity, accuracy, and precision of the method [211]. In other words, the electroluminescence method has good reproducibility and can be used in concurrency analysis. Moreover, the combination of the electroluminescence transducer with microfluidic structures enables us to determine the sample with an extremely small volume (1.8 microliter) and a low detection limit (~1.5 femtomoles) [212].

Among electrochemical methods, potentiometric and amperometric ones have received more attention from researchers. These methods are more useful for fabricating the point-of-care sensors. The most common electrochemical biosensors are the glucometer [213] and alcohol breath analyzer [214]. A drop of blood or a few seconds of exhalation is sufficient to cause a reaction. However, the accumulation of biological species on the electrode surface can reduce its transfer rate, leading to a negative effect on the sensor performance [215]. Today, electrochemical sensors are connected to a wireless control system, enabling us to monitor electroactive species remotely, while also transferring data from one device to another via Bluetooth [216].

9.2. Thermal Transducers

In thermal transducers, the changes in the heat produced or consumed in a chemical reaction are measured over a period of time [157]. Depending on the temperature range and the importance of the sensor resistance in physical and chemical environments, different thermal transducers can be used [217]. Notably, thermocouples are the most robust thermal sensors; they have good physical and chemical stability and operate in the temperature range of −200 °C to +3000 °C [218]. On the other hand, thermistors and semiconductors are employed in the temperature ranges of −50 °C to +300 °C and −40 °C to +100 °C, respectively [218]. These sensors have lower thermal resistance and stability than thermocouples [218]. Sometimes the thermal sensor information is not accurate and reliable, arising from factors such as heat loss (due to radiation or convection), sensor heating (after applying an electric field generated from an external power supply), and sensor deformation (due to mechanical pressure). Accordingly, thermal biosensors have rarely been considered in research studies [151].

9.3. Optical Transducers

Optical transducers are very popular among all transducers employed in the preparation of paper sensors because of their simplicity, availability, and capability [17]. The analyte-bioreceptor interaction leads to a change in the optical property of the detection element, which can be observed in the form of absorption, emission, scattering, and reflection of light. Depending on the type of change, colorimetric, and fluorometric methods can be used [17].

In the colorimetric method, changes taking place in the intensity of light absorbed or reflected are observed [219]. In fact, chromophores with different configurations cause the optical property to change after the reaction. In the case of nanoparticles, the change in the optical property arises from their aggregation or surface modification [219]. These changes are proportional to the amount of analyte in the sample. The most important advantage of the colorimetric methods is that the color changes can be seen with the naked eye [219]. It is also possible to obtain more accurate detection results by performing image analysis or spectrophotometric spectrum analysis. The absorption spectrum of the detection element is investigated before and after the interaction using the spectrophotometry analysis. Basically, the maximum wavelength is used for quantitative measurements [8]. The changes can decrease or increase the absorption at specified wavelength or can shift the peak to lower or higher wavelengths [8]. Scanners, cameras, or smartphones are used in the image analysis [220]. In this respect, the image of the detection element is obtained before and after the color changes, followed by calculating the values of the color (red, green, and blue) components for each image, pixel by pixel. The average values obtained for each color component are then calculated, and the difference between these values is determined before and after the interaction. The numerical average values are returned to the image format, creating a color-difference pattern [28]. The intensity of the resultant color difference is proportional to the concentration of analyte in the sample. Alternatively, the image analysis can certainly be a simpler and more appropriate approach for the development of a point-of-care method. In this case, the white background of the paper substrates does not interfere with the measurements [220]. However, the optical conditions of the medium can interfere with the detection of color changes. For this reason, the image analysis is performed under a controlled light condition, in a sealed cabinet [221]. The main problem of the image-analysis-based methods is their low sensitivity and long response time. Moreover, it might be necessary to use a signal processor to achieve a desired signal, thus increasing the time and cost of the analysis.

In the fluorometric method, the return of the excited species to the ground state results in the emission of photons with less energy and longer wavelengths [222]. Different fluorophore compounds, such as chemical dyes, semiconductor quantum dots, carbon quantum dots, and nanoclusters can be employed as detection elements in the fluorometric sensors [223]. The reaction between the analyte and the bioreceptor may increase or decrease the fluorescence intensity, which can be attributed to the effect of internal filtering, dynamic damping, static damping, or Förster resonance energy transfer (FRET) [224]. Unlike colorimetric methods, the fluorometric sensors are highly selective and sensitive when employing a fluorometer or a fluorescence-sensitive camera [225]. The sensor is placed on a UV lamp with an excitation wavelength of 365 nm monitoring the fluorescence emission via the camera [225]. In these methods, the fluorescence of the paper substrate interferes with the measurement, which can be removed by subtracting the sample fluorescence from the background fluorescence, by using a standard compound (thus verifying the measurement validity), or by embedding a filter in the signal receiving device [165].

10. Application of Paper Biosensors in the Detection of Toxic Materials

In continuance, we review the studies carried out on the development of paper biosensors in the detection of mycotoxins, organophosphates, bacteria, and metal ions. Of course, the instrumental and analytical information of these biosensors are summarized in Tables 1–4.

10.1. Mycotoxins Detection

Toxins can be of biological origin, such as plants, animals, bacteria, and fungi. According to the source of production, toxins can be classified into various categories, such as botulinum neurotoxin, conotoxins, diphtheria toxin, notexin, tetrodotoxin, phycotoxins, phytotoxins, and so on [226]. Among them, mycotoxins are secondary metabolites produced by fungi, such as *Fusarium*, *Aspergillus*, and *Penicillium*, when harvested or stored improperly [227]. These toxins can enter the human body either directly (through the consumption of contaminated agricultural materials) or indirectly (through the consumption of animal products), causing cancer in tissues, as well as gene, liver, and kidney poisoning [228]. Moreover, they may cause disorders in the nervous and reproductive systems [228]. So far, a wide range of mycotoxins have been identified. The best known of these are: aflatoxins (AF), ochratoxins (OTA), fumonisins, patulin, zearalenone (ZEA), and trichothecenes [229]. According to the International Agency for Research on Cancer (IARC), mycotoxins can be divided into two categories: potent carcinogens (e.g., aflatoxins) and substances that can be (but are not necessarily) carcinogenic (e.g., ochratoxins) [230].

Among mycotoxins, many studies have been carried out on aflatoxins [231]. In 1960, aflatoxin was found to be the cause of the turkey X disease [232]. This toxin is a metabolite produced by the growth of *Aspergillus flavus* and *Aspergillus parasiticus* [231]. These fungi can be found in agricultural products, such as wheat, peanuts, bran, sesame seeds, peppers, and a variety of spices [233]. However, the growth of fungi increases by keeping these species in certain conditions, such as humidity above 7% and temperatures between 13 °C and 40 °C [234]. According to the Food and Drug Administration (FDA), the permissible level of aflatoxin in food samples should be between 20 and 300 ppb [235]. The aflatoxins can be classified as aflatoxins (AF) B1, B2, G1, G2, M1, and M2 [230]. Among them, the most toxic type of aflatoxins is aflatoxin B1, resulting in poisoning through both swallowing and skin penetration [236]. The liver, the most important organ in the body, is severely affected by aflatoxin B1 [236]. Liver failure causes fatty infiltration, necrosis, hemorrhage, fibrosis, regeneration of nodules, and even bile duct proliferation/hyperplasia [236]. To prevent these problems, the presence of aflatoxin B1 and other types of mycotoxins should be detected, and its amount be determined using analytical methods. Traditional methods, such as chromatography (e.g., thin layer and high performance liquid chromatography), spectrophotometric methods (e.g., colorimetry or fluorimetry), and enzyme-linked immunoassay methods, are used for the qualitative and quantitative analysis of aflatoxins [227]. While these methods are accurate and sensitive, they are costly and time-consuming and require special laboratory conditions. Certainly, using a biosensor that can detect and measure the toxins at the sampling site, with the help of a dedicated receptor, could be a simpler and more user-friendly method.

As seen in Table 1, most biosensors employed to detect and measure mycotoxins have a 2D design and lateral flow structure. In research studies, mycotoxins are extracted from agricultural and food samples. The bioreceptor of the resulting sensors is primarily a monoclonal antibody. The antibodies are either conjugated to gold nanoparticles with a colorimetric transducer or coated on a fluorophore, in which the amount of mycotoxin is detected by fluorescence. The reasoning behind the use of a fluorescence probe is that, since the substrate is covered with a black background, due to the color of the analyte (thus being unsuitable for the colorimetric method), it does not properly show the changes in the color of the gold nanoparticles [237].

Most lateral flow-based methods use a competitive mechanism for the interaction between the antigen and antibody in the test line. This approach was used for the detection of AFB1, by using a 1-ethyl-3(3-dimethylaminopropyl) carbodiimide hydrochloride (EDC)-mediated method [237] or AFM1 with polystyrene microspheres, enclosing time-resolved fluorescent europium (III) [Eu(III)-TRFM] [238]. The TRFM immunoassay is a new method, based on lateral flow, in which fluorescence microspheres are used as probes, giving rise to good sensitivity and a high linear range [239]. Most TRFM lanthanide complexes are used to make these sensors. Lanthanide elements, such as Eu(III), Tb(III), Sm(III), and Dy(III), are involved in the formation of the complexes [238]. The fluorescence intensity of the resultant compounds is weak, reducing the sensor's performance in detecting extremely small amounts of analyte. Tang and Wang et al. proved, in two separate studies, that the fluorescence emission intensity of the complexes could be increased by encapsulating them in polystyrene or by chelating them with silica nanoparticles, respectively [240,241]. In a typical study, Tang et al. [242] used this strategy for detection of six types of hazardous chemical compounds, including AFB1, AFB2, AFG1, AFG2, carbaryl, and carbofuran. These compounds were measured for five corn samples, with detection limits of 0.03 $ng \cdot mL^{-1}$, 0.02 $ng \cdot mL^{-1}$, and 60.2 $ng \cdot mL^{-1}$ [242]. In lateral flow colorimetric methods, the detection element is mostly made of gold nanoparticles (mainly synthesized by sodium citrate). Unlike fluorescence, these methods do not interfere with the fluorescence emission of the substrate. The detection is also possible with the naked eye and does not require an excitatory stimulus, such as the UV lamp. Extensive studies have been performed based on lateral flow colorimetric methods. Notably, according to Wang et al. [243], the colorimetric method shows good performance in detecting AFM1 in raw milk samples in the presence of microorganisms, such as *Escherichia coli* O157:H7, having a detection limit of 50 $pgmL^{-1}$. In this regard, two test lines, one for the AFM1 competitive reaction and the other for the detection of *Escherichia coli*, were embedded in the detection pad [243]. This design has also been observed in the simultaneous detection of different species of mycotoxins. In fact, a wide range of mycotoxins exist in real samples. Instead of fabricating a separate sensor for each mycotoxin, it is possible to place bioreceptors of all toxins on a pad, thus reducing the cost and time of analysis. In simultaneous measurements, coating antigens of different toxins (e.g., aflatoxins, ochratoxins, deoxynivalenol, ochratoxin A, T-2 toxin, and zearalenone) are sprayed along each other on the detection pad [68,244,245]. The detection limits of these sensors are given in Table 1. As can be inferred, most of the sensors used in the simultaneous detection have the same design. Nevertheless, the visual detection limits for determining the amount of a particular toxin, obtained by the various methods, are different from each other because the detection and determination of toxins depends on factors such as the antibody concentration, nanoparticle concentration, nanoparticle size, pH value of Ab-GNP interaction, blocking buffer, and the order of the test lines [68]. For example, the use of BSA in the blocking buffer composition is suitable for detecting AFB1, whereas it is not effective for detecting OTA. This is due to the interaction of BSA with OTA, thereby reducing the sensitivity of the measurement. Therefore, ovalbumin (OVA) is used in the blocking buffer of a mixture containing AFB1 and OTA [246]. The pH, antibody concentration, and nanoparticle concentration should be such that a stable Ab-GNP conjugate is formed, avoiding the aggregation of nanoparticles [68]. The Ab-GNP conjugate instability can affect the formation of a stable test line, thus changing the sensitivity and performance of the sensor [68]. Basically, pH affects the electrostatic interaction between nanoparticles and antibodies, so that the antibody creates a large steric hindrance, preventing the nanoparticles from being aggregated [68]. On the other hand, experiments have shown that the greater the distance between the test line and the conjugate pad, the better the measurement sensitivity [68]. Of course, this is a relative phenomenon, which can be suitable for some toxins and not effective for other toxin types. In addition to antibodies, aptamer can be used as bioreceptor in lateral flow structure. This system was used by Zhang et al. [247] for the detection of OTA in corn samples through fluorescence techniques. Limited studies have been performed using microfluidic structures, likely due to the type of paper. In fact, it is difficult to fix the bioreceptor on the Whatman paper used in the fabrication of microfluidic devices. This contrasts with lateral flow sensors, employing a nitrocellulose paper as the substrate. However, these tools have been used to detect mycotoxins, establishing the detection mechanism, based on the interaction between toxins and aptamer. The aptamer of a specific toxin is initially absorbed on a gold nanoparticle and then separated from it by adding analyte to the solution. The high affinity of the aptamer for the analyte is responsible for this separation, followed by the interaction with the toxin. Moreover, the configuration of the aptamer changes to G-quadruplex [248]. If a salt, such as NaCl, is added to the solution, the charge repulsion between the nanoparticles is reduced, causing them to accumulate and change color from red to blue [248]. Kasoju et al. have employed this strategy to determine AFB1 and AFM1 in food and milk samples, respectively. Thus, the concentration of analytes can be determined with an accuracy of picomolar [249]. To perform simultaneous measurements in microfluidic structures, Sheini [250] proposed a nanoparticle-based color sensor array, without the use of bioreceptors. In this case, the interaction between the analyte and the organic compounds, coated on silver and gold nanoparticles, was responsible for the color changes. In fact, the accumulation of nanoparticles in the presence of analytes gives rise to the color changes. As shown in Figure 4, the sensor array creates a unique pattern for each toxin that can be used to determine the type of toxin (and its chemical structure) and to produce fungus, leading to a detection limit in the range of nanomolar [250]. Mycotoxins have also been evaluated by electrochemical methods. In this respect, Migliorini et al. have been able to detect AFB1 by an electrochemical sensor, in which the anti-AFB1 antibody was covalently immobilized on a film prepared by multi-walled carbon nanotubes and chitosan. The resulting sensor provided an AFB1 detection limit of 0.62 ng·mL^{-1} and excellent recovery in the corn sample [251].

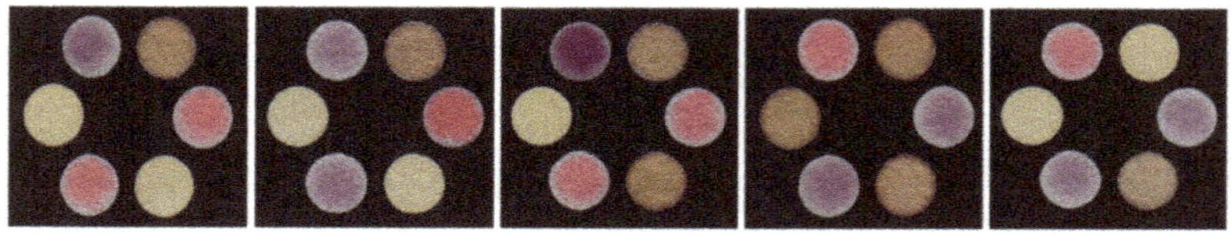

Figure 4. Colorimetric sensor array, based on gold and silver nanoparticles for the individual and simultaneous detection of alfatoxins and ochratoxin A, zearalenone (reprinted with permission from [250], copyright (2020) Springer Nature).

10.2. Organophosphates Detection

Pesticides are divided into three categories: organochlorines, organophosphates, and carbamates [215,256]. Among them, the tendency to produce organophosphates is higher, as their decomposition process is rapidly carried out, with the help of microorganisms or natural environmental processes [7]. However, organophosphates have a long half-life and accumulate in the environment for a certain period of time. Other types of organophosphates can act as nerve agents, which are used in chemical warfare, or enter into the river and groundwater cycle from the waste of industrial products, such as plastics, lubricants, refrigerants, fuels, solvents, dispersants, and surfactants [7,257]. The chemical structure of these compounds falls into the two following groups: oxon (P=O) and thion (P=S) [258]. Both groups have the ability to inhibit the activity of cholinesterase enzymes that act as neurotransmitters, giving rise to the acetylcholine accumulation in the intersynaptic space [258]. In turn, this results in neurological, respiratory and cardiovascular disorders and even death (in acute cases) [258]. The organophosphate toxicity, which has been proven since 1960 [259], poisons 3 million and kills 200,000 people annually [260]. Many efforts have been devoted to detecting and determining these compounds (quickly and in a timely manner) in real samples, one of which is the use of biosensors. As represented in Table 2, in organophosphate biosensors, most enzymes are used as bioreceptors, involving the mechanism of enzyme activity inhibition by organophosphate in the body. In this mechanism, acetylcholine is initially hydrolyzed to choline in the presence of acetylcholinesterase. The resultant product can react with the detection element (being made of gold or silver nanoparticles) used in the sensor structure, thus changing its color. In another mechanism, the choline produced in the presence of choline oxidase is oxidized to H_2O_2, followed by its participation in the oxidation reaction with a redox organic substance that changes the oxidized state of the colored substrate to a reduced form, resulting in a change in its color [261]. The colorimetric reagent used so far for these sensor systems are indophenol acetate, dithiobisnitrobenzoate, indoxyl acetate, and 3,3′,5,5′-tetramethylbenzidine. In one study, a three-layer sensor was designed to detect chlorpyrifos [262]. The configuration, mechanism of reaction, and response of the sensor are shown in Figure 5. Using double-sided adhesive, the layers were stacked on top of each other, so that the middle layer contained the ACHE and indoxyl acetate reacting together to produce indigo blue. The production process of indigo and, subsequently, the intensity of the blue color was reduced after the entrance of the analyte from the first layer and the inhibition of enzyme activity. The resulting color changes could be seen in the third layer [262].

Table 1. Different type of paper-based biosensor for mycotoxin detection.

Type of Mycotoxin	Device Structure	Bioreceptor	Detection Method	Sensing Element	Media	Linear Range	Detection Limit	Ref.
AFB1	Immunofiltration assay	Anti-AFB1	Colorimetric	AuNPs	Rice, corn, and wheat	0–4000 ng·mL^{-1}	2.0 ng·mL^{-1}	[252]
AFB1	Immunodipstick assay	Anti-AFB1	Colorimetric	Core-Shell AgAuNPs	Rice, wheat, sunflower, cotton, chillies, and almonds	0.1–10.0 ng·mL^{-1}	0.1 ng·mL^{-1}	[253]
AFB1	Microfluidic assay	Aptamer	Colorimetric	AuNPs	-	1 pM–1 mM	10 nM	[254]
AFB1	Lateral flow assay	Anti-AFB1	Fluorimetric	Fluorescent microsphere	Soybean sauce	-	2.5 µgL^{-1}	[237]
AFB1	Printed electrode	Anti-AFB1	Impedimetric	MWCNT/chitosan	Maize flour	1.0 to 30.0 ng·mL^{-1}	0.62 ng·mL^{-1}	[251]
AFM1	Lateral flow assay	Anti-AFM1	Colorimetric	AuNPs	Milk	-	50 pgmL^{-1}	[243]
AFM1	Microfluidic assay	Aptamer	Colorimetric	AuNPs	Milk	1.0 pM to 1.0 µM	10.0 nM	[249]
AFM1	Lateral flow assay	Anti-AFB1	Fluorimetric	Fluorescent microsphere	Milk powder, UHT, and pasteurized milk	0.05–2.0 ng·mL^{-1}	0.019 ng·mL^{-1}	[238]
OTA	Lateral flow assay	Aptamer	Fluorimetric	Fluorescent probe	Corn	1–1000 ng·mL^{-1}	0.40 ng·mL^{-1}	[247]
AFs	Lateral flow assay	Anti-AFs	Fluorimetric	Eu(III) NPs	Corn	0.03–3.90 ngg^{-1}	0.03 ngg^{-1}	[242]
AFB1, AFM1, DON, OTA, T-2, ZEN	Lateral flow assay	Anti-AFB1, Anti-AFM1, Anti-DON, Anti-OTA, Anti-T-2, Anti-ZEN	Fluorimetric	Protein microarrays	Water	0.04–1.69 ng·mL^{-1} 0.45–3.90 ng·mL^{-1} 20.20–69.23 ng·mL^{-1} 35.68–363.18 ng·mL^{-1} 0.11–1.81 ng·mL^{-1} 0.08–7.47 ng·mL^{-1}	0.01 ng·mL^{-1} 0.24 ng·mL^{-1} 15.45 ng·mL^{-1} 15.39 ng·mL^{-1} 0.05 ng·mL^{-1} 0.01 ng·mL^{-1}	[246]
AFB1, OTA, ZEN	Lateral flow assay	Anti-AFB1, Anti-OTA, Anti-ZEN	Colorimetric	AuNPs	Peanuts, maize, and rice		0.25 ng·mL^{-1} 0.5 ng·mL^{-1} 1.0 ng·mL^{-1}	[244]
AFB1, OTA, ZEN	Lateral flow assay	Anti-AFB1, Anti-OTA, Anti-ZEN	Colorimetric	AuNPs	Corn, rice and peanut		0.10–0.13 µg·kg^{-1} 0.42–0.46 µg·kg^{-1} 0.19–0.24 µg·kg^{-1}	[68]
ZEAs, DONs, T-2s, AFs, FBs	Lateral flow assay	Anti-ZEAs, Anti-DONs, Anti-T-2s Anti-AFs, Anti-FBs,	Colorimetric	AuNPs	Cereal		0.04–0.17 µg·kg^{-1} 0.06–49 µg·kg^{-1} 0.15–0.22 µg·kg^{-1} 0.056–0.49 µg·kg^{-1} 0.53–1.05 µg·kg^{-1}	[245]
AFB1 ZEA DON	Lateral flow assay	Anti-AFB1, Anti-ZEA, Anti-DON,	Colorimetric	AuNPs	Cereals		0.05 µg·kg^{-1} 1.0 µg·kg^{-1} 3.0 µg·kg^{-1}	[255]
AFB1 AFG1 AFM1 OTA ZEN	Microfluidic assay	Free	Colorimetric	AuNPs and AgNPs	Pistachio, wheat, coffee, and milk	3.1–7800 ng·mL^{-1} 8.2–8400 ng·mL^{-1} 2.5–8200 ng·mL^{-1} 4.0–3800 ng·mL^{-1} 8.0–7900 ng·mL^{-1}	2.7 ng·mL^{-1} 7.3 ng·mL^{-1} 2.1 ng·mL^{-1} 3.3 ng·mL^{-1} 7.0 ng·mL^{-1}	[250]

AF: aflatoxin; NP: nanoparticle; MWCNT: multi-walled carbon nanotubes; UHT: ultrahigh-temperature; OTA: ochratoxin A; fluorescent probe: 5′-biotin-(CH2)$_6$-ttt-ttt-ttt-ttt-ttt-ttt-3′; DON: deoxynivalenol; T-2: T-2 toxin; ZEN: zearalenone; FB: fumonisin.

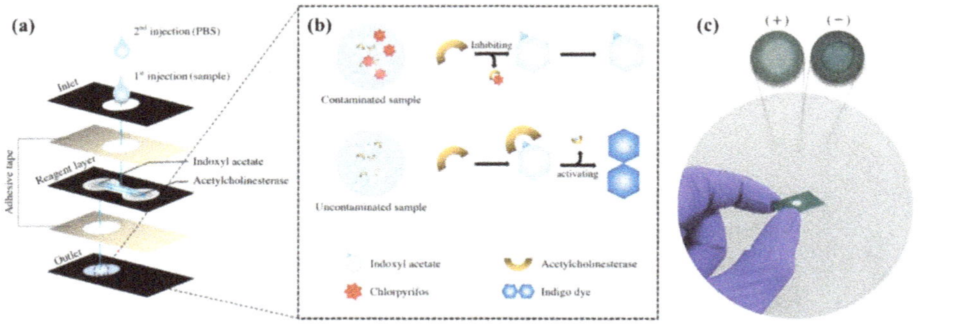

Figure 5. 3D sensor configuration, obtained via the stacking method. The sensor was used to detect chlorpyrifos through the enzymatic procedure. (**a**) The proposed procedure for creation of sensor and detection of pesticide; (**b**) the proposed mechanism for detection of pesticide: ACHE and indoxyl acetate enzymes reacting together to produce indigo blue (the intensity of blue color was reduced after the entrance of the sample from the first layer and inhibition of enzyme activity); and (**c**) the response of the sensor in the presence of contaminated (positive) and normal (negative) samples (reprinted with permission from [262], copyright (2018) Springer Nature).

The origami method can also replace the stacking one, due to its easier and more accurate fabrication process of a 3D system. By considering a bilayer sensor, comprising of the ACHE enzyme in the first layer and indophenol acetate in the second layer, the solution sample is injected into the enzyme-containing layer [263]. After folding the layers, the interaction between the enzyme and the colorimetric indicator occurred, and the color of the sensor changed from blue to colorless in the presence of phosphorus analytes (e.g., chlorpyrifos). The possibility of tracking the analyte amount in the sample using any type of smartphone has been proven, as well. In this way, ambient light sensors were used and installed on smartphones [263]. In these sensors, it is possible to control the ambient light conditions and adjust the backlight, preventing the sensitivity of the CMOS sensors or the quality of the camera lens from affecting the reception of high-resolution images [263]. The color indicators can be replaced by nanoparticles, whose central metal atom is capable of participating in the oxidation reaction [264]. The most common type of nanoparticles is the metal oxide, including nanoceria (CeO_2) nanoparticles [265]. The interaction of these nanoparticles with H_2O_2 changes the Ce^{3+}/Ce^{4+} ratio on the surface, thus varying their color from colorless to yellow [266]. In the presence of organophosphates, the activity of enzyme was inhibited thereby reducing the H_2O_2 amount produced, and turning the paper color from yellow to colorless [267]. This strategy was used for detection of methyl-paraoxon or chlorpyrifos [267]. The micro-spot structure is mostly used in fluorescence-based biosensors. In these sensors, the fluorescent probes, such as tetraphenylethene (TPE) functionalized with maleimide group [268] or ZnCdSe\CdTe\Zn-nanoporphyrin [269], were applied for the detection of diazinon and dimethoate, dichlorvos, and demeton, respectively. Studies have also been carried out on micro-spot sensors with the surface-enhanced Raman scattering mechanism, modifying the paper substrate with a solution of gold nanoparticles coating by Raman probe (e.g., 4-mercaptobenzoic acid) [270]. The prepared sensor is exposed to a sample containing methyl parathion, having a linear response at the analyte concentration range of 0.018–0.354 $\mu g \cdot cm^{-2}$, and detection limit of 0.011 $\mu g \cdot cm^2$ [270].

Enzyme biosensors have also been popular in the electrochemical studies of organophosphates [271]. Compared to colorimetric methods, the electrochemical biosensors are resistant to interference from the color or opacity of the sample matrix [271]. The cholinesterase enzymes are used in the structure of these biosensors. For example, butyrylcholinesterase was used for detection of methyl parathion with a potentiometric method [140] or simultaneous determination of various type of pesticide with an amprometric method [272]. In the later, the graphite working electrodes were modified by Prussian blue nanoparticle integrated with carbon black, preventing the precipitation of thiol products on the electrode surface while also increasing the oxidative current relative to unmodified electrodes. Additionally, the current is measured in the presence and absence of the analyte using a pad, reducing the analysis cost and time [272]. Despite their applicability in detecting organophosphates, cholinesterases are also prone to other interfering species in the medium [215]. Consequently, cholinesterases do not show good selectivity against enzymes such as phosphotriesterase (PET), and interact only with compounds containing specific ester bonds that are also found in organophosphates [273]. Hondred et al. have used PET in the fabrication of the paraoxon biosensor and employed the inkjet maskless lithography method for designing the first printed graphene sensor [274]. By designing an unmodified enzyme immunosorbent assay, one can use a graphene electrode to detect parathion, according to Mehta et al. [275]. In this respect, anti-parathion antibodies were attached to the electrode surface by means of amines-based material (e.g., 2-aminobenzyl amine) as a mediator. The sensor response was recorded by an impedance method, resulting in a detection limit of 0.052 ng·L^{-1} for tomato and carrot samples [275].

Enzyme-based colorimetric and electrochemical methods do not have good selectivity for similar organophosphates, preventing the possibility of their simultaneous determination in a real sample containing several species of organophosphates. In addition, despite the high specificity of immunoassay methods, it is difficult to immobilize multiple antibodies on a piece of paper. In order to simultaneously monitor organophosphates by paper sensors, Bagheri research group has proposed a nanoparticle-based color sensor array [276]. They used modified gold and silver nanoparticles, generating a fingerprint response for all pesticides studied. As seen in Figure 6, the sensor response show a discrimination between pesticides from non-pesticides, carbamates from organophosphate, oxones from thions, and aliphatic thions from non-aliphatic species, providing a good sensitivity in the order of nanogram per milliliter for separate quantitative and simultaneous analysis of pesticides [276]. Elsewhere, Bagheri et al. have used nanoparticle-based origami paper structures to analyze sulfur-containing organophosphates in the vapor phase, simultaneously [277].

10.3. Pathogen Bacteria Detection

Pathogenic bacteria is one of the causes of human disease leading to infections in the body or even death [277]. Bacterial infections can originate from food, water, or air after entering the body, infecting areas such as the lungs, stomach, intestines, skin, kidneys, bladder, and even the blood [281]. The bacterial infection in its mildest form leads to intoxication with fever, chills, and fatigue [281]. Annually, two million people are killed by waterborne pathogens alone [282]. Pathogenic bacteria can be classified into two categories: gram-positive and gram-negative [283]. Compared to gram-negative bacteria, the cell wall of gram-positive bacteria has a thick peptidoglycan layer, without the outer membrane. This causes the gram-positive and gram-negative bacteria to turn purple and pink in the gram staining test, respectively [283]. Bacteria such as *Staphylococcus aureus*, *Listeria monocytogenes*, *Streptococcus agalactiae*, and *Enterococcus faecalis* are in the gram-positive group, and bacteria such as *Escherichia coli*, *Klebsiella pneumoniae*, *Proteus mirabilis*, *Enterobacter aerogenes*, and *Pseudomonas aeruginosa* are in the gram-negative group [282].

The most common clinical methods used to detect bacterial infections are as follows: culturing, enzyme-linked immunosorbent assay, and polymerase chain reaction [284]. While the accuracy and sensitivity of these methods are very good and have been used as a gold standard approach in medical centers for many years, they take a long time (at least 48 h) to determine the infection and the type of bacteria [285]. The diagnosis of infectious diseases in a short time is of great importance. For this reason, many efforts have been made to develop rapid test sensors, a considerable part of which are based on paper biosensors. A sensor is highly efficient when it can detect fewer bacteria in the shortest possible time.

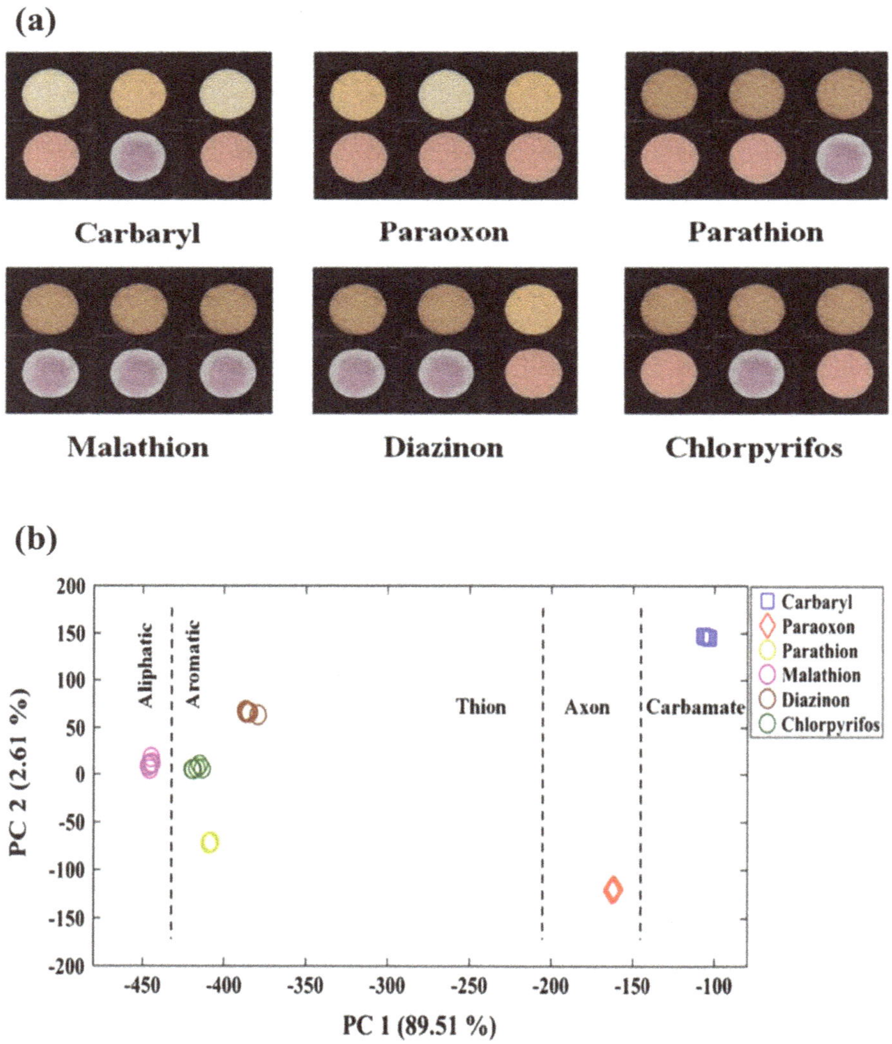

Figure 6. Paper-based electronic tongue for detection of pesticides. This sensor can discriminate carbamate, oxon organophosphate, and thion organophosphate from each other. It can also separate aliphatic thions from non-aliphatic species. This figure illustrates the actual sensor response (**a**) and PCA score plot (**b**) which are shown the discriminatory ability of proposed sensor (Reprinted with permission from [276], copyright (2020) Springer Nature).

Table 2. Different type of paper-based biosensor for organophosphate detection.

Type of OP	Device Structure	Bioreceptor	Detection Method	Sensing Element	Media	Linear Range	Detection Limit	Ref.
Parathion	Origami	Enzyme	Potentiometric	butyrylcholine-sensitive membrane		0.1–1.0 nm	0.06 nm	[140]
Parathion	Printed electrode	Anti-parathion	Impedimetric	Gr/ABA	Tomato and carrot	0.1–1000 ng·L^{-1}	52 pg·L^{-1}	[275]
Parathion	Micro spot	Free	SERS	Au NPs\4 MBA	apple	0.018–0.354 μg·cm^{-2}	0.011 μg·cm^{-2}	[270]
Paraoxon	Printed electrode	Enzyme	Amperometric	PtNP-IML-PGE/GA	Soil and water	0.1–1.0 nM	3 nM	[274]
Malathion	Printed electrode	Mitochondria	Voltammetric	Quinone			20 nM	[278]
Chlorpyrifos	3D paper	Enzyme	Colorimetric	Indoxyl acetate		0–25.0 ppm	8.60 ppm	[262]
Trichlorfon	Microfluidic assay	Free	Colorimetric	ammonium molybdate method	Pak choi, broccoli, swamp, cabbage		1.65 μg·mL^{-1}	[279]
Diazinon	Micro spot	Enzyme	Fluorimetric	TPE	Human serum	0.3–5.0 ng·mL^{-1}	0.23 ng·mL^{-1}	[268]
Paraoxon Chlorpyrifos	Micro spot	Enzyme	Colorimetric	Nanoceria	Human serum	0–100.0 ng·mL^{-1} 0–60 ng·mL^{-1}	18.0 ng·mL^{-1} 5.3 ng·mL^{-1}	[267]
Phorate, avermectin, imidacloprid	3D paper	Enzyme	Colorimetric	Indophenol acetate	Lettuce and rice			[280]
Chlorpyrifos parathion methyl-parathion malathion fenitrothion carbaryl	Origami	Enzyme	Colorimetric	indophenol acetate	Cabbage extracts	5.0–100.0 μg·mL^{-1} 1.0–8.0 μg·mL^{-1} 0.5–6.0 μg·mL^{-1} 0.5–6.0 μg·mL^{-1} 0.5–6.0 μg·mL^{-1} 1.0–8.0 μg·mL^{-1}	3.3 μg·mL^{-1} 0.52 μg·mL^{-1} 0.46 μg·mL^{-1} 0.45 μg·mL^{-1} 0.47 μg·mL^{-1} 0.51 μg·mL^{-1}	[263]
Dimethoate dichlorvos, demeton	Micro spot	Free	Fluorimetric	CdTe QDs/ZnCdSe QDs/Nano-ZnTPyP	Apple and cabbage			[269]
paraoxon 2.4-DCPA atrazine	Origami	Enzyme	Electrochemical	GP/CB/PBNPs GP/CB GP/CB	River water	2–20 ppb 100–600 ppb 10–100 ppb	2 ppb 50 ppb 10 ppb	[272]
Carbaryl, paraoxon, parathion, malathion, diazinon, chlorpyrifos	Origami	Array-based e-tongue	Colorimetric	AuNPs and AgNPs	Tap water, apple juice, rice	35.0–2500.0 ng·mL^{-1} 25.0–5000.0 ng·mL^{-1} 35.0–5000.0 ng·mL^{-1} 20.0–2500.0 ng·mL^{-1} 50.0–7500.0 ng·mL^{-1} 40.0–2500.0 ng·mL^{-1}	29.0 ng·mL^{-1} 22.0 ng·mL^{-1} 32.0 ng·mL^{-1} 17.0 ng·mL^{-1} 45.0 ng·mL^{-1} 36.0 ng·mL^{-1}	[276]

Table 2. *Cont.*

Type of OP	Device Structure	Bioreceptor	Detection Method	Sensing Element	Media	Linear Range	Detection Limit	Ref.
Parathion, malathion, diazinon, chlorpyrifos	Origami	Array-based e-nose	Colorimetric	AuNPs and AgNPs	Ambient air	70–1000 ng·mL^{-1} 110–810 ng·mL^{-1} 90–730 ng·mL^{-1} 130–730 ng·mL^{-1}	58.0 ng·mL^{-1} 103.0 ng·mL^{-1} 81.0 ng·mL^{-1} 117.0 ng·mL^{-1}	[277]

OP: Organophosphate, 2-ABA: 2-Aminobenzyl amine, Gr: Graphene, PtNPs: Platinum nanoparticles, IML: Inkjet maskless lithography, PGE: Patterned graphene electrodes, GA: Glutaraldehyde, SERS: Surface-enhanced Raman scattering (SERS), TPE: Tetraphenylethylene, e-tongue: electronic tongue, 2.4-DCPA: 2,4-dichlorophenoxyacetic acid, GP: Graphite, CB: Carbon black, PBNPs: Prussian blue nanoparticle.

Thus far, the use of paper biosensors has been reported in the study of food spoilage (cold cuts, sausages, beef, and pork), beverages (orange juice, milk, and drinking water), vegetables and fruits (cucumber and lettuce), and biological samples (serum and urine), as can be observed in Table 3. Enzymes and aptamers have been mainly used as bioreceptors in the colorimetric detection of bacteria (Table 3). For example, Creran et al. used β-galactosidase (β-gal) enzyme to determine *Escherichia coli* XL1 and *Bacillus subtilis* in drinking water samples [286]. This enzyme can catalyze the conversion reaction of yellow chlorophenolred-β-D-galactopyranoside (CPRG) to its purple state. In this respect, the anionic enzyme is first electrostatically attached to a gold nanoparticle with a positive electric surface. In the presence of analyte, the nanoparticle tends to interact with bacteria more, resulting in the separation of the enzyme from the complex structure, while also reacting with CPRG to change the sensor color from yellow to purple [286]. In addition to β-gal, Jokerst et al. have employed phosphatidylinositol-specific phospholipase C (PI-PLC) and esterase [287]. The PI-PLC catalyzes the reaction of 5-bromo-4-chloro-3-indolyl-myo-inositol phosphate (X-InP) to the indigo form, and the esterase catalyzes the reaction of 5-bromo-6-chloro-3-indolyl caprylate to its purple state. The β-gal, PI-PLC, and esterase enzymes have been used to detect *Escherichia coli*, *Listeria monocytogenes*, and *Salmonella enterica*, respectively [287]. According to Jokerst et al., by combining paper biosensors and polymerase chain reaction (PCR), it is possible to detect 10 Cfu/cm^2 in less than 12 s [287]. The detection mechanism of *Staphylococcus aureus*, with a sensor proposed by Suaifan et al., is based on the proteolytic activity of *S. aureus* proteases on a magnetic nanobeads-peptide probe being immobilized on the gold platform via an Au-S connection [288]. During the reaction, magnetic nanobeads detached from the surface are separated from the gold platform using an external magnet, resulting in the appearance of yellow color. The time required for this measurement has been reported to be 1 min [288]. The α-glucosidase enzyme is secreted by various species of *Cronobacter* spp., catalyzing the 5-bromo-4-chloro-3-indolyl-α-D-glucopyranoside (XαGlc) color species to the form of indigo, which can be used in a paper sensor. This can allow for the detection of the presence and amount of different species of *Cronobacter* spp., up to a concentration of 10 Cfu/cm^2 in 10 h [289].

The detection of *Staphylococcus aureus* has been investigated using the nanozyme nature of nanoclusters. In one study, Bagheri Pebdeni et al. synthesized Au/Pt bimetallic nanoclusters using cytosine-rich, single-strand DNA, catalyzing TMB in its oxidized form in the presence of H_2O_2, due to the peroxidizing properties [290]. This causes the sensor to change from colorless to dark blue. The catalytic activity of the nanoclusters changes in the presence of bacteria, leading to the production of less oxidized TMB and, consequently, reducing the intensity of the blue color [290]. The schematic diagram for the working mechanism of this sensor is represented in Figure 7.

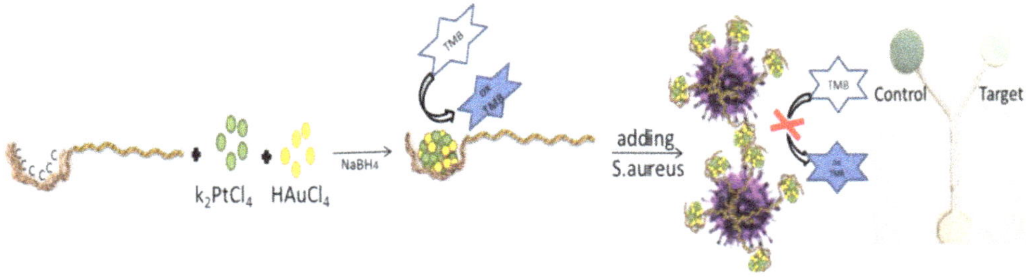

Figure 7. The nanozyme-based sensor for detection of *Staphylococcus aureus*. Au/Pt bimetallic nanoclusters synthesized by cytosine-rich, single-strand DNA show enzyme-like activity, to catalyze TMB to its oxidized form in the presence of H_2O_2; the color of sensor changes from colorless to dark blue. In the presence of bacteria strain, the nanoclusters interact with the bacteria, and the redox reaction of TMB is inhibited, leading to a reduction in the intensity of the blue color (reprinted with permission from [290], copyright (2020) Elsevier).

Unlike colorimetric methods, electrochemical techniques mostly employ antibodies and aptamers as bioreceptors (Table 3). In most cases, the detection element is a graphene electrode (or its derivatives) coated on paper by screen printing. To increase sensor performance, the graphene electrode can be modified by some compounds, such as gold nanoparticles and polymers. In this direction, using an electrodeposition technique, Wang et al. coated gold nanoparticles on an electrode surface made of reduced graphene oxide (rGO). After immobilizing the antibody on the electrode surface using the biotin-streptavidin system, they used *Escherichia coli O157: H7*, which was detected by an amperometric biosensor with a detection limit of 1.5×10^2 Cfu/mL [291]. Neutral red (NR)-modified graphene electrodes have been used in the design of sandwich immunoassay systems, according to Mo et al. [292]. In this respect, polyaniline (PANI) was precipitated on a carbon electrode, and the primary antibody (Ab1) was bound to the polymer via gold nanoparticles. On the other hand, rGO-NR was electrostatically bound to Au@Pt nanoparticles with a negative surface charge. The secondary antibody (Ab2) was bound to Au@Pt nanoparticles. Bacteria were sandwiched between these two layers, thereby increasing the sensitivity of the electrochemical signal [292].

Graphene electrodes can be functionalized with thermoresponsive polymers, such as poly (N-isopropylacrylamide) (PNIPAm). These polymers are sensitive to the physiological temperature (37 °C). Khan et al. have used the PNIPAm polymer to fabricate the PNIPAm-graphene nanoplatelet electrode nanocomposite, being precipitated on a gold substrate coated on paper [293]. The fibrous structure of the nanocomposite is such that it receives bacterial cells without the need for the presence of antibodies, thus changing the resistance of the gold platform. The resulting nanocomposite sensor was used to detect *S. mutans*, *B. subtilis*, and *E. coli* belonging to gram-positive and gram-negative bacteria groups, respectively. Moreover, the detection time for this sensor was reduced to 10 min [293].

The electrochemical behavior of graphene electrodes, modified by a specific bioreceptor, can be different from each other in the detection of bacteria. For example, Hernández et al. immobilized an aptamer on GO and rGO [294]. A covalent approach was used to immobilize the bioreceptor on GO, whereby amide bonds formed between the carboxyl group of GO and the amino groups of the aptamer. Alternatively, the rGO is non-covalently modified by the pyrenyl aptamer head. The aptamer has a strong tendency to interact with bacteria, dominating π–π interactions to separate its negatively ionized phosphodiester groups from the electrode surface in the presence of analyte. In turn, this changes the electrode potential. The experimental observations have shown that the modified rGO electrode has less noise and a higher detection limit than the GO. Hernández et al. have used this potentiometric system to detect *Staphylococcus aureus*, with a detection limit of 1 Cfu/mL in less than 2 min [294].

Sometimes antibodies can be covalently attached to carbon nanotubes with the help of a mediator, in order to detect bacteria. The covalent bonding reduces the time of the functionalization process of the paper surface and increases the stability of the sensor. In a study by Bhardwaj et al. [295], bacterial antibodies were immobilized on single-walled carbon nanotubes, using N-(3-dimethylaminopropyl)-N′-ethylcarbodiimide hydrochloride/N-hydroxysuccinimide mediator, allowing for the detection of *S. aureus* in milk samples. Accordingly, bacteria could be detected in the linear range of $10-10^7$ Cfu/mL in less than 30 min. The resulting sensor can determine *S. aureus* in the presence of other bacterial species, including *Escherichia coli B*, *Bacillus subtilis*, and *S. epidermidis*, providing excellent selectivity, due to the use of the specific antibody [295].

In recent years, with the development of sensor arrays, it has become possible to detect and determine bacteria simultaneously [296–300]. The detection mechanism is based on changes in the optical or electrical properties of the indicator after the interaction with bacteria or metabolites emitted from the bacterial growth. The detection elements can be organic semiconductors, conductive polymers, polymer carbon black composites, and dyes [301–306]. In a study by Bordbar et al. [282], the sensor array was composed of sixteen nanoparticles (eight gold and eight silver nanoparticles), each of which was synthesized by 8 coating agents, consisting of protein, polymer, surfactant, carbohydrate, and amino acid exposed to ten strains of bacterial. The detection process was evaluated in ambient water samples and 300 urine samples of ill and healthy persons. It was reported that the sensor array was capable of determining 10^2 Cfu/mL during 10 min in the culture media, 50 min in the aqueous sample, and 30 min in the urine sample. This sensor can be used to detect urinary tract infections with high accuracy [282]. Moreover, it is possible to use a paper sensor array in the diagnosis of sepsis in children. This has been designed by Sheini, through immobilizing gold and silver metal nanoclusters on Whatman paper [307]. The resulting sensor scan determine sepsis-causing bacteria, including *Staphylococcus aureus*, *Streptococcus pyogenes*, *Escherichia coli*, and *Pseudomonas aeruginosa*, in pediatric serum samples and provide a detection limit of less than 100 Cfu/mL. In this case, changes in fluorescence emission intensity were recorded in less than 15 s, with the help of a hand-held UV lamp and a smartphone [307]. One can see the results of this study in Figure 8.

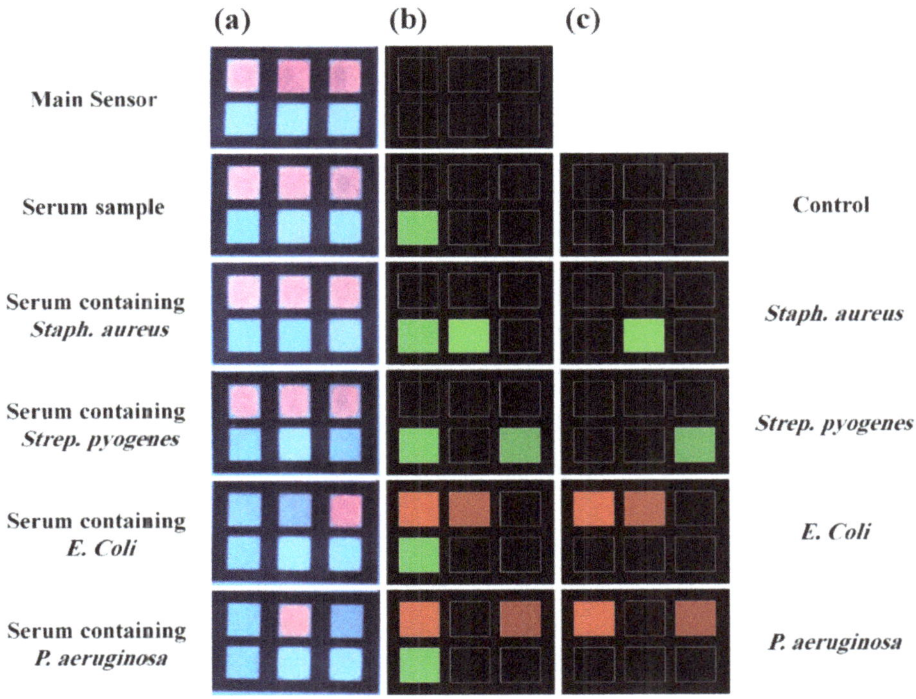

Figure 8. A fluorimetric, paper-based sensor for detecting of bacteria-caused sepsis. Array was constructed by gold and copper nanoclusters, which are synthesized by proteins. The interaction between detection elements and serum contaminated with bacteria leads to turning off fluorescence emission. This figure represents the actual response of sensor (**a**) and respective difference maps (**b**) for the control and each bacteria strain. (**c**) The response of sensor after eliminating the effect of the pure serum sample (reprinted with permission from [307], copyright (2021) Elsevier).

Table 3. Different type of paper-based biosensor for bacteria detection.

Type of Bacteria	Device Structure	Bioreceptor	Detection Method	Sensing Element	Media	Detection Limit	Time	Ref.
E. coli O157:H7	paper electrode	Antibody	Impedimetric	rGOPE/AuNPs	Ground beef and cucumber	1.5×10^{-2} Cfu mL^{-1}		[291]
E. coli XL1	Inkjet-Printed test strip	Enzyme	Colorimetric	CPRG	Drinking water	10^2 bacteria mL^{-1}	5 min	[286]
E. coli	Micro spot	Aptamer	Fluorimetric	Fluorogenic DNAzyme probe	Milk, apple juice, and drinking water	100 cells mL^{-1}		[308]
E. coli K12	Origami	Aptamer	Colorimetric	TMB, Hemin, H_2O_2	Juice and milk	10^3 Cfu mL^{-1}	35 min	[309]
E. coli O157:H7	Printed electrode	Antibody	Electrochemical	SPCE-PANI-AuNPs-Ab1 and PANI-rGO-NR-Au@Pt-Ab2	Milk and pork	2.84×10^3 Cfu mL^{-1}	60 min	[292]
S. aureus	Paper electrode	Aptamer	Potentiometric	GO or rGO		1 Cfu mL^{-1}	1–2 min	[294]
S. aureus	Test strip	Enzyme	Colorimetric	Magnetic nanobeads/Peptide	Ground beef, turkey sausage, lettuce, milk, and dust samples	7 Cfu mL^{-1} (Pure broth culture) 40 Cfu mL^{-1} (food products) 100 Cfu mL^{-1} (environmental samples)	1 min	[288]
S. aureus	Paper electrode	Antibody	Electrochemical	SWCNT	Milk	13 Cfu mL^{-1}	30 min	[295]
S. aureus	Microfluidic assay	Nanozym	Colorimetric	DNA-Au/Pt BMNCs, H_2O_2, and TMB	Milk, orange juice, and human serum	80 Cfu mL^{-1}	60 min	[290]
E. faecalis	Paper strip	Free	Colorimetric	Resazurin			10 min	[310]
Salmonella typh.	Paper strip	Antibody	Potentiometric	PAMAM$(NH_2)_{64}$/GA/Ab	Apple juice	5 cells mL^{-1}	<1 h	[311]
Each type	Printed electrode	Protein	Impedimetric	Con A	Water	1.9×10^3 Cfu mL^{-1}		[312]
Cronobacter spp.	Micro spot	Enzyme	Colorimetric	XαGlc		10 Cfu cm^{-2}	10 h	[289]
E. coli O157:H7, *Salmonella*, and *Listeria*	Micro spot	Enzyme	Colorimetric	CPRG, Magenta caprylate, and X-InP	Bologna	10 Cfu cm^{-2}	8 h	[287]
E. coli, *S. mutans*, and *B. subtilis*	Paper electrode	Polymer	Thermoelectrochemical	Gr-PNIPAm-Au	Autoclave, tap and lake waters, and milk	5 cells mL^{-1}	less than 10 min	[293]
S. aureus, *Listeria*, *E. coli*, proteus, *klebsiella*, *E. aerogenes*, *P. aeruginosa*, *E. faecalis*, *S. agalactiae*, MRSA	Micro spot	Array-based E.nose	Colorimetric	AuNPs and AgNPs	Tap and mineral water, and human urine	1.0×10^2 Cfu mL^{-1}	50 min (water) 30 min (urine)	[282]

Table 3. *Cont.*

Type of Bacteria	Device Structure	Bioreceptor	Detection Method	Sensing Element	Media	Detection Limit	Time	Ref.
S. aureus *S. pyogenes* *E. coli* *P. aeruginosa*	Microfluidic assay	Array-based E.tongue	Fluorimetric	Protein based Au and CuNCs	Serum (for detecting sepsis)	43.0 Cfu mL^{-1} 63.5 Cfu mL^{-1} 26.0 Cfu mL^{-1} 47.0 Cfu mL^{-1}	15s	[307]

E. coli: *Escherichia coli*; rGOPE: reduced graphene oxide paper electrode; CPRG: chlorophenol red-β-D-galactopyranoside; NR: neutral red; PANI: polyaniline; SPCE: screen-printed carbon electrode; Ab: antibody; *S. aureus*: *Staphylococcus aureus*; GO: graphene oxide; rGO: reduced graphene oxide; SWCNT: single walled carbon nanotube; BMNCs: bimetallic nanoclusters; TMB: 3,3′,5,5′-tetramethylbenzidine; *E. faecalis*: *Enterococcus faecalis*; *Salmonella typh.*: *Salmonella typhimurium*; GA: glutaraldehyde; PAMAM: poly(amidoa-mine); Con A: lectin concanavalin A; XαGlc: 5-bromo-4-chloro-3-indolyl-α-D-glucopyranoside; *Listeria*: *listeria monocytogenes*; X-InP: 5-bromo-4-chloro-3-indolyl-myo-inositol phosphate; Magenta caprylate: 5-bromo-6-chloro-3-indolyl caprylate; PNIPAm: poly(N-isopropylacrylamide); *B. subtilis*: *Bacillus subtilis*; *S. Mutans*: *Streptococcus mutans*; MRSA: methicillin-resistant Staphylococcus aureus; *S. agalactiae*: *Streptococcus agalactiae*; *klebsiella*: *klebsiella pneumonia*; *E. aerogenes*: *Enterobacter aerogenes*; *P. aeruginosa*: *Pseudomonas aeruginosa*.

10.4. Heavy Metal Ions Detection

The waste from mines and other industrial centers, fertilizers, and pesticides, as well as fuels and pollutants from vehicles, ships, and heavy machinery, can be sources of heavy metal ions [313]. The entry of the metal ions into the environment reduces the quality of water, air, and soil, which in turn leads to the extinction of various plant and animal species [314]. When the human body is exposed to the heavy metal ions, they disrupt the body's immune system, resulting in respiratory, skin, digestive, kidney, and liver problems and even cancer, in acute cases [315]. As mentioned in Table 4, most studies on heavy metal poisoning have been performed for drinking and ambient water samples. According to the protocol of the World Health Organization, the permissible limit of heavy metals in aqueous samples should be in the range of $0.01-0.05\ mgL^{-1}$ [315].

Mercury (II) ions (Hg(II)) is a heavy metal ion with a limit of 6 μgL^{-1} in aqueous samples [316]. To determine Hg(II) ions by paper sensors, the following three structures have been used: micro-spot, microfluidic, and distance quantitation (Table 4). By preparing a paper aptasensor-based micro spot, Chen et al. determined Hg(II) ions with a detection limit of 50 nm in river and pond water samples [317]. A single-stranded DNA is physically bound to Au nanoparticles. In the presence of the analyte, the structure of the single-stranded DNA changes to the hairpin structure, due to the binding of thymine–Hg(II)–thymine, giving rise to the separation of the aptamer from the nanoparticles. By adding salt to the reaction medium, the electron repulsion of the bare nanoparticles is reduced, causing the nanoparticles aggregation and leading to a change in the color of the sensor from red to blue [317]. Pt nanoparticles can have peroxidase-like properties, catalyzing the TMB oxidation reaction to its oxidized form and changing the color of the sensor from colorless to blue [318]. Furthermore, Pt nanoparticle can be compounded with the Hg(II) ion, producing Hg-Pt alloy [318]. As a result, the presence of Hg(II) in the reaction medium inhibits the enzymatic activity and reduces the intensity of the blue color of the sensor [318]. Chen et al. used this mechanism to determine Hg(II) ions in drinking and pond water samples, having a detection limit of 0.01 μM, performed on the micro-spot paper [318]. Some mercury sensors have been fabricated based on the length measurement. The sensing element in these sensors can be a colorimetric reagent (e.g., dithizone) [319] or a fluorescent compound (e.g., nitrogen-doped carbon dots (NCDs)) [316]. In the former, the color change (from yellow to purple) is due to the formation of dithizone-Hg precipitate on the surface of the paper, detecting Hg(II) ions with a detection limit of 0.93 μgmL^{-1} [319]. In the latter, the analyte causes the fluorescence of NCDs to be quenched. This fluorescence quench is due to the formation of a non-fluorescent Hg-NCDs complex, arising from the covalent interaction of NCDs electrons with empty orbitals of the Hg(II) ion. The electrons of NCDs are supplied by the C=N, C=O, and C−OH functional groups in the carbon dot structure. On the other hand, the surface of CDs has been modified by ethylenediamine, so that the mercury tends to interact with the nitrogen existing in this compound [316]. Based on the observations by Ninwong et al., the detection limit of the latter method is 0.005 μgmL^{-1}, being considerably smaller than that of the colorimetric method [316]. Nashukha et al. have designed a membraneless gas separation μPAD for determining Hg (II) ions [320]. The proposed sensor consists of donor and acceptor layers together with an interlayer space. In the donor layer, analyte with iodide added in large quantities to the medium forms a water-soluble $HgI_4{}^{2-}$ complex. The remaining iodide reacts with the iodate (already immobilized on the paper), resulting in the production of volatile iodine. In turn, the iodine vapors pass through the interlayer space and penetrate the acceptor layer, so that they can react with the iodide-starch indicator to form a tri-iodide starch complex. Accordingly, the color of the acceptor layer changes to purple, corresponding to the concentration of Hg(II) ions [320].

Based on the US Environmental Protection Agency (EPA), the permissible level of copper (II) (Cu(II)) ions in a drinking water sample is 1.3 ppm [321]. Numerous paper sensors have been designed to detect Cu(II) ions, based on colorimetric, fluorometric, and electrochemical methods. In the colorimetric method, the paper sensor can have 2D microfluidic, lateral flow, or 3D origami structure (Table 4).

Ratnarathorn et al. have used a 2D microfluidic structure to measure Cu(II) ions in tap and pond water samples [322]. The measurement was performed based on the accumulation of silver nanoparticles, whose surface was simultaneously modified by homocysteine and dithiothreitol. The detection limit of this method was reported to be 7.8 nM [322]. Quinn et al. introduced distance-based μPADs, whose detection zone was coated with dithiooxamide [321]. In this regard, the metal ion is separated from the sample texture using the solid-phase extraction method, eliminating the interfering effect of the texture, while also causing the metal ion to be pre-concentrated. This improves the detection limit of the method up to 20 ppb [321]. One of the disadvantages of microfluidic methods is that the sensing element in the sensor moves around the detection zone with the sample stream, making the monitoring of the sensor response erroneous [276]. Moreover, a part of the sensor response disappears in the microfluidic methods. Polyvinylchloride can be used to stabilize the color on the paper uniformly [62]. Sharifi et al. used this mechanism to quantitatively determine Cu(II) ions with pyrocatechol violet and chrome azurol S indicators [62]. In the origami structure, a waste layer is designed to collect the ions unreacted with the color indicator, improving the detection limit of Cu(II) ion measurements with organic color indicators to less than 2 mgL^{-1} [62]. To electrochemically study Cu(II) ions, Wang et al. designed a three-electrode paper device, using a magnetron sputtering technology (MST) on a nitrocellulose paper substrate [323]. The MST creates a uniform porous structure, without the need for additional surface modification. The measurement was performed by square-wave stripping voltammetry, with a detection limit of 2 μgL^{-1} [323].

The fluorescence detection was performed with sensing elements, such as quantum dots and nanoclusters. At Fang's suggestion, a distance-based μPAD was designed with the help of BSA-Au nanocrystals [324]. The fluorescence of the nanocluster is quenched in the presence of metal ions, allowing for adjusting the detection limit by controlling the water absorbed on the pad. Therefore, it can easily be used to detect Cu(II) ions in complex samples such as blood, soil, and sewage. The sensor response is not affected by a mixture of other ions, providing sensitivities in the range of 5 μM [324].

The poisoning by cadmium (II) (Cd(II)) ions can cause serious damage to the liver and kidneys [325]. The permissible limit of Cd(II) in the aqueous sample has been reported to be between 3 and 5 ppb [326]. It is possible to specifically determine Cd(II) ions by a lateral flow immunoassay method. López Marzo et al. [327] proposed a method based on a competitive reaction between Cd−EDTA complex and Cd−EDTA−BSA−AuNP to interact with the same active site of antibody immobilized on the test line. In the absence of cadmium (II) ion, a distinct red color is detected in the test line whose intensity decreases with increasing thee cadmium concentration and replacement of Cd−EDTA with conjugated nanoparticles [327]. In another study, the lock-and-key theory has been employed to design the sensor. The receptor was an ion-imprinted polymer that reacted specifically with Cd(II) ions. Increasing the concentration of metal ions changed the color of the sensor with an origami paper structure from yellow to red. The detection limit of the measurement was 0.4 $ng\cdot mL^{-1}$ [328].

Since the permissible level of lead (II) (Pb(II)) ions in ambient samples is 15 ppb [329], POCT has been designed to detect them in a variety of paper samples. In this design, carbon dots can act as sensing elements. Gupta et al. have synthesized CDs by heating a biological medium such as potato-dextrose agar (PDA) in the microwave [330]. The chemical structure of these CDs comprises hydroxyl and carboxylic groups, forming a stable complex with Pb(II) ions which results in fluorescence quenching of CDs. The detection limit of this micro-spot paper sensor was 110 pM [330]. Wang et al. employed a combination of two CDs with blue and red fluorescence emission as the sensing element [331]. The blue CDs were synthesized by a combination of sodium citrate and polyacrylamide, having amine and carboxylic groups. The red CDs contained amine groups related to p-phenylenediamine (p-PDA). In the presence of Pb (II) ions, the fluorescence of blue CDs is quenched due to the interaction of lead with the carboxylic groups, whereas the emission intensity of red CDs increases, providing a detection limit of 2.8 nM [331].

Compared to other metal ions, the presence of chromium (III) (Cr(III)) ions is recommended in the human diet at concentrations of 50–200 mgdL^{-1}, influencing the glucose, lipid, and protein metabolism efficiently [332] However, at high concentrations, Cr(III) ion can damage the cell by binding to DNA [332]. In order to determine Cr(III) ions, Elavarasi et al. proposed a paper sensor based on gold nanoparticles, synthesized by citrate, without the coating of any other chemical compound on their surface [333]. In the presence of metal ions, gold nanoparticles accumulate and change color from red to blue. This accumulation is due to the tendency of chromium (III) ions to interact with citrate oxygen groups, demonstrating a high selectivity to the ions [333].

The paper sensors are capable of detecting multiple metal ions in ambient samples simultaneously. To this end, Feng et al. [334] have designed a sensor array consisting of nine indicators from the derivatives of 4,4-difluoro-4-bora-3a,4a-diaza-s-indacen, allowing for qualitatively and quantitatively determining seven metal ions, including cobalt(II), mercury(II), copper(II), cadmium(II), nickel(II), zinc(II), and silver(I) in aqueous waste samples with a detection limit of 10^{-7} M. The designed sensor comprised an absorber layer located below the detection area, absorbing 800 μL of the solution containing metal ions to be in contact with fluorescence indicators [334]. In another study, Feng et al. have used a microfluidic structure to simultaneously detect metal ions. The detection process was performed by changing the color of pyridylazo derivatives. As an advantage, this detection method does not require tools to inject the sample onto the paper surface as it is in constant contact with the indicator. In the aforementioned studies, cross reactive indicators were used to design the sensor [335]. Nevertheless, in some cases, it is possible to use indicators that specifically interact with a metal species. For example, nickel(II), chromium(VI), mercury(II), and iron(II) ions can be detected using Dimethylglyoxim [336–338], 1,5-diphenylcarbazide [336], Michler's thioketone [336], bathophenanthroline [337], or 10-phenanthroline [338], respectively. To detect copper (II) ions, the following materials can be used: diethyldithiocarbamate [337], bathocuproine [338], and cuprizone [339]. Sensors with strip or micro-spot structure based on these indicators are employed to evaluate the contamination of drinking and environmental water samples [340,341]. Most of the sensor designs use image analysis software with the capability of being installed on a smartphone, thus analyzing the corresponding color changes.

Table 4. Different type of paper-based sensor for heavy metal detection.

Type of Metal Ions	Device Structure	Detection Method	Sensing Element	Media	Linear Range	Detection Limit	Ref.
Hg(II)	Microfluidic assay	Colorimetric	KI, KIO_3, Starch	Soil	50–350 mg L^{-1}	20 mg L^{-1}	[320]
Hg(II)	Distance-based sensor	Fluorimetric	NCD	Drinking, pond, and tap waters	0.5–25 mg L^{-1}	0.5 mg L^{-1}	[316]
Hg(II)	Distance-based sensor	Colorimetric	Dithizone	whitening cream	1–30 µg mL^{-1}	0.93 µg mL^{-1}	[319]
Hg(II)	Micro spot	Colorimetric	PtNPs-TMB	Pond and tap waters	0.025–0.5 µM	0.01 µM	[318]
Hg(II)	Micro spot	Colorimetric	ssDNA-AuNPs	Water	0–2 µM	50 nM	[317]
Cu(II)	Microfluidic assay	Colorimetric	Hcy-DTT-AgNP	Water	7.8–62.8 µM	7.8 nM	[322]
Cu(II)	Microfluidic assay	Fluorometric	CdTe QDs-Cu-IIP	Sea and lake waters	0.032-3.2 mg L^{-1}	0.012 mg L^{-1}	[342]
Cu(II)	Distance-based sensor	Fluorometric	BSA-AuNCs		5–500 µM	5 µM	[324]
Cu(II)	Distance-based sensor	Colorimetric	Dithiooxamide	Drinking water	20–500,000 ppb	20 ppb	[321]
Cu(II)	Paper electrode	SWSV		Lake waters	5–1000 µg L^{-1}	2 µg·L^{-1}	[323]
Cu(II)	Micro paper	Fluorometric	CdTe QDs/GCNNs	Tea soup, orange juice, and red wine	0.01~5.0 µg·mL^{-1}	0.47 ng·mL^{-1}	[343]
Cu(II)	Origami	Colorimetric	Chrome azurol S, Pyrocatechol violet	Rain and Tab waters	5.0–1400.0 mg L^{-1} 5.0–200.0 mg L^{-1}	1.7 mg L^{-1} 1.9 mg L^{-1}	[62]
Cr(III)	Paper strip	Colorimetric	Citrate-AuNPs		10^{-3}–10^{-6} M	1.06×10^{-7} M	[333]
Cd(II)	Lateral flow	Colorimetric	Antibody/modified AuNPs	Drinking and tap waters	0.4–10 ppb	0.1 ppb	[327]
Cd(II)	Origami	Colorimetric	Ion imprinted polymer	Water	1–100 ng mL^{-1}	0.4 ng mL^{-1}	[328]
Pb(II)	Paper strip	Fluorometric	CDs (potato-dextrose agar)	Human cells	Up to 1 µM	106 pM	[330]
Pb(II)	Paper strip	Colorimetric	Mixture of blue CDs and red CDs	Tap water and lake water	15−80 nM	2.89 nM	[331]
Hg(II) Cu(II)	Origami	Fluorometric	CdTe QDs-IIP	Lake and sea waters	0.26–34.0 µgL^{-1} 0.11–58.0 µgL^{-1}	0.056 µg·L^{-1} 0.035 µg·L^{-1}	[340]
Hg(II) Pb(II)	Origami	ECL	Si@CNCs and Ru@AuNPs	Lake water and human serum	5.0×10^{-10} to 1.0×10^{-6} M 3.0×10^{-11}–3.0×10^{-6} M	0.2 nM 10 pM	[344]
Cd(II) Pb(II)	Printed electrode	SWSV		Salty soda and dirty ground waters	10–100 ppb 10–100 ppb	2.3 ppb 2.0 ppb	[345]
Ni(II), Cr(VI), Hg(II)	Microfluidic assay	Colorimetric	DMG DPC MT	Lake water		0.24 ppm 0.18 ppm 0.19 ppm	[336]

Table 4. *Cont.*

Type of Metal Ions	Device Structure	Detection Method	Sensing Element	Media	Linear Range	Detection Limit	Ref.
Fe(II) Ni(II) Cu(II)	Barrier-free patterned paper	Colorimetric	BP DMG DDC	Pond water	0.5–20 ppm 0.4–20 ppm 0.5–20 ppm	0.25 ppm 0.4 ppm 0.5 ppm	[337]
Fe(II) Cu(II) Ni(II)	Micro spot	Colorimetric	BC Phen DMG	Tap and lake water and papermaking wastewater	0.5–500 mg L^{-1} 0.5–500 mg L^{-1} 2–500 mg L^{-1}	0.5 mg L^{-1} 0.5 mg L^{-1} 2 mg L^{-1}	[338]
Zn(II) Cr(II) Cu(II) Pb(II) Mn(II)	Filter paper	Colorimetric	ZI, cyanide and cyclohexanone DPC CPZ ALS PAN and cyanide	Wastewater	2.00–6.00 mg L^{-1} 0.10–0.50 mg L^{-1} 0.30–8.00 mg L^{-1} 0.08–0.60 mg L^{-1} 0.20–1.00 mg L^{-1}	0.63 mg L^{-1} 0.07 mg L^{-1} 0.17 mg L^{-1} 0.03 mg L^{-1} 0.11 mg L^{-1}	[339]
Hg(II) Ag(I) Cu(II) Cd(II) Pb(II), Cr(VI) Ni(II)	Microfluidic assay	Colorimetric	CPRG	Distilled, tap, lake, and fall water		0.001 ppm 0.002 ppm 0.020 ppm 0.020 ppm 0.140 ppm 0.150 ppm 0.230 ppm	[341]
Hg(II), Cd(II), Co(II), Cu(II), Ni(II), Zn(II), and Ag(I)	Array-based e-tongue	Fluorometric	DPA derivatives	Wastewater		10^{-7} M	[334]
Hg(II), Cd(II), Co(II), Cu(II), Ni(II), Zn(II), and Pb(II)	Array-based e-tongue	Colorimetric	Pyridylazo compounds	Sewage water		50 µM	[335]

NCD: Nitrogen-doped carbon dots; TMB: 3,3′,5,5′-Tetramethylbenzidine; Hcy: homocysteine; DTT: Dithiothreitol; Cu-IIP: Cu(II) imprinted polymers; BSA: bovine serum albumi; SWSV: square-wave stripping voltammetry (SWSV); GCNNs: graphite carbon nitride; CDs: carbon dots; IIP: ion imprinted polymer; ECL: electrochemiluminescence; Ru@AuNPs: $Ru(bpy)_3{}^{2+}$-gold nanoparticles (AuNPs); Si@CNCs: carbon nanocrystals (CNCs) capped silica nanoparticles; ZI: zincon; DPC: 1,5-Diphenylcarbazide; CPZ: cuprizone; ALS: alizarin red S; PAN: 1-(2-pyridylazo)-2-naphthol; MT: Michler's thioketone; DMG: dimethylglyoxim; BP: Bathophenanthroline; DDC: diethyldithiocarbamate; BC: bathocuproine; Phen: 1,10-phenanthroline; CPRG: chlorophenol red β-galactopyrano-side; DPA: Di-2-picolyamine.

11. Conclusions

Due to their simplicity, low cost, easy fabrication, ease of use, and reliable performance, PPOCTs have been welcomed by a large number of research groups. They consume extremely small volumes of analytes, indicators, and receptors. PPOCTs are diverse in design structure and device dimension, having a high ability to detect the analyte in different physical states. However, the performance of these sensors is limited by some disadvantages, such as sample evaporation on the surface of paper, sample entrapment between the paper fibers, low mechanical stability, the infiltration of samples or reagent into the paper layers, appearing coffee effect in the detection zones, variations in the flow rate of samples due to their different viscosity, using readers with low sensitivity, interference of the environmental factors in sensor stability and the optical factors in collection of sensor responses. Limitations can be reduced by some effective solutions, such as sealing paper substrates, modifying the surface of channels, and detecting zones with polymers or proteins to block the paper pores or hydrophobicize the reagents. This can be achieved by using a sealed box to eliminate ambient light (resulting in a reproducible response) and connecting the readers to the wireless system, enabling us to receive the information without the presence in the infected area or to transfer the information to a strategic system. Most of these sensors are commercially available and are used by medical, food, pharmaceutical, and forensics centers, as well as by the environment and the general public. Overall, a variety of paper biosensors have been reported to detect the hazardous contaminations mentioned in this review. However, the development of paper tools for the simultaneous detection of aflatoxins, organophosphates, bacteria, and metal ions (using bioreceptors and paper sensor arrays) could attract the further attention of researchers.

Author Contributions: M.M.B.; conceptualization, methodology, writing–original draft preparation, A.S.; investigation, writing–original draft preparation, P.H.; conceptualization, methodology, writing–original draft preparation, A.H.; conceptualization, methodology, writing–review, and editing, H.B. All authors have read and agreed to the published version of the manuscript.

Funding: This study has not received any funding.

Institutional Review Board Statement: Not applicable.

Informed Consent Statement: Not applicable.

Data Availability Statement: Not applicable.

Acknowledgments: The authors gratefully acknowledge Research Council of Baqiyatallah University of Medical Sciences.

Conflicts of Interest: The authors declare no conflict of interest.

References

1. Vallero, D.A. Fundamentals of environmental chemistry. In *Environmental Contaminants: Assessment and Control*; Elsevier: Amsterdam, The Netherlands, 2010; pp. 289–332.
2. Kudr, J.; Zitka, O.; Klimanek, M.; Vrba, R.; Adam, V. Microfluidic electrochemical devices for pollution analysis–A review. *Sens. Actuators B Chem.* **2017**, *246*, 578–590. [CrossRef]
3. WHO. *7 Million Premature Deaths Annually Linked to Air Pollution*; World Health Organization: Geneva, Switzerland, 2019.
4. Landrigan, P.J.; Fuller, R. Global health and environmental pollution. *Int. J. Public Health* **2015**, *60*, 761–762. [CrossRef] [PubMed]
5. Lead poisoning and health. *Saudi Med. J.* **2013**, *34*, 1090–1091.
6. WHO. *Arsenic*; World Health Organization: Geneva, Switzerland, 2016.
7. Sidhu, G.K.; Singh, S.; Kumar, V.; Dhanjal, D.S.; Datta, S.; Singh, J. Toxicity, monitoring and biodegradation of organophosphate pesticides: A review. *Crit. Rev. Environ. Sci. Technol.* **2019**, *49*, 1135–1187. [CrossRef]
8. Bordbar, M.M.; Hemmateenejad, B.; Tashkhourian, J.; Nami-Ana, S.F. An optoelectronic tongue based on an array of gold and silver nanoparticles for analysis of natural, synthetic and biological antioxidants. *Microchim. Acta* **2018**, *185*, 493. [CrossRef]
9. George, M. The origins and the future of microfluidics. *Nature* **2006**, *442*, 368.
10. Ren, K.; Zhou, J.; Wu, H. Materials for microfluidic chip fabrication. *Acc. Chem. Res.* **2013**, *46*, 2396–2406. [CrossRef]
11. Olanrewaju, A.; Beaugrand, M.; Yafia, M.; Juncker, D. Capillary microfluidics in microchannels: From microfluidic networks to capillaric circuits. *Lab Chip* **2018**, *18*, 2323–2347. [CrossRef] [PubMed]

12. Hansen, S.; Wahed, A.A. El Point-of-care or point-of-need diagnostic tests: Time to change outbreak investigation and pathogen detection. *Trop. Med. Infect. Dis.* **2020**, *5*, 151. [CrossRef] [PubMed]
13. Niculescu, A.G.; Chircov, C.; Bîrcă, A.C.; Grumezescu, A.M. Fabrication and applications of microfluidic devices: A review. *Int. J. Mol. Sci.* **2021**, *22*, 2011. [CrossRef]
14. Vashist, S.K. Point-of-care diagnostics: Recent advances and trends. *Biosensors* **2017**, *7*, 62. [CrossRef] [PubMed]
15. Kumar, S.; Nehra, M.; Khurana, S.; Dilbaghi, N.; Kumar, V.; Kaushik, A.; Kim, K.H. Aspects of point-of-care diagnostics for personalized health wellness. *Int. J. Nanomed.* **2021**, *16*, 383–402. [CrossRef]
16. Mohankumar, P.; Ajayan, J.; Mohanraj, T.; Yasodharan, R. Recent developments in biosensors for healthcare and biomedical applications: A review. *Meas. J. Int. Meas. Confed.* **2021**, *167*, 108293. [CrossRef]
17. Chen, C.; Wang, J. Optical biosensors: An exhaustive and comprehensive review. *Analyst* **2020**, *145*, 1605–1628. [CrossRef]
18. Chen, S.; Shamsi, M.H. Biosensors-on-chip: A topical review. *J. Micromech. Microeng.* **2017**, *27*, 083001. [CrossRef]
19. Azmi, A.; Azman, A.A.; Ibrahim, S.; Yunus, M.A.M. Techniques in advancing the capabilities of various nitrate detection methods: A review. *Int. J. Smart Sens. Intell. Syst.* **2017**, *10*, 223–261. [CrossRef]
20. Nery, E.W.; Kubota, L.T. Sensing approaches on paper-based devices: A review. *Anal. Bioanal. Chem.* **2013**, *405*, 7573–7595. [CrossRef] [PubMed]
21. Nguyen, T.; Chidambara, V.A.; Andreasen, S.Z.; Golabi, M.; Huynh, V.N.; Linh, Q.T.; Bang, D.D.; Wolff, A. Point-of-care devices for pathogen detections: The three most important factors to realise towards commercialization. *TrAC Trends Anal. Chem.* **2020**, *131*, 116004. [CrossRef]
22. Cate, D.M.; Adkins, J.A.; Mettakoonpitak, J.; Henry, C.S. Recent developments in paper-based microfluidic devices. *Anal. Chem.* **2015**, *87*, 19–41. [CrossRef]
23. Gong, M.M.; Sinton, D. Turning the Page: Advancing Paper-Based Microfluidics for Broad Diagnostic Application. *Chem. Rev.* **2017**, *117*, 8447–8480. [CrossRef]
24. Bracher, P.J.; Gupta, M.; Whitesides, G.M. Shaped films of ionotropic hydrogels fabricated using templates of patterned paper. *Adv. Mater.* **2009**, *21*, 445–450. [CrossRef]
25. Tseng, S.C.; Yu, C.C.; Wan, D.; Chen, H.L.; Wang, L.A.; Wu, M.C.; Su, W.F.; Han, H.C.; Chen, L.C. Eco-friendly plasmonic sensors: Using the photothermal effect to prepare metal nanoparticle-containing test papers for highly sensitive colorimetric detection. *Anal. Chem.* **2012**, *84*, 5140–5145. [CrossRef]
26. Ngo, Y.H.; Li, D.; Simon, G.P.; Garnier, G. Gold nanoparticle-paper as a three-dimensional surface enhanced raman scattering substrate. *Langmuir* **2012**, *28*, 8782–8790. [CrossRef]
27. Jarujamrus, P.; Tian, J.; Li, X.; Siripinyanond, A.; Shiowatana, J.; Shen, W. Mechanisms of red blood cells agglutination in antibody-treated paper. *Analyst* **2012**, *137*, 2205–2210. [CrossRef] [PubMed]
28. Bordbar, M.M.; Tashkhourian, J.; Hemmateenejad, B. Structural Elucidation and Ultrasensitive Analyses of Volatile Organic Compounds by Paper-Based Nano-Optoelectronic Noses. *ACS Sens.* **2019**, *4*, 1442–1451. [CrossRef]
29. Singh, A.T.; Lantigua, D.; Meka, A.; Taing, S.; Pandher, M.; Camci-Unal, G. Paper-based sensors: Emerging themes and applications. *Sensors* **2018**, *18*, 2838. [CrossRef] [PubMed]
30. Kuswandi, B.; Ensafi, A.A. Perspective—Paper-Based Biosensors: Trending Topic in Clinical Diagnostics Developments and Commercialization. *J. Electrochem. Soc.* **2020**, *167*, 037509. [CrossRef]
31. Ratajczak, K.; Stobiecka, M. High-performance modified cellulose paper-based biosensors for medical diagnostics and early cancer screening: A concise review. *Carbohydr. Polym.* **2020**, *229*, 115463. [CrossRef] [PubMed]
32. Kouisni, L.; Rochefort, D. Confocal microscopy study of polymer microcapsules for enzyme immobilisation in paper substrates. *J. Appl. Polym. Sci.* **2009**, *111*, 1–10. [CrossRef]
33. Akyazi, T.; Basabe-Desmonts, L.; Benito-Lopez, F. Review on microfluidic paper-based analytical devices towards commercialisation. *Anal. Chim. Acta* **2018**, *1001*, 1–17. [CrossRef] [PubMed]
34. Fu, L.M.; Wang, Y.N. Detection methods and applications of microfluidic paper-based analytical devices. *TrAC Trends Anal. Chem.* **2018**, *107*, 196–211. [CrossRef]
35. Tang, R.H.; Yang, H.; Choi, J.R.; Gong, Y.; Feng, S.S.; Pingguan-Murphy, B.; Huang, Q.S.; Shi, J.L.; Mei, Q.B.; Xu, F. Advances in paper-based sample pretreatment for point-of-care testing. *Crit. Rev. Biotechnol.* **2017**, *37*, 411–428. [CrossRef] [PubMed]
36. Bracher, P.J.; Gupta, M.; Whitesides, G.M. Patterned paper as a template for the delivery of reactants in the fabrication of planar materials. *Soft Matter* **2010**, *6*, 4303–4309. [CrossRef]
37. Tang, R.H.; Liu, L.N.; Zhang, S.F.; He, X.C.; Li, X.J.; Xu, F.; Ni, Y.H.; Li, F. A review on advances in methods for modification of paper supports for use in point-of-care testing. *Microchim. Acta* **2019**, *186*, 521. [CrossRef] [PubMed]
38. Bongiovanni, R.; Zeno, E.; Pollicino, A.; Serafini, P.M.; Tonelli, C. UV light-induced grafting of fluorinated monomer onto cellulose sheets. *Cellulose* **2011**, *18*, 117–126. [CrossRef]
39. Princi, E.; Vicini, S. Graft polymerisation of ethyl acrylate/methyl methacrylate copolymers: A tool for the consolidation of paper-based materials. *Eur. Polym. J.* **2008**, *44*, 2392–2403. [CrossRef]
40. Liana, D.D.; Raguse, B.; Justin Gooding, J.; Chow, E. Recent advances in paper-based sensors. *Sensors* **2012**, *12*, 11505–11526. [CrossRef]
41. Kühl, S.; Krummenauer, F.; Dagassan-Berndt, D.; Lambrecht, T.J.; d'Hoedt, B.; Schulze, R.K.W. Ink-jet printout of radiographs on transparent film and glossy paper versus monitor display: An ROC analysis. *Clin. Oral Investig.* **2011**, *15*, 351–356. [CrossRef]

42. Lin, Y.; Gritsenko, D.; Liu, Q.; Lu, X.; Xu, J. Recent Advancements in Functionalized Paper-Based Electronics. *ACS Appl. Mater. Interfaces* **2016**, *8*, 20501–20515. [CrossRef] [PubMed]
43. Kamel, S.; Khattab, T.A. Recent advances in cellulose-based biosensors for medical diagnosis. *Biosensors* **2020**, *10*, 67. [CrossRef]
44. Li, Z.; Askim, J.R.; Suslick, K.S. The Optoelectronic Nose: Colorimetric and Fluorometric Sensor Arrays. *Chem. Rev.* **2019**, *119*, 231–292. [CrossRef]
45. Li, Z.; Suslick, K.S. The Optoelectronic Nose. *Acc. Chem. Res.* **2021**, *54*, 950–960. [CrossRef]
46. Sola Martínez, R.A.; Pastor Hernández, J.M.; Yanes Torrado, Ó.; Cánovas Díaz, M.; de Diego Puente, T.; Vinaixa Crevillent, M. Exhaled volatile organic compounds analysis in clinical pediatrics: A systematic review. *Pediatr. Res.* **2021**, *89*, 1352–1363. [CrossRef] [PubMed]
47. Oakley-Girvan, I.; Davis, S.W. Breath based volatile organic compounds in the detection of breast, lung, and colorectal cancers: A systematic review. *Cancer Biomark.* **2017**, *21*, 29–39. [CrossRef]
48. Bel'skaya, L.V.; Sarf, E.A.; Shalygin, S.P.; Postnova, T.V.; Kosenok, V.K. Identification of salivary volatile organic compounds as potential markers of stomach and colorectal cancer: A pilot study. *J. Oral Biosci.* **2020**, *62*, 212–221. [CrossRef]
49. Mirzaei, Y.; Gholami, A.; Bordbar, M.M. A distance-based paper sensor for rapid detection of blood lactate concentration using gold nanoparticles synthesized by Satureja hortensis. *Sens. Actuators B Chem.* **2021**, *345*, 130445. [CrossRef]
50. Khalid, T.; Aggio, R.; White, P.; De Lacy Costello, B.; Persad, R.; Al-Kateb, H.; Jones, P.; Probert, C.S.; Ratcliffe, N. Urinary volatile organic compounds for the detection of prostate cancer. *PLoS ONE* **2015**, *10*, e0143283. [CrossRef] [PubMed]
51. Roy, M.; Yadav, B.K. Electronic nose for detection of food adulteration: A review. *J. Food Sci. Technol.* **2021**, 1–13. [CrossRef]
52. Bordbar, M.M.; Tashkhourian, J.; Hemmateenejad, B. Qualitative and quantitative analysis of toxic materials in adulterated fruit pickle samples by a colorimetric sensor array. *Sens. Actuators B Chem.* **2018**, *257*, 783–791. [CrossRef]
53. Hemmateenejad, B.; Tashkhourian, J.; Bordbar, M.M.; Mobaraki, N. Development of colorimetric sensor array for discrimination of herbal medicine. *J. Iran. Chem. Soc.* **2017**, *14*, 595–604. [CrossRef]
54. Shi, H.; Zhang, M.; Adhikari, B. Advances of electronic nose and its application in fresh foods: A review. *Crit. Rev. Food Sci. Nutr.* **2018**, *58*, 2700–2710. [CrossRef]
55. Zhang, Y.; Askim, J.R.; Zhong, W.; Orlean, P.; Suslick, K.S. Identification of pathogenic fungi with an optoelectronic nose. *Analyst* **2014**, *139*, 1922–1928. [CrossRef] [PubMed]
56. Lough, F.; Perry, J.D.; Stanforth, S.P.; Dean, J.R. Detection of exogenous VOCs as a novel in vitro diagnostic technique for the detection of pathogenic bacteria. *TrAC Trends Anal. Chem.* **2017**, *87*, 71–81. [CrossRef]
57. Tai, H.; Duan, Z.; Wang, Y.; Wang, S.; Jiang, Y. Paper-Based Sensors for Gas, Humidity, and Strain Detections: A Review. *ACS Appl. Mater. Interfaces* **2020**, *12*, 31037–31053. [CrossRef]
58. Boeker, P. On 'electronic nose'methodology. *Sens. Actuators B Chem.* **2014**, *204*, 2–17. [CrossRef]
59. Morbioli, G.G.; Mazzu-Nascimento, T.; Stockton, A.M.; Carrilho, E. Technical aspects and challenges of colorimetric detection with microfluidic paper-based analytical devices (μPADs)—A review. *Anal. Chim. Acta* **2017**, *970*, 1–22. [CrossRef] [PubMed]
60. He, Y.; Wu, Y.; Fu, J.Z.; Wu, W. Bin Fabrication of paper-based microfluidic analysis devices: A review. *RSC Adv.* **2015**, *5*, 78109–78127. [CrossRef]
61. Podrazka, M.; Báczyńska, E.; Kundys, M.; Jeleń, P.S.; Nery, E.W. Electronic tongue-A tool for all tastes? *Biosensors* **2017**, *8*, 3. [CrossRef]
62. Sharifi, H.; Tashkhourian, J.; Hemmateenejad, B. A 3D origami paper-based analytical device combined with PVC membrane for colorimetric assay of heavy metal ions: Application to determination of Cu(II) in water samples. *Anal. Chim. Acta* **2020**, *1126*, 114–123. [CrossRef]
63. Zhang, Y.; Li, X.; Li, H.; Song, M.; Feng, L.; Guan, Y. Postage stamp-sized array sensor for the sensitive screening test of heavy-metal ions. *Analyst* **2014**, *139*, 4887–4893. [CrossRef]
64. Yetisen, A.K.; Akram, M.S.; Lowe, C.R. Paper-based microfluidic point-of-care diagnostic devices. *Lab Chip* **2013**, *13*, 2210–2251. [CrossRef]
65. Nguyen, Q.H.; Kim, M.I. Nanomaterial-mediated paper-based biosensors for colorimetric pathogen detection. *TrAC Trends Anal. Chem.* **2020**, *132*, 116038. [CrossRef]
66. Quesada-González, D.; Merkoçi, A. Nanoparticle-based lateral flow biosensors. *Biosens. Bioelectron.* **2015**, *73*, 47–63. [CrossRef]
67. Parolo, C.; Sena-Torralba, A.; Bergua, J.F.; Calucho, E.; Fuentes-Chust, C.; Hu, L.; Rivas, L.; Álvarez-Diduk, R.; Nguyen, E.P.; Cinti, S.; et al. Tutorial: Design and fabrication of nanoparticle-based lateral-flow immunoassays. *Nat. Protoc.* **2020**, *15*, 3788–3816. [CrossRef] [PubMed]
68. Chen, Y.; Chen, Q.; Han, M.; Zhou, J.; Gong, L.; Niu, Y.; Zhang, Y.; He, L.; Zhang, L. Development and optimization of a multiplex lateral flow immunoassay for the simultaneous determination of three mycotoxins in corn, rice and peanut. *Food Chem.* **2016**, *213*, 478–484. [CrossRef] [PubMed]
69. Quesada-González, D.; Baiocco, A.; Martos, A.A.; de la Escosura-Muñiz, A.; Palleschi, G.; Merkoçi, A. Iridium oxide (IV) nanoparticle-based electrocatalytic detection of PBDE. *Biosens. Bioelectron.* **2019**, *127*, 150–154. [CrossRef] [PubMed]
70. Tsai, T.T.; Huang, T.H.; Chen, C.A.; Ho, N.Y.J.; Chou, Y.J.; Chen, C.F. Development a stacking pad design for enhancing the sensitivity of lateral flow immunoassay. *Sci. Rep.* **2018**, *8*, 17319. [CrossRef]
71. Jia, L.; David, M.; Joanne, M. Enhancing the signal of lateral flow immunoassays by using different developing methods. *Sens. Mater.* **2015**, *27*, 549–561.

72. Bahadır, E.B.; Sezgintürk, M.K. Lateral flow assays: Principles, designs and labels. *TrAC Trends Anal. Chem.* **2016**, *82*, 286–306. [CrossRef]
73. Luo, K.; Kim, H.Y.; Oh, M.H.; Kim, Y.R. Paper-based lateral flow strip assay for the detection of foodborne pathogens: Principles, applications, technological challenges and opportunities. *Crit. Rev. Food Sci. Nutr.* **2020**, *60*, 157–170. [CrossRef]
74. Huang, Y.; Xu, T.; Wang, W.; Wen, Y.; Li, K.; Qian, L.; Zhang, X.; Liu, G. Lateral flow biosensors based on the use of micro- and nanomaterials: A review on recent developments. *Microchim. Acta* **2020**, *187*, 70. [CrossRef]
75. Sharma, R.; Gautam, P.B.; Rajput, Y.S.; Mann, B.; Gandhi, K. Identification of analyte of interest through lateral flow assay. In Nano-Technological and Biochemical Techniques for Assessing the Quality and Safety of Milk and Milk Products. 2019. Available online: https://www.academia.edu/download/58315915/CAFT_Compendium.pdf#page=111 (accessed on 29 August 2021).
76. Byzova, N.A.; Zherdev, A.V.; Khlebtsov, B.N.; Burov, A.M.; Khlebtsov, N.G.; Dzantiev, B.B. Advantages of highly spherical gold nanoparticles as labels for lateral flow immunoassay. *Sensors* **2020**, *20*, 3608. [CrossRef] [PubMed]
77. Nguyen, V.T.; Song, S.; Park, S.; Joo, C. Recent advances in high-sensitivity detection methods for paper-based lateral-flow assay. *Biosens. Bioelectron.* **2020**, *152*, 112015. [CrossRef]
78. Apilux, A.; Rengpipat, S.; Suwanjang, W.; Chailapakul, O. Development of competitive lateral flow immunoassay coupled with silver enhancement for simple and sensitive salivary cortisol detection. *EXCLI J.* **2018**, *17*, 1198–1209. [CrossRef]
79. Rey, E.G.; O'Dell, D.; Mehta, S.; Erickson, D. Mitigating the Hook Effect in Lateral Flow Sandwich Immunoassays Using Real-Time Reaction Kinetics. *Anal. Chem.* **2017**, *89*, 5095–5100. [CrossRef]
80. Ge, X.; Asiri, A.M.; Du, D.; Wen, W.; Wang, S.; Lin, Y. Nanomaterial-enhanced paper-based biosensors. *TrAC Trends Anal. Chem.* **2014**, *58*, 31–39. [CrossRef]
81. Ross, G.M.S.; Salentijn, G.I.; Nielen, M.W.F. A critical comparison between flow-through and lateral flow immunoassay formats for visual and smartphone-based multiplex allergen detection. *Biosensors* **2019**, *9*, 143. [CrossRef]
82. Taghizadeh-Behbahani, M.; Hemmateenejad, B.; Shamsipur, M.; Tavassoli, A. A paper-based length of stain analytical device for naked eye (readout-free) detection of cystic fibrosis. *Anal. Chim. Acta* **2019**, *1080*, 138–145. [CrossRef]
83. Alsaeed, B.; Mansour, F.R. Distance-based paper microfluidics; principle, technical aspects and applications. *Microchem. J.* **2020**, *155*, 104664. [CrossRef]
84. Zhang, J.X.J.; Hoshino, K. Microfluidics and micro total analytical systems. In *Molecular Sensors and Nanodevices: Principles, Designs and Applications in Biomedical Engineering*, 2nd ed.; Elsevier: Amsterdam, The Netherlands, 2019; pp. 113–179. ISBN 978-0-12-814862-4.
85. Jiang, L.; Korivi, N.S. Microfluidics: Technologies and applications. *Nanolithogr. Art Fabr. Nanoelectron. Nanophotonic Devices Syst.* **2013**, 424–443. [CrossRef]
86. Melin, J.; Van Der Wijngaart, W.; Stemme, G. Behaviour and design considerations for continuous flow closed-open-closed liquid microchannels. *Lab Chip* **2005**, *5*, 682–686. [CrossRef]
87. Konda, A.; Morin, S.A. Flow-directed synthesis of spatially variant arrays of branched zinc oxide mesostructures. *Nanoscale* **2017**, *9*, 8393–8400. [CrossRef]
88. Chokkalingam, V.; Tel, J.; Wimmers, F.; Liu, X.; Semenov, S.; Thiele, J.; Figdor, C.G.; Huck, W.T.S. Probing cellular heterogeneity in cytokine-secreting immune cells using droplet-based microfluidics. *Lab Chip* **2013**, *13*, 4740–4744. [CrossRef] [PubMed]
89. Pesant, J.; Hareng, M.; Mourey, B.; Perbet, J. Electrodes for a Device Operating by Electrically Controlled Fluid Displacement. U.S. Patent 4569575, 11 February 1986.
90. Martinez, A.W.; Phillips, S.T.; Butte, M.J.; Whitesides, G.M. Patterned Paper as a Platform for Inexpensive, Low-Volume, Portable Bioassays. *Angew. Chem.* **2007**, *119*, 1340–1342. [CrossRef]
91. DeBlois, R.W.; Bean, C.P. Counting and sizing of submicron particles by the resistive pulse technique. *Rev. Sci. Instrum.* **1970**, *41*, 909–916. [CrossRef]
92. Kauffman, P.; Fu, E.; Lutz, B.; Yager, P. Visualization and measurement of flow in two-dimensional paper networks. *Lab Chip* **2010**, *10*, 2614–2617. [CrossRef] [PubMed]
93. Martinez, A.W.; Phillips, S.T.; Whitesides, G.M. Three-dimensional microfluidic devices fabricated in layered paper and tape. *Proc. Natl. Acad. Sci. USA* **2008**, *105*, 19606–19611. [CrossRef]
94. Liu, H.; Crooks, R.M. Three-dimensional paper microfluidic devices assembled using the principles of origami. *J. Am. Chem. Soc.* **2011**, *133*, 17564–17566. [CrossRef]
95. Santhiago, M.; Nery, E.W.; Santos, G.P.; Kubota, L.T. Microfluidic paper-based devices for bioanalytical applications. *Bioanalysis* **2014**, *6*, 89–106. [CrossRef]
96. Lim, H.; Jafry, A.T.; Lee, J. Fabrication, flow control, and applications of microfluidic paper-based analytical devices. *Molecules* **2019**, *24*, 2869. [CrossRef]
97. Fenton, E.M.; Mascarenas, M.R.; López, G.P.; Sibbett, S.S. Multiplex lateral-flow test strips fabricated by two-dimensional shaping. *ACS Appl. Mater. Interfaces* **2009**, *1*, 124–129. [CrossRef]
98. Sadri, B.; Goswami, D.; Martinez, R.V. Rapid fabrication of epidermal paper-based electronic devices using razor printing. *Micromachines* **2018**, *9*, 420. [CrossRef]
99. Cassano, C.L.; Fan, Z.H. Laminated paper-based analytical devices (LPAD): Fabrication, characterization, and assays. *Microfluid. Nanofluidics* **2013**, *15*, 173–181. [CrossRef]

100. Jafry, A.T.; Lim, H.; Sung, W.K.; Lee, J. Flexible time–temperature indicator: A versatile platform for laminated paper-based analytical devices. *Microfluid. Nanofluidics* **2017**, *21*, 57. [CrossRef]
101. Glavan, A.C.; Martinez, R.V.; Maxwell, E.J.; Subramaniam, A.B.; Nunes, R.M.D.; Soh, S.; Whitesides, G.M. Rapid fabrication of pressure-driven open-channel microfluidic devices in omniphobic RF paper *Lab Chip* **2013**, *13*, 2922–2930. [CrossRef]
102. Thuo, M.M.; Martinez, R.V.; Lan, W.J.; Liu, X.; Barber, J.; Atkinson, M.B.J.; Bandarage, D.; Bloch, J.F.; Whitesides, G.M. Fabrication of low-cost paper-based microfluidic devices by embossing or cut-and-stack methods. *Chem. Mater.* **2014**, *26*, 4230–4237. [CrossRef]
103. Theillet, G.; Rubens, A.; Foucault, F.; Dalbon, P.; Rozand, C.; Leparc-Goffart, I.; Bedin, F. Laser-cut paper-based device for the detection of dengue non-structural NS1 protein and specific IgM in human samples. *Arch. Virol.* **2018**, *163*, 1757–1767. [CrossRef]
104. Spicar-Mihalic, P.; Houghtaling, J.; Fu, E.; Yager, P.; Liang, T.; Toley, B. CO_2 laser cutting and ablative etching for the fabrication of paper-based devices. *J. Micromech. Microeng.* **2013**, *23*, 067003. [CrossRef]
105. Xia, Y.; Si, J.; Li, Z. Fabrication techniques for microfluidic paper-based analytical devices and their applications for biological testing: A review. *Biosens. Bioelectron.* **2016**, *77*, 774–789. [CrossRef]
106. Jeong, S.G.; Kim, J.; Nam, J.O.; Song, Y.S.; Lee, C.S. Paper-based analytical device for quantitative urinalysis. *Int. Neurourol. J.* **2013**, *17*, 155–161. [CrossRef]
107. Songok, J.; Tuominen, M.; Teisala, H.; Haapanen, J.; Mäkelä, J.M.; Kuusipalo, J.; Toivakka, M. Paper-Based Microfluidics: Fabrication Technique and Dynamics of Capillary-Driven Surface Flow. *ACS Appl. Mater. Interfaces* **2016**, *6*, 20060–20066. [CrossRef] [PubMed]
108. Sones, C.L.; Katis, I.N.; He, P.J.W.; Mills, B.; Namiq, M.F.; Shardlow, P.; Ibsen, M.; Eason, R.W. Laser-induced photo-polymerisation for creation of paper-based fluidic devices. *Lab Chip* **2014**, *14*, 4567–4574. [CrossRef] [PubMed]
109. Nargang, T.M.; Dierkes, R.; Bruchmann, J.; Keller, N.; Sachsenheimer, K.; Lee-Thedieck, C.; Kotz, F.; Helmer, D.; Rapp, B.E. Photolithographic structuring of soft, extremely foldable and autoclavable hydrophobic barriers in paper. *Anal. Methods* **2018**, *10*, 4028–4035. [CrossRef]
110. Carrilho, E.; Martinez, A.W.; Whitesides, G.M. Understanding wax printing A simple micropatterning process for paper-based microfluidics. *Anal. Chem.* **2009**, *81*, 7091–7095. [CrossRef] [PubMed]
111. Suresh, V.; Qunya, O.; Kanta, B.L.; Yuh, L.Y.; Chong, K.S.L. Non-invasive paper-based microfluidic device for ultra-low detection of urea through enzyme catalysis. *R. Soc. Open Sci.* **2018**, *5*, 171980. [CrossRef]
112. Songjaroen, T.; Dungchai, W.; Chailapakul, O.; Laiwattanapaisal, W. Novel, simple and low-cost alternative method for fabrication of paper-based microfluidics by wax dipping. *Talanta* **2011**, *85*, 2587–2593. [CrossRef]
113. Zhang, A.L.; Zha, Y. Fabrication of paper-based microfluidic device using printed circuit technology. *AIP Adv.* **2012**, *2*, 022171. [CrossRef]
114. Noh, H.; Phillips, S.T. Metering the capillary-driven flow of fluids in paper-based microfluidic devices. *Anal. Chem.* **2010**, *82*, 4181–4187. [CrossRef]
115. Maejima, K.; Tomikawa, S.; Suzuki, K.; Citterio, D. Inkjet printing: An integrated and green chemical approach to microfluidic paper-based analytical devices. *RSC Adv.* **2013**, *3*, 9258–9263. [CrossRef]
116. Li, X.; Tian, J.; Garnier, G.; Shen, W. Fabrication of paper-based microfluidic sensors by printing. *Colloids Surf. B Biointerfaces* **2010**, *76*, 564–570. [CrossRef] [PubMed]
117. Wang, J.; Monton, M.R.N.; Zhang, X.; Filipe, C.D.M.; Pelton, R.; Brennan, J.D. Hydrophobic sol-gel channel patterning strategies for paper-based microfluidics. *Lab Chip* **2014**, *14*, 691–695. [CrossRef]
118. Hiltunen, J.; Liedert, C.; Hiltunen, M.; Huttunen, O.H.; Hiitola-Keinänen, J.; Aikio, S.; Harjanne, M.; Kurkinen, M.; Hakalahti, L.; Lee, L.P. Roll-to-roll fabrication of integrated PDMS-paper microfluidics for nucleic acid amplification. *Lab Chip* **2018**, *18*, 1552–1559. [CrossRef]
119. Olkkonen, J.; Lehtinen, K.; Erho, T. Flexographically printed fluidic structures in paper. *Anal. Chem.* **2010**, *82*, 10246–10250. [CrossRef]
120. Bracher, P.J.; Gupta, M.; MacK, E.T.; Whitesides, G.M. Heterogeneous films of ionotropic hydrogels fabricated from delivery templates of patterned paper. *ACS Appl. Mater. Interfaces* **2009**, *1*, 1807–1812. [CrossRef]
121. Ghosh, R.; Gopalakrishnan, S.; Savitha, R.; Renganathan, T.; Pushpavanam, S. Fabrication of laser printed microfluidic paper-based analytical devices (LP-µPADs) for point-of-care applications. *Sci. Rep.* **2019**, *9*, 7896. [CrossRef]
122. Mohamed, H.M. Screen-printed disposable electrodes: Pharmaceutical applications and recent developments. *TrAC Trends Anal. Chem.* **2016**, *82*, 1–11. [CrossRef]
123. Lamas-Ardisana, P.J.; Martínez-Paredes, G.; Añorga, L.; Grande, H.J. Glucose biosensor based on disposable electrochemical paper-based transducers fully fabricated by screen-printing. *Biosens. Bioelectron.* **2018**, *109*, 8–12. [CrossRef] [PubMed]
124. Beitollahi, H.; Mohammadi, S.Z.; Safaei, M.; Tajik, S. Applications of electrochemical sensors and biosensors based on modified screen-printed electrodes: A review. *Anal. Methods* **2020**, *12*, 1547–1560. [CrossRef]
125. He, Y.; Wu, Y.; Fu, J.Z.; Gao, Q.; Qiu, J.J. Developments of 3D Printing Microfluidics and Applications in Chemistry and Biology: A Review. *Electroanalysis* **2016**, *28*, 1658–1678. [CrossRef]
126. He, Y.; Gao, Q.; Wu, W.B.; Nie, J.; Fu, J.Z. 3D printed paper-based microfluidic analytical devices. *Micromachines* **2016**, *7*, 108. [CrossRef] [PubMed]

127. Curto, V.F.; Lopez-Ruiz, N.; Capitan-Vallvey, L.F.; Palma, A.J.; Benito-Lopez, F.; Diamond, D. Fast prototyping of paper-based microfluidic devices by contact stamping using indelible ink. *RSC Adv.* **2013**, *3*, 18811–18816. [CrossRef]
128. Yao, X.H.; Jia, T.; Xie, C.Q.; Fu, J.Z.; He, Y. Facial fabrication of paper-based flexible electronics with flash foam stamp lithography. *Microsyst. Technol.* **2017**, *23*, 4419–4426. [CrossRef]
129. Liu, N.; Xu, J.; An, H.J.; Phan, D.T.; Hashimoto, M.; Lew, W.S. Direct spraying method for fabrication of paper-based microfluidic devices. *J. Micromech. Microeng.* **2017**, *27*, 104001. [CrossRef]
130. Nurak, T.; Praphairaksit, N.; Chailapakul, O. Fabrication of paper-based devices by lacquer spraying method for the determination of nickel (II) ion in waste water. *Talanta* **2013**, *114*, 291–296. [CrossRef] [PubMed]
131. Kwong, P.; Gupta, M. Vapor phase deposition of functional polymers onto paper-based microfluidic devices for advanced unit operations. *Anal. Chem.* **2012**, *84*, 10129–10135. [CrossRef] [PubMed]
132. Lawal, A.T.; Wallace, G.G. Vapour phase polymerisation of conducting and non-conducting polymers: A review. *Talanta* **2014**, *119*, 133–143. [CrossRef]
133. Demirel, G.; Babur, E. Vapor-phase deposition of polymers as a simple and versatile technique to generate paper-based microfluidic platforms for bioassay applications. *Analyst* **2014**, *139*, 2326–2331. [CrossRef]
134. Obeso, C.G.; Sousa, M.P.; Song, W.; Rodriguez-Pérez, M.A.; Bhushan, B.; Mano, J.F. Modification of paper using polyhydroxybutyrate to obtain biomimetic superhydrophobic substrates. *Colloids Surf. A Physicochem. Eng. Asp.* **2013**, *416*, 51–55. [CrossRef]
135. Sheini, A.; Aseman, M.D.; Bordbar, M.M. Origami paper analytical assay based on metal complex sensor for rapid determination of blood cyanide concentration in fire survivors. *Sci. Rep.* **2021**, *11*, 3521. [CrossRef]
136. Sheini, A. A paper-based device for the colorimetric determination of ammonia and carbon dioxide using thiomalic acid and maltol functionalized silver nanoparticles: Application to the enzymatic determination of urea in saliva and blood. *Microchim. Acta* **2020**, *187*, 565. [CrossRef] [PubMed]
137. Jeong, S.G.; Lee, S.H.; Choi, C.H.; Kim, J.; Lee, C.S. Toward instrument-free digital measurements: A three-dimensional microfluidic device fabricated in a single sheet of paper by double-sided printing and lamination. *Lab Chip* **2015**, *15*, 1188–1194. [CrossRef] [PubMed]
138. Mohammadifar, M.; Zhang, J.; Yazgan, I.; Sadik, O.; Choi, S. Power-on-paper: Origami-inspired fabrication of 3-D microbial fuel cells. *Renew. Energy* **2018**, *118*, 695–700. [CrossRef]
139. Park, C.; Han, Y.D.; Kim, H.V.; Lee, J.; Yoon, H.C.; Park, S. Double-sided 3D printing on paper towards mass production of three-dimensional paper-based microfluidic analytical devices (3D-μPADs). *Lab Chip* **2018**, *18*, 1533–1538. [CrossRef]
140. Ding, J.; Li, B.; Chen, L.; Qin, W. A Three-Dimensional Origami Paper-Based Device for Potentiometric Biosensing. *Angew. Chem. Int. Ed.* **2016**, *55*, 13033–13037. [CrossRef]
141. Razmi, N.; Baradaran, B.; Hejazi, M.; Hasanzadeh, M.; Mosafer, J.; Mokhtarzadeh, A.; de la Guardia, M. Recent advances on aptamer-based biosensors to detection of platelet-derived growth factor. *Biosens. Bioelectron.* **2018**, *113*, 58–71. [CrossRef] [PubMed]
142. Juzgado, A.; Soldà, A.; Ostric, A.; Criado, A.; Valenti, G.; Rapino, S.; Conti, G.; Fracasso, G.; Paolucci, F.; Prato, M. Highly sensitive electrochemiluminescence detection of a prostate cancer biomarker. *J. Mater. Chem. B* **2017**, *5*, 6681–6687. [CrossRef] [PubMed]
143. Vo-Dinh, T.; Cullum, B. Biosensors and biochips: Advances in biological and medical diagnostics. *Fresenius. J. Anal. Chem.* **2000**, *366*, 540–551. [CrossRef]
144. Valenti, G.; Rampazzo, E.; Biavardi, E.; Villani, E.; Fracasso, G.; Marcaccio, M.; Bertani, F.; Ramarli, D.; Dalcanale, E.; Paolucci, F.; et al. An electrochemiluminescence-supramolecular approach to sarcosine detection for early diagnosis of prostate cancer. *Faraday Discuss.* **2015**, *185*, 299–309. [CrossRef]
145. Holford, T.R.J.; Davis, F.; Higson, S.P.J. Recent trends in antibody based sensors. *Biosens. Bioelectron.* **2012**, *34*, 12–24. [CrossRef]
146. Aleman, J.; Kilic, T.; Mille, L.S.; Shin, S.R.; Zhang, Y.S. Microfluidic integration of regeneratable electrochemical affinity-based biosensors for continual monitoring of organ-on-a-chip devices. *Nat. Protoc.* **2021**, *16*, 2564–2593. [CrossRef]
147. JD, B.; BG, B.; WG, V. Measurement of monoclonal antibody affinity by non-competitive enzyme immunoassay. *J. Immunol. Methods* **1987**, *100*, 173–179.
148. Nistor, C.; Emnéus, J. Chapter 9 Immunoassay: Potentials and limitations. *Compr. Anal. Chem.* **2005**, *44*, 375–427. [CrossRef]
149. Jost, C.; Plückthun, A. Engineered proteins with desired specificity: DARPins, other alternative scaffolds and bispecific IgGs. *Curr. Opin. Struct. Biol.* **2014**, *27*, 102–112. [CrossRef] [PubMed]
150. Mehlhorn, A.; Rahimi, P.; Joseph, Y. Aptamer-based biosensors for antibiotic detection: A review. *Biosensors* **2018**, *8*, 54. [CrossRef]
151. Mello, L.D.; Kubota, L.T. Review of the use of biosensors as analytical tools in the food and drink industries. *Food Chem.* **2002**, *77*, 237–256. [CrossRef]
152. Du, Y.; Dong, S. Nucleic acid biosensors: Recent advances and perspectives. *Anal. Chem.* **2017**, *89*, 189–215. [CrossRef]
153. Bollella, P.; Gorton, L. Enzyme based amperometric biosensors. *Curr. Opin. Electrochem.* **2018**, *10*, 157–173. [CrossRef]
154. Mross, S.; Pierrat, S.; Zimmermann, T.; Kraft, M. Microfluidic enzymatic biosensing systems: A review. *Biosens. Bioelectron.* **2015**, *70*, 376–391. [CrossRef] [PubMed]
155. Marazuela, M.D.; Moreno-Bondi, M.C. Fiber-optic biosensors—An overview. *Anal. Bioanal. Chem.* **2002**, *372*, 664–682. [CrossRef] [PubMed]
156. Upadhyay, L.; Verm, N. Enzyme Inhibition Based Biosensors: A Review. *Anal. Lett.* **2013**, *46*, 225–241. [CrossRef]

157. Mehrotra, P. Biosensors and their applications—A review. *J. Oral Biol. Craniofacial Res.* **2016**, *6*, 153–159. [CrossRef]
158. Kazemi-Darsanaki, R.; Azizzadeh, A.; Nourbakhsh, M.; Raeisi, G.; AzizollahiAliabadi, M. Biosensors: Functions and Applications. *J. Biol. Today's World* **2013**, *2*, 20–23. [CrossRef]
159. Nakamura, H. Current status of water environment and their microbial biosensor techniques—Part II: Recent trends in microbial biosensor development. *Anal. Bioanal. Chem.* **2018**, *410*, 3967–3989. [CrossRef]
160. Varzakas, T.; Nikoleli, G.-P.; Nikolelis, D. *Tissue, Microorganisms, Organelles, and Cell-Based Biosensors*, 1st ed.; Taylor & Francis Group: Abingdon, UK, 2013.
161. Campàs, M.; Carpentier, R.; Rouillon, R. Plant tissue-and photosynthesis-based biosensors. *Biotechnol. Adv.* **2008**, *26*, 370–378. [CrossRef] [PubMed]
162. Jung, Y.; Jeong, J.Y.; Chung, B.H. Recent advances in immobilization methods of antibodies on solid supports. *Analyst* **2008**, *133*, 697–701. [CrossRef]
163. Nery, E.W.; Kubota, L.T. Evaluation of enzyme immobilization methods for paper-based devices—A glucose oxidase study. *J. Pharm. Biomed. Anal.* **2016**, *117*, 551–559. [CrossRef] [PubMed]
164. Kong, F.; Hu, Y.F. Biomolecule immobilization techniques for bioactive paper fabrication. *Anal. Bioanal. Chem.* **2012**, *403*, 7–13. [CrossRef]
165. Narsaiah, K.; Jha, S.N.; Bhardwaj, R.; Sharma, R.; Kumar, R. Optical biosensors for food quality and safety assurance—A review. *J. Food Sci. Technol.* **2012**, *49*, 383–406. [CrossRef]
166. Kaur, K.; Kaushal, P. Enzymes as analytical tools for the assessment of food quality and food safety. *Biomass Biofuels Biochem. Adv. Enzym. Technol.* **2019**, 273–292. [CrossRef]
167. Yamaguchi, H.; Kiyota, Y.; Miyazaki, M. Techniques for preparation of cross-linked enzyme aggregates and their applications in bioconversions. *Catalysts* **2018**, *8*, 174. [CrossRef]
168. Liebich, V.J.; Avrutina, O.: Habermann, J.; Hillscher, L.M.; Langhans, M.; Meckel, T.; Biesalski, M.; Kolmar, H. Toward Fabrication of Bioactive Papers: Covalent Immobilization of Peptides and Proteins. *Biomacromolecules* **2021**. [CrossRef]
169. Kasoju, N.; Nguyen, L.T.B.; Padalhin, A.R.; Dye, J.F.; Cui, Z.; Ye, H. Techniques for modifying biomaterials to improve hemocompatibility. *Hemocompat. Biomater. Clin. Appl. Blood-Biomater. Interact.* **2018**, 191–220. [CrossRef]
170. Al-Husseini, Z.N.O. A Literature Review on the Indicators in Precipitation. *Am. Int. J. Sci. Eng. Res.* **2019**, *2*, 1–8. [CrossRef]
171. Martinez, A.W.; Phillips, S.T.; Whitesides, G.M.; Carrilho, E. Diagnostics for the developing world: Microfluidic paper-based analytical devices. *Anal. Chem.* **2010**, *82*, 3–10. [CrossRef]
172. Liu, M.M.; Lian, X.; Liu, H.; Guo, Z.Z.; Huang, H.H.; Lei, Y.; Peng, H.P.; Chen, W.; Lin, X.H.; Liu, A.L.; et al. A colorimetric assay for sensitive detection of hydrogen peroxide and glucose in microfluidic paper-based analytical devices integrated with starch-iodide-gelatin system. *Talanta* **2019**, *200*, 511–517. [CrossRef]
173. Almeida, L.C.; Correia, J.P.; Viana, A.S. Electrochemical and optical characterization of thin polydopamine films on carbon surfaces for enzymatic sensors. *Electrochim. Acta* **2018**, *263*, 480–489. [CrossRef]
174. Mu, C.; Lu, H.; Bao, J.; Zhang, Q. Visual colorimetric 'turn-off' biosensor for ascorbic acid detection based on hypochlorite–3,3′,5,5′,-Tetramethylbenzidine system. *Spectrochim. Acta Part A Mol. Biomol. Spectrosc.* **2018**, *201*, 61–66. [CrossRef]
175. Gao, S.; Zheng, X.; Hu, B.; Sun, M.; Wu, J.; Jiao, B.; Wang, L. Enzyme-linked, aptamer-based, competitive biolayer interferometry biosensor for palytoxin. *Biosens. Bioelectron.* **2017**, *89*, 952–958. [CrossRef]
176. Ding, L.; Gong, Z.; Yan, M.; Yu, J.; Song, X. Determination of glucose by using fluorescent silicon nanoparticles and an inner filter caused by peroxidase-induced oxidation of o-phenylenediamine by hydrogen peroxide. *Microchim. Acta* **2017**, *184*, 4531–4536. [CrossRef]
177. Fan, Y.; Shi, S.; Ma, J.; Guo, Y. A paper-based electrochemical immunosensor with reduced graphene oxide/thionine/gold nanoparticles nanocomposites modification for the detection of cancer antigen 125. *Biosens. Bioelectron.* **2019**, *135*, 1–7. [CrossRef]
178. Dutta, G.; Lillehoj, P.B. An ultrasensitive enzyme-free electrochemical immunosensor based on redox cycling amplification using methylene blue. *Analyst* **2017**, *142*, 3492–3499. [CrossRef]
179. Hamidi-Asl, E.; Raoof, J.B.; Ojani, R.; Hejazi, M.S. Indigo carmine as new label in PNA biosensor for detection of short sequence of p53 tumor suppressor gene. *Electroanalysis* **2013**, *25*, 2075–2083. [CrossRef]
180. Chen, Q.; Huang, F.; Cai, G.; Wang, M.; Lin, J. An optical biosensor using immunomagnetic separation, urease catalysis and pH indication for rapid and sensitive detection of Listeria monocytogenes. *Sens. Actuators B Chem.* **2018**, *258*, 447–453. [CrossRef]
181. Van der Schueren, L.; de Clerck, K. Coloration and application of pH-sensitive dyes on textile materials. *Color. Technol.* **2012**, *128*, 82–90. [CrossRef]
182. Crosland, M.; Hannaway, O. Gay-Lussac, Scientist and Bourgeois. *Phys. Today* **1981**, *34*, 84–86. [CrossRef]
183. Mahmoudi, M.; Lohse, S.E.; Murphy, C.J.; Suslick, K.S. Identification of Nanoparticles with a Colorimetric Sensor Array. *ACS Sens.* **2016**, *1*, 17–21. [CrossRef]
184. Askim, J.R.; Mahmoudi, M.; Suslick, K.S. Optical sensor arrays for chemical sensing: The optoelectronic nose. *Chem. Soc. Rev.* **2013**, *42*, 8649–8682. [CrossRef]
185. Sedgwick, A.C.; Brewster, J.T.; Wu, T.; Feng, X.; Bull, S.D.; Qian, X.; Sessler, J.L.; James, T.D.; Anslyn, E.V.; Sun, X. Indicator displacement assays (IDAs): The past, present and future. *Chem. Soc. Rev.* **2021**, *50*, 9–38. [CrossRef]
186. Khajehsharifi, H.; Bordbar, M.M. A highly selective chemosensor for detection and determination of cyanide by using an indicator displacement assay and PC-ANN and its logic gate behavior. *Sens. Actuators B Chem.* **2015**, *209*, 1015–1022. [CrossRef]

187. Ealia, S.A.M.; Saravanakumar, M.P. A review on the classification, characterisation, synthesis of nanoparticles and their application. *IOP Conf. Ser. Mater. Sci.* **2017**, *263*, 032019.
188. Cuenya, B.R. Synthesis and catalytic properties of metal nanoparticles: Size, shape, support, composition, and oxidation state effects. *Thin Solid Film.* **2010**, *518*, 3127–3150. [CrossRef]
189. Lan, L.; Yao, Y.; Ping, J.; Ying, Y. Recent advances in nanomaterial-based biosensors for antibiotics detection. *Biosens. Bioelectron.* **2017**, *91*, 504–514. [CrossRef] [PubMed]
190. Zeng, Y.; Zhu, Z.; Du, D.; Lin, Y. Nanomaterial-based electrochemical biosensors for food safety. *J. Electroanal. Chem.* **2016**, *781*, 147–154. [CrossRef]
191. Srikar, S.K.; Giri, D.D.; Pal, D.B.; Mishra, P.K.; Upadhyay, S.N. Green Synthesis of Silver Nanoparticles: A Review. *Green Sustain. Chem.* **2016**, *6*, 34–56. [CrossRef]
192. Srinoi, P.; Chen, Y.T.; Vittur, V.; Marquez, M.D.; Lee, T.R. Bimetallic nanoparticles: Enhanced magnetic and optical properties for emerging biological applications. *Appl. Sci.* **2018**, *8*, 1106. [CrossRef]
193. Sharma, G.; Kumar, A.; Sharma, S.; Naushad, M.; Prakash Dwivedi, R.; ALOthman, Z.A.; Mola, G.T. Novel development of nanoparticles to bimetallic nanoparticles and their composites: A review. *J. King Saud Univ. Sci.* **2019**, *31*, 257–269. [CrossRef]
194. Wu, K.; Su, D.; Liu, J.; Saha, R.; Wang, J.P. Magnetic nanoparticles in nanomedicine: A review of recent advances. *Nanotechnology* **2019**, *30*, 502003. [CrossRef]
195. Pastucha, M.; Farka, Z.; Lacina, K.; Mikušová, Z.; Skládal, P. Magnetic nanoparticles for smart electrochemical immunoassays: A review on recent developments. *Microchim. Acta* **2019**, *186*, 1–26. [CrossRef]
196. Mohammed, L.; Gomaa, H.G.; Ragab, D.; Zhu, J. Magnetic nanoparticles for environmental and biomedical applications: A review. *Particuology* **2017**, *30*, 1–14. [CrossRef]
197. Asadian, E.; Ghalkhani, M.; Shahrokhian, S. Electrochemical sensing based on carbon nanoparticles: A review. *Sens. Actuators B Chem.* **2019**, *293*, 183–209. [CrossRef]
198. LeCroy, G.E.; Yang, S.T.; Yang, F.; Liu, Y.; Fernando, K.A.S.; Bunker, C.E.; Hu, Y.; Luo, P.G.; Sun, Y.P. Functionalized carbon nanoparticles: Syntheses and applications in optical bioimaging and energy conversion. *Coord. Chem. Rev.* **2016**, *320*, 66–81. [CrossRef]
199. Balarastaghi, M.; Ahmadi, V. Formulation of atomic positions and carbon–carbon bond length in armchair graphene nanoribbons: An ab initio study. *J. Theor. Appl. Phys.* **2017**, *11*, 191–199. [CrossRef]
200. Ji, K.; Han, J.; Hirata, A.; Fujita, T.; Shen, Y.; Ning, S.; Liu, P.; Kashani, H.; Tian, Y.; Ito, Y.; et al. Lithium intercalation into bilayer graphene. *Nat. Commun.* **2019**, *10*, 68–70. [CrossRef] [PubMed]
201. Kokorina, A.A.; Ermakov, A.V.; Abramova, A.M.; Goryacheva, I.Y.; Sukhorukov, G.B. Carbon nanoparticles and materials on their basis. *Colloids Interfaces* **2020**, *4*, 42. [CrossRef]
202. Anzar, N.; Hasan, R.; Tyagi, M.; Yadav, N.; Narang, J. Carbon nanotube—A review on Synthesis, Properties and plethora of applications in the field of biomedical science. *Sens. Int.* **2020**, *1*, 100003. [CrossRef]
203. Campuzano, S.; Yáñez-Sedeño, P.; Pingarrón, J.M. Carbon dots and graphene quantum dots in electrochemical biosensing. *Nanomaterials* **2019**, *9*, 634. [CrossRef]
204. Zhang, Y.; Zhang, C.; Xu, C.; Wang, X.; Liu, C.; Waterhouse, G.I.N.; Wang, Y.; Yin, H. Ultrasmall Au nanoclusters for biomedical and biosensing applications: A mini-review. *Talanta* **2019**, *200*, 432–442. [CrossRef]
205. Cui, H.; Shao, Z.S.; Song, Z.; Wang, Y.B.; Wang, H.S. Development of gold nanoclusters: From preparation to applications in the field of biomedicine. *J. Mater. Chem. C* **2020**, *8*, 14312–14333. [CrossRef]
206. Kimmel, D.W.; Leblanc, G.; Meschievitz, M.E.; Cliffel, D.E. Electrochemical sensors and biosensors. *Anal. Chem.* **2012**, *84*, 685–707. [CrossRef]
207. Kanyong, P.; Krampa, F.D.; Aniweh, Y.; Awandare, G.A. Enzyme-based amperometric galactose biosensors: A review. *Microchim. Acta* **2017**, *184*, 3663–3671. [CrossRef]
208. Ding, J.; Qin, W. Recent advances in potentiometric biosensors. *TrAC Trends Anal. Chem.* **2020**, *124*, 115803. [CrossRef]
209. Bahadir, E.B.; Sezgintürk, M.K. A review on impedimetric biosensors. *Artif. Cells Nanomed. Biotechnol.* **2016**, *44*, 248–262. [CrossRef]
210. Adley, C.C.; Ryan, M.P. Conductometric biosensors for high throughput screening of pathogens in food. *High Throughput Screen. Food Saf. Assess. Biosens. Technol. Hyperspectr. Imaging Pract. Appl.* **2015**, 315–326. [CrossRef]
211. Chen, D.; Zhang, M.; Zhou, F.; Hai, H.; Li, J. Ultrasensitive electroluminescence biosensor for a breast cancer marker microRNA based on target cyclic regeneration and multi-labeled magnetized nanoparticles. *Microchim. Acta* **2019**, *186*, 1–9. [CrossRef] [PubMed]
212. Shamsi, M.H.; Choi, K.; Ng, A.H.C.; Dean Chamberlain, M.; Wheeler, A.R. Electrochemiluminescence on digital microfluidics for microRNA analysis. *Biosens. Bioelectron.* **2016**, *77*, 845–852. [CrossRef] [PubMed]
213. Nie, Z.; Deiss, F.; Liu, X.; Akbulut, O.; Whitesides, G.M. Integration of paper-based microfluidic devices with commercial electrochemical readers. *Lab Chip* **2010**, *10*, 3163–3169. [CrossRef]
214. Millo, T.; Jaiswal, A.K.; Prasad, Y.S.; Murty, O.P. Breath alcohol analyzer and its forensic applications. *J. Forensic Med. Toxicol.* **2010**, *27*, 55–59.
215. Van Dyk, J.S.; Pletschke, B. Review on the use of enzymes for the detection of organochlorine, organophosphate and carbamate pesticides in the environment. *Chemosphere* **2011**, *82*, 291–307. [CrossRef]

216. Mercer, C.; Bennett, R.; Conghaile, P.; Rusling, J.F.; Leech, D. Glucose biosensor based on open-source wireless microfluidic potentiostat. *Sens. Actuators B Chem.* **2019**, *290*, 616–624. [CrossRef]
217. Ramanathan, K.; Danielsson, B. Principles and applications of thermal biosensors. *Biosens. Bioelectron.* **2001**, *16*, 417–423. [CrossRef]
218. Alan, T. Improving the accuracy of temperature measurements. *Sens. Rev.* **2001**, *21*, 193–198.
219. Yu, L.; Li, N. Noble metal nanoparticles-based colorimetric biosensor for visual quantification: A mini review. *Chemosensors* **2019**, *7*, 53. [CrossRef]
220. Huang, X.; Xu, D.; Chen, J.; Liu, J.; Li, Y.; Song, J.; Ma, X.; Guo, J. Smartphone-based analytical biosensors. *Analyst* **2018**, *143*, 5339–5351. [CrossRef] [PubMed]
221. Hemmateenejad, B.; Mobaraki, N.; Shakerizadeh-Shirazi, F.; Miri, R. Multivariate image analysis-thin layer chromatography (MIA-TLC) for simultaneous determination of co-eluting components. *Analyst* **2010**, *135*, 1747–1758. [CrossRef] [PubMed]
222. Leopold, A.V.; Shcherbakova, D.M.; Verkhusha, V.V. Fluorescent Biosensors for Neurotransmission and Neuromodulation: Engineering and Applications. *Front. Cell. Neurosci.* **2019**, *13*, 474. [CrossRef] [PubMed]
223. Gaviria-Arroyave, M.I.; Cano, J.B.; Peñuela, G.A. Nanomaterial-based fluorescent biosensors for monitoring environmental pollutants: A critical review. *Talanta Open* **2020**, *2*, 100006. [CrossRef]
224. Van De Weert, M.; Stella, L. Fluorescence quenching and ligand binding: A critical discussion of a popular methodology. *J. Mol. Struct.* **2011**, *998*, 144–150. [CrossRef]
225. Ulep, T.H.; Yoon, J.Y. Challenges in paper-based fluorogenic optical sensing with smartphones. *Nano Converg.* **2018**, *5*, 1–11. [CrossRef]
226. Biological Toxin Safe Work Practices. Available online: https://www.ehs.washington.edu/resource/biological-toxin-safe-work-practices-65 (accessed on 29 August 2021).
227. Singh, J.; Mehta, A. Rapid and sensitive detection of mycotoxins by advanced and emerging analytical methods: A review. *Food Sci. Nutr.* **2020**, *8*, 2183–2204. [CrossRef]
228. Haque, M.A.; Wang, Y.; Shen, Z.; Li, X.; Saleemi, M.K.; He, C. Mycotoxin contamination and control strategy in human, domestic animal and poultry: A review. *Microb. Pathog.* **2020**, *142*, 104095. [CrossRef]
229. Al-Jaal, B.A.; Jaganjac, M.; Barcaru, A.; Horvatovich, P.; Latiff, A. Aflatoxin, fumonisin, ochratoxin, zearalenone and deoxynivalenol biomarkers in human biological fluids: A systematic literature review, 2001–2018. *Food Chem. Toxicol.* **2019**, *129*, 211–228. [CrossRef] [PubMed]
230. Chauhan, R.; Singh, J.; Sachdev, T.; Basu, T.; Malhotra, B.D. Recent advances in mycotoxins detection. *Biosens. Bioelectron.* **2016**, *81*, 532–545. [CrossRef]
231. Liu, D.; Li, W.; Zhu, C.; Li, Y.; Shen, X.; Li, L.; Yan, X.; You, T. Recent progress on electrochemical biosensing of aflatoxins: A review. *TrAC Trends Anal. Chem.* **2020**, *133*, 115966. [CrossRef]
232. Wannop, C.C. The Histopathology of Turkey "X" Disease in Great Britain. *Avian Dis.* **1961**, *5*, 371. [CrossRef]
233. Negash, D. A Review of Aflatoxin: Occurrence, Prevention, and Gaps in Both Food and Feed Safety. *J. Appl. Microb. Res.* **2018**, *1*, 35–43. [CrossRef]
234. Robertson, A. Risk of Aflatoxin Contamination Increases with Hot and Dry Growing Conditions. Available online: https://lib.dr.iastate.edu/cropnews/1383/ (accessed on 29 August 2021).
235. FDA Guidance for Industry: Action Levels for Poisonous or Deleterious Substances in Human Food and Animal Feed. Available online: https://www.fda.gov/media/121202/download (accessed on 29 August 2021).
236. Xue, Z.; Zhang, Y.; Yu, W.; Zhang, J.; Wang, J.; Wan, F.; Kim, Y.; Liu, Y.; Kou, X. Recent advances in aflatoxin B1 detection based on nanotechnology and nanomaterials-A review. *Anal. Chim. Acta* **2019**, *1069*, 1–27. [CrossRef] [PubMed]
237. Liu, D.; Huang, Y.; Chen, M.; Wang, S.; Liu, K.; Lai, W. Rapid detection method for aflatoxin B1 in soybean sauce based on fluorescent microspheres probe. *Food Control* **2015**, *50*, 659–662. [CrossRef]
238. Li, M.; Wang, H.; Sun, J.; Ji, J.; Ye, Y.; Lu, X.; Zhang, Y.; Sun, X. Rapid, on-site, and sensitive detection of aflatoxin M1 in milk products by using time-resolved fluorescence microsphere test strip. *Food Control* **2021**, *121*, 107616. [CrossRef]
239. Yang, Q.; Zhu, J.; Ma, F.; Li, P.; Zhang, L.; Zhang, W.; Ding, X.; Zhang, Q. Quantitative determination of major capsaicinoids in serum by ELISA and time-resolved fluorescent immunoassay based on monoclonal antibodies. *Biosens. Bioelectron.* **2016**, *81*, 229–235. [CrossRef]
240. Tang, X.; Li, P.; Zhang, Q.; Zhang, Z.; Zhang, W.; Jiang, J. Time-Resolved Fluorescence Immunochromatographic Assay Developed Using Two Idiotypic Nanobodies for Rapid, Quantitative, and Simultaneous Detection of Aflatoxin and Zearalenone in Maize and Its Products. *Anal. Chem.* **2017**, *89*, 11520–11528. [CrossRef] [PubMed]
241. Wang, Q.X.; Xue, S.F.; Chen, Z.H.; Ma, S.H.; Zhang, S.; Shi, G.; Zhang, M. Dual lanthanide-doped complexes: The development of a time-resolved ratiometric fluorescent probe for anthrax biomarker and a paper-based visual sensor. *Biosens. Bioelectron.* **2017**, *94*, 388–393. [CrossRef]
242. Tang, X.; Zhang, Q.; Zhang, Z.; Ding, X.; Jiang, J.; Zhang, W.; Li, P. Rapid, on-site and quantitative paper-based immunoassay platform for concurrent determination of pesticide residues and mycotoxins. *Anal. Chim. Acta* **2019**, *1078*, 142–150. [CrossRef] [PubMed]

243. Wang, C.; Peng, J.; Liu, D.F.; Xing, K.Y.; Zhang, G.G.; Huang, Z.; Cheng, S.; Zhu, F.F.; Duan, M.L.; Zhang, K.Y.; et al. Lateral flow immunoassay integrated with competitive and sandwich models for the detection of aflatoxin M1 and Escherichia coli O157:H7 in milk. *J. Dairy Sci.* **2018**, *101*, 8767–8777. [CrossRef]
244. Li, X.; Li, P.; Zhang, Q.; Li, R.; Zhang, W.; Zhang, Z.; Ding, X.; Tang, X. Multi-component immunochromatographic assay for simultaneous detection of aflatoxin B1, ochratoxin A and zearalenone in agro-food. *Biosens. Bioelectron.* **2013**, *49*, 426–432. [CrossRef] [PubMed]
245. Kong, D.; Liu, L.; Song, S.; Suryoprabowo, S.; Li, A.; Kuang, H.; Wang, L.; Xu, C. A gold nanoparticle-based semi-quantitative and quantitative ultrasensitive paper sensor for the detection of twenty mycotoxins. *Nanoscale* **2016**, *8*, 5245–5253. [CrossRef]
246. Wang, Y.; Liu, N.; Ning, B.; Liu, M.; Lv, Z.; Sun, Z.; Peng, Y.; Chen, C.; Li, J.; Gao, Z. Simultaneous and rapid detection of six different mycotoxins using an immunochip. *Biosens. Bioelectron.* **2012**, *34*, 44–50. [CrossRef]
247. Zhang, G.; Zhu, C.; Huang, Y.; Yan, J.; Chen, A. A lateral flow strip based aptasensor for detection of Ochratoxin a in corn samples. *Molecules* **2018**, *23*, 291. [CrossRef] [PubMed]
248. Kyung-Mi, S.; Seonghwan, L.; Changill, B. Aptamers and Their Biological Applications. *Sensors* **2012**, *12*, 612–631.
249. Kasoju, A.; Shahdeo, D.; Khan, A.A.; Shrikrishna, N.S.; Mahari, S.; Alanazi, A.M.; Bhat, M.A.; Giri, J.; Gandhi, S. Fabrication of microfluidic device for Aflatoxin M1 detection in milk samples with specific aptamers. *Sci. Rep.* **2020**, *10*, 1–8. [CrossRef]
250. Sheini, A. Colorimetric aggregation assay based on array of gold and silver nanoparticles for simultaneous analysis of aflatoxins, ochratoxin and zearalenone by using chemometric analysis and paper based analytical devices. *Microchim. Acta* **2020**, *187*, 1–11. [CrossRef]
251. Migliorini, F.L.; dos Santos, D.M.; Soares, A.C.; Mattoso, L.H.C.; Oliveira, O.N.; Correa, D.S. Design of a low-cost and disposable paper-based immunosensor for the rapid and sensitive detection of aflatoxin B1. *Chemosensors* **2020**, *8*, 87. [CrossRef]
252. Ye, Y.; Zhou, Y.; Mo, Z.; Cheng, W.; Yang, S.; Wang, X.; Chen, F. Rapid detection of aflatoxin B1 on membrane by dot-immunogold filtration assay. *Talanta* **2010**, *81*, 792–798. [CrossRef]
253. Liao, J.Y.; Li, H. Lateral flow immunodipstick for visual detection of aflatoxin B1 in food using immuno-nanoparticles composed of a silver core and a gold shell. *Microchim. Acta* **2010**, *171*, 289–295. [CrossRef]
254. Kasoju, A.; Shrikrishna, N.S.; Shahdeo, D.; Khan, A.A.; Alanazi, A.M.; Gandhi, S. Microfluidic paper device for rapid detection of aflatoxin B1 using an aptamer based colorimetric assay. *RSC Adv.* **2020**, *10*, 11843–11850. [CrossRef]
255. Song, S.; Liu, N.; Zhao, Z.; Njumbe Ediage, E.; Wu, S.; Sun, C.; De Saeger, S.; Wu, A. Multiplex lateral flow immunoassay for mycotoxin determination. *Anal. Chem.* **2014**, *86*, 4995–5001. [CrossRef] [PubMed]
256. Pirsaheb, M.; Hossini, H.; Asadi, F.; Janjani, H. A systematic review on organochlorine and organophosphorus pesticides content in water resources. *Toxin Rev.* **2017**, *36*, 210–221. [CrossRef]
257. Diauudin, F.N.; Rashid, J.I.A.; Knight, V.F.; Wan Yunus, W.M.Z.; Ong, K.K.; Kasim, N.A.M.; Abdul Halim, N.; Noor, S.A.M. A review of current advances in the detection of organophosphorus chemical warfare agents based biosensor approaches. *Sens. Bio-Sens. Res.* **2019**, *26*, 100305. [CrossRef]
258. Obare, S.O.; De, C.; Guo, W.; Haywood, T.L.; Samuels, T.A.; Adams, C.P.; Masika, N.O.; Murray, D.H.; Anderson, G.A.; Campbell, K.; et al. Fluorescent chemosensors for toxic organophosphorus pesticides: A review. *Sensors* **2010**, *10*, 7018–7043. [CrossRef] [PubMed]
259. Gourie-Devi, M. Neurological practice: An Indian perspective. *Ann. Indian Acad. Neurol.* **2006**, *9*, 129. [CrossRef]
260. Hmouda, H.; Salem, C.B.; Bouraoui, K. Management of acute organophosphorus pesticide poisoning. *Lancet* **2008**, *371*, 2169–2170. [CrossRef]
261. Pundir, C.S.; Chauhan, N. Acetylcholinesterase inhibition-based biosensors for pesticide determination: A review. *Anal. Biochem.* **2012**, *429*, 19–31. [CrossRef] [PubMed]
262. Kim, H.J.; Kim, Y.; Park, S.J.; Kwon, C.; Noh, H. Development of Colorimetric Paper Sensor for Pesticide Detection Using Competitive-inhibiting Reaction. *Biochip J.* **2018**, *12*, 326–331. [CrossRef]
263. Fu, Q.; Zhang, C.; Xie, J.; Li, Z.; Qu, L.; Cai, X.; Ouyang, H.; Song, Y.; Du, D.; Lin, Y.; et al. Ambient light sensor based colorimetric dipstick reader for rapid monitoring organophosphate pesticides on a smart phone. *Anal. Chim. Acta* **2019**, *1092*, 126–131. [CrossRef]
264. George, J.M.; Antony, A.; Mathew, B. Metal oxide nanoparticles in electrochemical sensing and biosensing: A review. *Microchim. Acta* **2018**, *185*, 1–26. [CrossRef]
265. Kargozar, S.; Baino, F.; Hoseini, S.J.; Hamzehlou, S.; Darroudi, M.; Verdi, J.; Hasanzadeh, L.; Kim, H.W.; Mozafari, M. Biomedical applications of nanoceria: New roles for an old player. *Nanomedicine* **2018**, *13*, 3051–3069. [CrossRef]
266. Liu, B.; Sun, Z.; Huang, P.J.J.; Liu, J. Hydrogen peroxide displacing DNA from nanoceria: Mechanism and detection of glucose in serum. *J. Am. Chem. Soc.* **2015**, *137*, 1290–1295. [CrossRef] [PubMed]
267. Nouanthavong, S.; Nacapricha, D.; Henry, C.S.; Sameenoi, Y. Pesticide analysis using nanoceria-coated paper-based devices as a detection platform. *Analyst* **2016**, *141*, 1837–1846. [CrossRef]
268. Chang, J.; Li, H.; Hou, T.; Li, F. Paper-based fluorescent sensor for rapid naked-eye detection of acetylcholinesterase activity and organophosphorus pesticides with high sensitivity and selectivity. *Biosens. Bioelectron.* **2016**, *86*, 971–977. [CrossRef] [PubMed]
269. Wang, Q.; Yin, Q.; Fan, Y.; Zhang, L.; Xu, Y.; Hu, O.; Guo, X.; Shi, Q.; Fu, H.; She, Y. Double quantum dots-nanoporphyrin fluorescence-visualized paper-based sensors for detecting organophosphorus pesticides. *Talanta* **2019**, *199*, 46–53. [CrossRef]

270. Xie, J.; Li, L.; Khan, I.M.; Wang, Z.; Ma, X. Flexible paper-based SERS substrate strategy for rapid detection of methyl parathion on the surface of fruit. *Spectrochim. Acta Part A Mol. Biomol. Spectrosc.* **2020**, *231*, 118104. [CrossRef] [PubMed]
271. Xiong, S.; Deng, Y.; Zhou, Y.; Gong, D.; Xu, Y.; Yang, L.; Chen, H.; Chen, L.; Song, T.; Luo, A.; et al. Current progress in biosensors for organophosphorus pesticides based on enzyme functionalized nanostructures: A review. *Anal. Methods* **2018**, *10*, 5468–5479. [CrossRef]
272. Arduini, F.; Cinti, S.; Caratelli, V.; Amendola, L.; Palleschi, G.; Moscone, D. Origami multiple paper-based electrochemical biosensors for pesticide detection. *Biosens. Bioelectron.* **2019**, *126*, 346–354. [CrossRef] [PubMed]
273. Bigley, A.N.; Raushel, F.M. Catalytic mechanisms for phosphotriesterases. *Biochim. Biophys. Acta Proteins Proteom.* **2013**, *1834*, 443–453. [CrossRef] [PubMed]
274. Hondred, J.A.; Breger, J.C.; Alves, N.J.; Trammell, S.A.; Walper, S.A.; Medintz, I.L.; Claussen, J.C. Printed Graphene Electrochemical Biosensors Fabricated by Inkjet Maskless Lithography for Rapid and Sensitive Detection of Organophosphates. *ACS Appl. Mater. Interfaces* **2018**, *10*, 11125–11134. [CrossRef]
275. Mehta, J.; Vinayak, P.; Tuteja, S.K.; Chhabra, V.A.; Bhardwaj, N.; Paul, A.K.; Kim, K.H.; Deep, A. Graphene modified screen printed immunosensor for highly sensitive detection of parathion. *Biosens. Bioelectron* **2016**, *83*, 339–346. [CrossRef]
276. Bordbar, M.M.; Nguyen, T.A.; Arduini, F.; Bagheri, H. A paper-based colorimetric sensor array for discrimination and simultaneous determination of organophosphate and carbamate pesticides in tap water, apple juice, and rice. *Microchim. Acta* **2020**, *187*, 1–13. [CrossRef]
277. Bordbar, M.M.; Nguyen, T.A.; Tran, A.Q.; Bagheri, H. Optoelectronic nose based on an origami paper sensor for selective detection of pesticide aerosols. *Sci. Rep.* **2020**, *10*, 1–12. [CrossRef]
278. Wang, T.; Reid, R.C.; Minteer, S.D. A Paper-based Mitochondrial Electrochemical Biosensor for Pesticide Detection. *Electroanalysis* **2016**, *28*, 854–859. [CrossRef]
279. Deng, S.; Yang, T.; Zhang, W.; Ren, C.; Zhang, J.; Zhang, Y.; Cui, T.; Yue, W. Rapid detection of trichlorfon residues by a microfluidic paper-based phosphorus-detection chip (μPPC). *New J. Chem.* **2019**, *43*, 7194–7197. [CrossRef]
280. Yang, N.; Shaheen, N.; Xie, L.; Yu, J.; Ahmad, H.; Mao, H. Pesticide residues identification by optical spectrum in the time-sequence of enzyme inhibitors performed on microfluidic paper-based analytical devices (μPADs). *Molecules* **2019**, *24*, 2428. [CrossRef]
281. Bridle, H.; Miller, B.; Desmulliez, M.P.Y. Application of microfluidics in waterborne pathogen monitoring: A review. *Water Res.* **2014**, *55*, 256–271. [CrossRef] [PubMed]
282. Bordbar, M.M.; Tashkhourian, J.; Tavassoli, A.; Bahramali, E.; Hemmateenejad, B. Ultrafast detection of infectious bacteria using optoelectronic nose based on metallic nanoparticles. *Sens. Actuators B Chem.* **2020**, *319*, 128262. [CrossRef]
283. Gregersen, T. Rapid method for distinction of gram-negative from gram-positive bacteria. *Eur. J. Appl. Microbiol. Biotechnol.* **1978**, *5*, 123–127. [CrossRef]
284. Saravanan, A.; Kumar, P.S.; Hemavathy, R.V.; Jeevanantham, S.; Kamalesh, R.; Sneha, S.; Yaashikaa, P.R. Methods of detection of food-borne pathogens: A review. *Environ. Chem. Lett.* **2021**, *19*, 189–207. [CrossRef]
285. Rajapaksha, P.; Elbourne, A.; Gangadoo, S.; Brown, R.; Cozzolino, D.; Chapman, J. A review of methods for the detection of pathogenic microorganisms. *Analyst* **2019**, *144*, 396–411. [CrossRef]
286. Creran, B.; Li, X.; Duncan, B.; Kim, C.S.; Moyano, D.F.; Rotello, V.M. Detection of bacteria using inkjet-printed enzymatic test strips. *ACS Appl. Mater. Interfaces* **2014**, *6*, 19525–19530. [CrossRef] [PubMed]
287. Jokerst, J.C.; Adkins, J.A.; Bisha, B.; Mentele, M.M.; Goodridge, L.D.; Henry, C.S. Development of a paper-based analytical device for colorimetric detection of select foodborne pathogens. *Anal. Chem.* **2012**, *84*, 2900–2907. [CrossRef]
288. Suaifan, G.A.R.Y.; Alhogail, S.; Zourob, M. Rapid and low-cost biosensor for the detection of *Staphylococcus aureus*. *Biosens. Bioelectron.* **2017**, *90*, 230–237. [CrossRef] [PubMed]
289. Sun, L.; Jiang, Y.; Pan, R.; Li, M.; Wang, R.; Chen, S.; Fu, S.; Man, C. A novel, simple and low-cost paper-based analytical device for colorimetric detection of *Cronobacter* spp. *Anal. Chim. Acta* **2018**, *1036*, 80–88. [CrossRef] [PubMed]
290. Bagheri Pebdeni, A.; Hosseini, M. Fast and selective whole cell detection of *Staphylococcus aureus* bacteria in food samples by paper based colorimetric nanobiosensor using peroxidase-like catalytic activity of DNA-Au/Pt bimetallic nanoclusters. *Microchem. J.* **2020**, *159*, 105475. [CrossRef]
291. Wang, Y.; Ping, J.; Ye, Z.; Wu, J.; Ying, Y. Impedimetric immunosensor based on gold nanoparticles modified graphene paper for label-free detection of Escherichia coli O157: H7. *Biosens. Bioelectron.* **2013**, *49*, 492–498. [CrossRef]
292. Mo, X.; Wu, Z.; Huang, J.; Zhao, G.; Dou, W. A sensitive and regenerative electrochemical immunosensor for quantitative detection of: *Escherichia coli* O157:H7 based on stable polyaniline coated screen-printed carbon electrode and rGO-NR-Au@Pt. *Anal. Methods* **2019**, *11*, 1475–1482. [CrossRef]
293. Khan, M.S.; Misra, S.K.; Dighe, K.; Wang, Z.; Schwartz-Duval, A.S.; Sar, D.; Pan, D. Electrically-receptive and thermally-responsive paper-based sensor chip for rapid detection of bacterial cells. *Biosens. Bioelectron.* **2018**, *110*, 132–140. [CrossRef] [PubMed]
294. Hernández, R.; Vallés, C.; Benito, A.M.; Maser, W.K.; Xavier Rius, F.; Riu, J. Graphene-based potentiometric biosensor for the immediate detection of living bacteria. *Biosens. Bioelectron.* **2014**, *54*, 553–557. [CrossRef] [PubMed]
295. Bhardwaj, J.; Devarakonda, S.; Kumar, S.; Jang, J. Development of a paper-based electrochemical immunosensor using an antibody-single walled carbon nanotubes bio-conjugate modified electrode for label-free detection of foodborne pathogens. *Sens. Actuators B Chem.* **2017**, *253*, 115–123. [CrossRef]

296. Han, J.; Cheng, H.; Wang, B.; Braun, M.S.; Fan, X.; Bender, M.; Huang, W.; Domhan, C.; Mier, W.; Lindner, T.; et al. A Polymer/Peptide Complex-Based Sensor Array That Discriminates Bacteria in Urine. *Angew. Chem.* **2017**, *129*, 15448–15453. [CrossRef]
297. Yan, P.; Ding, Z.; Li, X.; Dong, Y.; Fu, T.; Wu, Y. Colorimetric Sensor Array Based on Wulff-Type Boronate Functionalized AgNPs at Various pH for Bacteria Identification. *Anal. Chem.* **2019**, *91*, 12134–12137. [CrossRef]
298. Sun, H.; Tian, F.; Liang, Z.; Sun, T.; Yu, B.; Yang, S.X.; He, Q.; Zhang, L.; Liu, X. Sensor Array Optimization of Electronic Nose for Detection of Bacteria in Wound Infection. *IEEE Trans. Ind. Electron.* **2017**, *64*, 7350–7358. [CrossRef]
299. Wu, Y.; Wang, B.; Wang, K.; Yan, P. Identification of proteins and bacteria based on a metal ion-gold nanocluster sensor array. *Anal. Methods* **2018**, *10*, 3939–3944. [CrossRef]
300. Svechkarev, D.; Sadykov, M.R.; Bayles, K.W.; Mohs, A.M. Ratiometric Fluorescent Sensor Array as a Versatile Tool for Bacterial Pathogen Identification and Analysis. *ACS Sens.* **2018**, *3*, 700–708. [CrossRef] [PubMed]
301. Lai, S.Y.; Deffenderfer, O.F.; Hanson, W.; Phillips, M.P.; Thaler, E.R. Identification of upper respiratory bacterial pathogens with the electronic nose. *Laryngoscope* **2002**, *112*, 975–979. [CrossRef] [PubMed]
302. Carey, J.R.; Suslick, K.S.; Hulkower, K.I.; Imlay, J.A.; Imlay, K.R.C.; Ingison, C.K.; Ponder, J.B.; Sen, A.; Wittrig, A.E. Rapid identification of bacteria with a disposable colorimetric sensing array. *J. Am. Chem. Soc.* **2011**, *133*, 7571–7576. [CrossRef] [PubMed]
303. Lim, S.H.; Mix, S.; Xu, Z.; Taba, B.; Budvytiene, I.; Berliner, A.N.; Queralto, N.; Churi, Y.S.; Huang, R.S.; Eiden, M.; et al. Colorimetric sensor array allows fast detection and simultaneous identification of sepsis-causing bacteria in spiked blood culture. *J. Clin. Microbiol.* **2014**, *52*, 592–598. [CrossRef] [PubMed]
304. Canhoto, O.; Magan, N. Electronic nose technology for the detection of microbial and chemical contamination of potable water. *Sens. Actuators B Chem.* **2005**, *106*, 3–6. [CrossRef]
305. Dutta, R.; Das, A.; Stocks, N.G.; Morgan, D. Stochastic resonance-based electronic nose: A novel way to classify bacteria. *Sens. Actuators B Chem.* **2006**, *115*, 17–27. [CrossRef]
306. Chen, Q.; Li, H.; Ouyang, Q.; Zhao, J. Identification of spoilage bacteria using a simple colorimetric sensor array. *Sens. Actuators B Chem.* **2014**, *205*, 1–8. [CrossRef]
307. Sheini, A. A point-of-care testing sensor based on fluorescent nanoclusters for rapid detection of septicemia in children. *Sens. Actuators B Chem.* **2021**, *328*, 129029. [CrossRef]
308. Ali, M.M.; Brown, C.L.; Jahanshahi-Anbuhi, S.; Kannan, B.; Li, Y.; Filipe, C.D.M.; Brennan, J.D. A Printed Multicomponent Paper Sensor for Bacterial Detection. *Sci. Rep.* **2017**, *7*, 1–10. [CrossRef]
309. Sun, Y.; Chang, Y.; Zhang, Q.; Liu, M. An origami paper-based device printed with DNAzyme-containing DNA superstructures for Escherichia Coli detection. *Micromachines* **2019**, *10*, 531. [CrossRef]
310. Sun, Q.; Tam, N.F.Y.; Han, J.; Yung-Kang Peng, W.; Zhu, Z.; Chen, J.L. A simple paper-based colorimetric analytical device for rapid detection of Enterococcus faecalis under the stress of chlorophenols. *Talanta* **2021**, *225*, 121966. [CrossRef]
311. Silva, N.F.D.; Almeida, C.M.R.; Magalhães, J.M.C.S.; Gonçalves, M.P.; Freire, C.; Delerue-Matos, C. Development of a disposable paper-based potentiometric immunosensor for real-time detection of a foodborne pathogen. *Biosens. Bioelectron.* **2019**, *141*, 111317. [CrossRef]
312. Rengaraj, S.; Cruz-Izquierdo, Á.; Scott, J.L.; Di Lorenzo, M. Impedimetric paper-based biosensor for the detection of bacterial contamination in water. *Sens. Actuators B Chem.* **2018**, *265*, 50–58. [CrossRef]
313. Unnikrishnan, B.; Lien, C.-W.; Chu, H.-W.; Huang, C.-C. A review on metal nanozyme-based sensing of heavy metal ions: Challenges and future perspectives. *J. Hazard. Mater.* **2021**, *401*, 123397. [CrossRef]
314. Malik, L.A.; Bashir, A.; Qureashi, A.; Pandith, A.H. Detection and removal of heavy metal ions: A review. *Environ. Chem. Lett.* **2019**, *17*, 1495–1521. [CrossRef]
315. Sall, M.L.; Diaw, A.K.D.; Gningue-Sall, D.; Efremova Aaron, S.; Aaron, J.J. Toxic heavy metals: Impact on the environment and human health, and treatment with conducting organic polymers, a review. *Environ. Sci. Pollut. Res.* **2020**, *27*, 29927–29942. [CrossRef]
316. Ninwong, B.; Sangkaew, P.; Hapa, P.; Ratnarathorn, N.; Menger, R.F.; Henry, C.S.; Dungchai, W. Sensitive distance-based paper-based quantification of mercury ions using carbon nanodots and heating-based preconcentration. *RSC Adv.* **2020**, *10*, 9884–9893. [CrossRef]
317. Chen, G.H.; Chen, W.Y.; Yen, Y.C.; Wang, C.W.; Chang, H.T.; Chen, C.F. Detection of mercury(II) ions using colorimetric gold nanoparticles on paper-based analytical devices. *Anal. Chem.* **2014**, *86*, 6843–6849. [CrossRef] [PubMed]
318. Chen, W.; Fang, X.; Li, H.; Cao, H.; Kong, J. A Simple Paper-Based Colorimetric Device for Rapid Mercury(II) Assay. *Sci. Rep.* **2016**, *6*, 1–7. [CrossRef] [PubMed]
319. Cai, L.; Fang, Y.; Mo, Y.; Huang, Y.; Xu, C.; Zhang, Z.; Wang, M. Visual quantification of Hg on a microfluidic paper-based analytical device using distance-based detection technique. *AIP Adv.* **2017**, *7*, 85214. [CrossRef]
320. Nashukha, H.L.; Sitanurak, J.; Sulistyarti, H.; Nacapricha, D.; Uraisin, K. Simple and equipment-free paper-based device for determination of mercury in contaminated soil. *Molecules* **2021**, *26*, 2004. [CrossRef] [PubMed]
321. Quinn, C.W.; Cate, D.M.; Miller-Lionberg, D.D.; Reilly, T.; Volckens, J.; Henry, C.S. Solid-Phase Extraction Coupled to a Paper-Based Technique for Trace Copper Detection in Drinking Water. *Environ. Sci. Technol.* **2018**, *52*, 3567–3573. [CrossRef]

322. Ratnarathorn, N.; Chailapakul, O.; Henry, C.S.; Dungchai, W. Simple silver nanoparticle colorimetric sensing for copper by paper-based devices. *Talanta* **2012**, *99*, 552–557. [CrossRef]
323. Wang, X.; Sun, J.; Tong, J.; Guan, X.; Bian, C.; Xia, S. Paper-based sensor chip for heavy metal ion detection by SWSV. *Micromachines* **2018**, *9*, 150. [CrossRef]
324. Fang, X.; Zhao, Q.; Cao, H.; Liu, J.; Guan, M.; Kong, J. Rapid detection of Cu^{2+} by a paper-based microfluidic device coated with bovine serum albumin (BSA)-Au nanoclusters. *Analyst* **2015**, *140*, 7823–7826. [CrossRef]
325. Wang, H.; Da, L.; Yang, L.; Chu, S.; Yang, F.; Yu, S.; Jiang, C. Colorimetric fluorescent paper strip with smartphone platform for quantitative detection of cadmium ions in real samples. *J. Hazard. Mater.* **2020**, *392*, 122506. [CrossRef]
326. WHO. *Guidelines for Drinking-water Quality*, 3rd ed.; World Health Organization: Geneva, Switzerland, 2008. Available online: https://www.who.int/water_sanitation_health/dwq/fulltext.pdf (accessed on 29 August 2021).
327. López Marzo, A.M.; Pons, J.; Blake, D.A.; Merkoçi, A. All-integrated and highly sensitive paper based device with sample treatment platform for Cd^{2+} immunodetection in drinking/tap waters. *Anal. Chem.* **2013**, *85*, 3532–3538. [CrossRef]
328. Huang, K.; Chen, Y.; Zhou, F.; Zhao, X.; Liu, J.; Mei, S.; Zhou, Y.; Jing, T. Integrated ion imprinted polymers-paper composites for selective and sensitive detection of Cd(II) ions. *J. Hazard. Mater.* **2017**, *333*, 137–143. [CrossRef]
329. EPA 816-F-09-0004, United States Environmental Protection Agency. Available online: https://www.nrc.gov/docs/ML1307/ML13078A040.pdf (accessed on 29 August 2021).
330. Gupta, A.; Verma, N.C.; Khan, S.; Tiwari, S.; Chaudhary, A.; Nandi, C.K. Paper strip based and live cell ultrasensitive lead sensor using carbon dots synthesized from biological media. *Sens. Actuators B Chem.* **2016**, *232*, 107–114. [CrossRef]
331. Wang, H.; Yang, L.; Chu, S.; Liu, B.; Zhang, Q.; Zou, L.; Yu, S.; Jiang, C. Semiquantitative Visual Detection of Lead Ions with a Smartphone via a Colorimetric Paper-Based Analytical Device. *Anal. Chem.* **2019**, *91*, 9292–9299. [CrossRef] [PubMed]
332. Pechova, A.; Pavlata, L. Chromium as an essential nutrient: A review. *Vet. Med.* **2007**, *52*, 1–18. [CrossRef]
333. Elavarasi, M.; Rajeshwari, A.; Chandrasekaran, N.; Mukherjee, A. Simple colorimetric detection of Cr(III) in aqueous solutions by as synthesized citrate capped gold nanoparticles and development of a paper based assay. *Anal. Methods* **2013**, *5*, 6211–6218. [CrossRef]
334. Feng, L.; Li, H.; Niu, L.Y.; Guan, Y.S.; Duan, C.F.; Guan, Y.F.; Tung, C.H.; Yang, Q.Z. A fluorometric paper-based sensor array for the discrimination of heavy-metal ions. *Talanta* **2013**, *108*, 103–108. [CrossRef] [PubMed]
335. Feng, L.; Li, X.; Li, H.; Yang, W.; Chen, L.; Guan, Y. Enhancement of sensitivity of paper-based sensor array for the identification of heavy-metal ions. *Anal. Chim. Acta* **2013**, *780*, 74–80. [CrossRef] [PubMed]
336. Devadhasan, J.P.; Kim, J. A chemically functionalized paper-based microfluidic platform for multiplex heavy metal detection. *Sens. Actuators B Chem.* **2018**, *273*, 18–24. [CrossRef]
337. Zhang, L.; Guan, L.; Lu, Z.; Li, M.; Wu, J.; Cao, R.; Tian, J. Barrier-free patterned paper sensors for multiplexed heavy metal detection. *Talanta* **2019**, *196*, 408–414. [CrossRef] [PubMed]
338. Wu, J.; Li, M.; Tang, H.; Su, J.; He, M.; Chen, G.; Guan, L.; Tian, J. Portable paper sensors for the detection of heavy metals based on light transmission-improved quantification of colorimetric assays. *Analyst* **2019**, *144*, 6382–6390. [CrossRef] [PubMed]
339. Muhammad-aree, S.; Teepoo, S. On-site detection of heavy metals in wastewater using a single paper strip integrated with a smartphone. *Anal. Bioanal. Chem.* **2020**, *412*, 1395–1405. [CrossRef]
340. Qi, J.; Li, B.; Wang, X.; Zhang, Z.; Wang, Z.; Han, J.; Chen, L. Three-dimensional paper-based microfluidic chip device for multiplexed fluorescence detection of Cu^{2+} and Hg^{2+} ions based on ion imprinting technology. *Sens. Actuators B Chem.* **2017**, *251*, 224–233. [CrossRef]
341. Hossain, S.M.Z.; Brennan, J.D. β-Galactosidase-based colorimetric paper sensor for determination of heavy metals. *Anal. Chem.* **2011**, *83*, 8772–8778. [CrossRef]
342. Wang, X.R.; Li, B.W.; You, H.Y.; Chen, L.X. An ion imprinted polymers grafted paper-based fluorescent sensor based on quantum dots for detection of Cu^{2+} ions. *Chin. J. Anal. Chem.* **2015**, *43*, 1499–1504. [CrossRef]
343. He, K.; Zhan, X.; Liu, L.; Ruan, X.; Wu, Y. Ratiometric Fluorescent Paper-Based Sensor Based on CdTe Quantum Dots and Graphite Carbon Nitride Hybrid for Visual and Rapid Determination of Cu^{2+} in Drinks. *Photochem. Photobiol.* **2020**, *96*, 1154–1160. [CrossRef] [PubMed]
344. Zhang, M.; Ge, L.; Ge, S.; Yan, M.; Yu, J.; Huang, J.; Liu, S. Three-dimensional paper-based electrochemiluminescence device for simultaneous detection of Pb^{2+} and Hg^{2+} based on potential-control technique. *Biosens. Bioelectron.* **2013**, *41*, 544–550. [CrossRef] [PubMed]
345. Shi, J.; Tang, F.; Xing, H.; Zheng, H.; Bi, L.; Wang, W. Electrochemical detection of Pb and Cd in paper-based microfluidic devices. *J. Braz. Chem. Soc.* **2012**, *23*, 1124–1130. [CrossRef]

biosensors

MDPI

Review

Origami Paper-Based Electrochemical (Bio)Sensors: State of the Art and Perspective

Noemi Colozza [1,†], Veronica Caratelli [1,†], Danila Moscone [1] and Fabiana Arduini [1,2,*]

1 Department of Chemical Science and Technologies, University of Rome "Tor Vergata", Via della Ricerca Scientifica, 00133 Rome, Italy; noemi.colozza@uniroma2.it (N.C.); veronica.caratelli@uniroma2.it (V.C.); moscone@uniroma2.it (D.M.)
2 SENSE4MED, Via Renato Rascel 30, 00128 Rome, Italy
* Correspondence: fabiana.arduini@uniroma2.it
† These authors contributed equally.

Abstract: In the last 10 years, paper-based electrochemical biosensors have gathered attention from the scientific community for their unique advantages and sustainability vision. The use of papers in the design the electrochemical biosensors confers to these analytical tools several interesting features such as the management of the solution flow without external equipment, the fabrication of reagent-free devices exploiting the porosity of the paper to store the reagents, and the unprecedented capability to detect the target analyte in gas phase without any sampling system. Furthermore, cost-effective fabrication using printing technologies, including wax and screen-printing, combined with the use of this eco-friendly substrate and the possibility of reducing waste management after measuring by the incineration of the sensor, designate these type of sensors as eco-designed analytical tools. Additionally, the foldability feature of the paper has been recently exploited to design and fabricate 3D multifarious biosensors, which are able to detect different target analytes by using enzymes, antibodies, DNA, molecularly imprinted polymers, and cells as biocomponents. Interestingly, the 3D structure has recently boosted the self-powered paper-based biosensors, opening new frontiers in origami devices. This review aims to give an overview of the current state origami paper-based biosensors, pointing out how the foldability of the paper allows for the development of sensitive, selective, and easy-to-use smart and sustainable analytical devices.

Keywords: DNA; enzyme; antibody; cell; molecularly imprinted polymers

Citation: Colozza, N.; Caratelli, V.; Moscone, D.; Arduini, F. Origami Paper-Based Electrochemical (Bio)Sensors: State of the Art and Perspective. *Biosensors* **2021**, *11*, 328. https://doi.org/10.3390/bios11090328

Received: 1 August 2021
Accepted: 1 September 2021
Published: 10 September 2021

1. Introduction

In 2015, all United Nations Member States adopted the 2030 Agenda for Sustainable Development to provide a blueprint for peace and prosperity for people and the planet, both now and in the future. Analytical chemistry could carry out many activities to achieve the different sustainable development goals (SDGs), starting by ensuring healthy lives and promoting well-being for all at all ages (SDG3), through sustainable management of water and sanitation for all (SDG6), and conserving and sustainably using the oceans, seas, and marine resources, for sustainable development (SDG14), since the detection of biomarkers, pollutants, food quality indicators, among others are needed to achieve SDGs in a sustainable way. Nevertheless, if detection is important, it is also relevant how it must be carried out. If detection is accomplished by generating other pollutants, the detection avoids some sustainable key elements. The 11th Principle of Green Chemistry boosts the in situ analysis [1], but this feature is not enough. Thus, Green Analytical Chemistry has recently opened the concept of White Analytical Chemistry which extends and complements the Green Analytical Chemistry vision for giving coherence and synergy of the analytical, ecological, and practical attributes [2].

In this context, electrochemical (bio)sensors match several features such as the avoiding of organic solvents, the reduction of reagent consumption, the capability to measure on-site, and the reduced sample treatment, to name a few. A giant step was achieved by paper-based devices which have opened a new route in the sensing field, being at the state of the art the most eco-designed sensors.

Indeed, besides being plastic free, paper sensors are able to:

- Reduce the use of energy consumption (for instance, using the capillarity for the microfluidics instead of an external pump) [3];
- Reduce chemicals (since the reaction can happen in the few μL solution layer within the cellulose network) [4];
- Reduce or avoid the sample treatment by exploiting the porosity of the paper [5];
- Allow for the paper network as reservoir for modifying the sensors with nanomaterials [6,7];
- Deliver reagent-free analytical tools by exploiting the porosity of the paper [8,9];
- Carry out easily the multiplex analyses [10];
- Increase the sensitivity by multistep of sample loading and waiting to become dry before exploiting the 3D network of the paper [11];
- Overcome the limitation of alumina or polyester-based sensors (by measuring the analytes in gas phase of surface, without any additional external sampling system) [12–14].

In the paper-based devices field, if the Whiteside group was the pioneer in the colorimetric analysis [15–17], in the electrochemical field this primacy can be attributed to the Henry group [18], who combined the previously reported outstanding features of the paper with unrevealed features of the electrochemical detection, this type of transduction being characterized by high sensitivity, the capability to work in colored samples, and a connection with miniaturized commercially available transducers.

In the last 10 years, the research activity in this sector has had a rapid grown, attested by several reviews [19–28]. In 2020, a number of reviews appeared demonstrating that this is a hot topic, and if the first reviews are more general, the last ones are more particular such as potentiometric paper-based devices [28], attesting to the higher numbers of the articles published in this field. Regarding the configurations, if the first configuration of the paper-based devices is mainly based on the horizontal flow, recently the foldability has boosted the vertical microfluidics, opening for a further route in this paper-based device field and leading to the origami paper-based devices. The capability of paper to be cut and easily folded has motivated scientists to design several configurations allowing for multifarious and polyhedric devices for smart detection of several analytes. Herein, we reported, to our knowledge, the first review focused on electrochemical origami-paper based biosensors furnishing an overview of how the foldability features can generate interesting configurations of enzymatic, immuno-, DNA, cell, and molecular imprinting polymers (MIPs) biosensors, enhancing the sensitivity, selectivity, and easiness to carry out measurements (Table 1).

Our research covers the publication produced in the last ten years, going back to the first origami-like paper-based biosensor that has been reported. The papers have been selected by choosing the keywords "origami", "paper", "electrochemical biosensor" on the google scholar web search.

Table 1. Summary of the principal features of the origami paper-based electrochemical biosensors.

Type of Origami	Analyte	Sensing (Bio)Component	Electrochemical Technique	Matrix	Linear Range LOD	Ref.
Origami Paper-Based Electrochemical Enzymatic Biosensors						
Single folding	Glucose	Glucose oxidase preloaded on paper	Chronoamperometry	Human serum	0–24 mM	[29]
Single folding	Glucose	Glucose oxidase + ferrocenecarboxylic acid	Chronoamperometry	Human blood	1–12 mM 0.05 mM	[30]
Double folding	(a) Glucose (b) Lactate	(a) Glucose oxidase (b) Lactate oxidase	Chronoamperometry	Sweat	(a) 0.08–1.25 mM, 17.05 µM (b) 0.3–20.3 mM, 3.73 µM	[31]
Pop-up	Beta-Hydroxybutyrate	3-hydrozybutyrate dehydrogenase	Chronoamperometry	Human blood	0.1–6.0 mM 0.3 mM	[32]
Multiple folding	Organophosphorus pesticides	Inhibition of Butyrylcholinesterase, with a solid-state butyrylcholine-sensitive membrane	Potentiometry	Standard samples	0.1 and 1.0 nM 0.06 nM	[33]
Multiple folding	(a) Paraoxon (b) 2,4-dichlorophenoxyacetic acid (c) atrazine	Inhibition of: (a) butyrylcholinesterase (b) alkaline phosphatase (c) tyrosinase	Chronoamperometry	River water	(a) 2 ppb (b) 50 ppb (c) 10–100 ppb	[10]
Single folding	Sulfur mustard	Inhibition of choline oxidase	Chronoamperometry	(a) liquid and (b) aerosol sulfur mustard	(a) 1 mM (b) 0.019 g/m^3	[12]
Origami paper-based electrochemical DNA sensors						
Single folding	Mycobacterium tuberculosis ssDNA	Pyrrolidinyl peptide nucleic acid (acpcPNA) + $[Fe(CN)_6]^{3-/4-}$	Electrochemical impedance spectroscopy (EIS)	PCR-amplified DNA from blood samples	2–200 nM 1.24 nM	[34]
Double folding	Peanut allergen Ara h1	Synthetic aptamer against Ara h1 immobilized on graphene	Differential pulse voltammetry (DPV)	Cookie dough sample	50–1000 ng/mL 21.6 ng/mL	[35]
Multilayer (slipping layer)	ssDNA from hepatitis B virus	Complementary ssDNA #1 immobilized on MBs + Complementary ssDNA #2 labelled with AgNPs	Anodic stripping voltammetry	Standard solutions	85 pM	[36]

Table 1. *Cont.*

Type of Origami	Analyte	Sensing (Bio)Component	Electrochemical Technique	Matrix	Linear Range LOD	Ref.
Multiple folding	Complementary ssDNA	Complex formed by Biotin-DNA, avidin-HRP, and tetrahedral DNA	Chronoamperometry	Human serum	1–10 pM 1 pM	[37]
Origami paper-based electrochemical immunosensors						
Multilayer	Influenza virus H1N1	Horseradish peroxidase tagged antibody- H1N1 -primary antibody complex	Electrochemical impedance spectroscopy	Saliva	0–10,000 PFU/mL 4.7 PFU/mL	[38]
Multilayer	17β-estradiol	Monoclonal antibody immobilized on a MWCNTs/THI/AuNPs coated working electrode	Differential pulse voltammetry	Serum	0.01–100 ng/mL 10 pg/mL	[39]
Single folding	Prostate protein antigen	Carbon nanosperes-glucose oxidase-monoclonal antibody label	Differential pulse voltammetry	Serum	0.005–100 ng/mL 0.0012 ng/mL	[40]
Single folding	HIV p24 antigen	p24 antibody immobilized on zinc oxide nanowires coated working electrode	Electrochemical impedance spectroscopy	Serum	300 fg/mL	[41]
Single folding	Human interferon-gamma	Monoclonal human antibody was immobilized on polyaniline modified graphene electrode	Electrochemical impedance spectroscopy	Serum	5–1000 pg/mL 3.4 pg/mL	[42]
Multiple folding	(a) Human chorionic gonadotropin (b) prostate-specific antigen (c) carcinoembryonic antigen	Reduced graphene oxide/Ag@BSA/secondary antibody as signal label.	Cyclic voltammetry	Serum	(a) 0.002–120 mIU/mL, 0.0007 mIU/mL. (b) 0.001–110 pg/mL, 0.35 pg/mL. (c) 0.001–100 pg/mL, 0.33 pg/mL	[43]
Multiple folding	Human serum albumin and human immunoglobin G	Monoclonal anti-human antibodies immobilized onto pyrene carboxylic acid-modified single-walled carbon nanotubes	Chemiresistor	Urine, saliva, serum and whole blood	1.5 pM	[44]
Multilayer (slipping layer)	Ricin *a* chain	MB/anti-ricin *a* chain antibody/ricin *a* chain/monoclonal mouse anti-ricin *a* chain antibody/AgNP immunocomposite	Anodic stripping voltammetry	Standard solution	34 pM	[45]

Table 1. *Cont.*

Type of Origami	Analyte	Sensing (Bio)Component	Electrochemical Technique	Matrix	Linear Range LOD	Ref.
Multilayer (slipping layer)	Trefoil Factor 3	MB/spAb-TFF3-mpAb/2°mpAb/AgNP immunocomplex	Anodic stripping voltammetry	Urine	0.03–7.0 μg/mL	[46]
Origami paper-based electrochemical MIP sensors						
Single folding	Serotonin	Fe_3O_4@Au@SiO_2-MIP nanocomposite	Linear sweep voltammetry (LSV)	pharmaceutical and urine samples	0.5–1000 μM 0.08 μM	[47]
Double folding	Glycoproteins	SiO_2@Au/dsDNA/CeO_2 nanocomposite	Differential pulse voltammetry (DPV)	Egg white samples	$1–10^7$ pg/mL 0.87 pg/mL	[48]
Single folding + rotating elements	Carcino-embryonic antigen (CEA)	GO/chitosan/glutaraldehyde/dopamine	Differential pulse voltammetry (DPV)	Human serum	1.0–500.0 ng/mL 0.32 ng/mL	[49]
Origami paper-based electrochemical cell-based biosensors						
Single folding	Human acute promyelocytic leukemia cells (HL-60)	Aptamer KH1C12	Differential pulse voltammetry (DPV)	Human serum	$5.0 \times 10^2–7.5 \times 10^7$ cells/mL 4 cells/10 μL	[50]
Single folding	Human chronic myelogenous leukemia cells (K-562)	Concanavalin A immobilized on IL/3D-AuNPs/GN/composite	Differential pulse voltammetry (DPV)	Standard solutions	$1.0 \times 10^3–5.0 \times 10^6$ cells/mL 200 cells/mL	[51]
Single folding	Casein allergen	Rat basophilic leukemia (RBL-2H3) mast cells	Differential pulse voltammetry (DPV)	Standard solutions	$10^{-7}–10^{-5}$ g/mL 32 ng/mL	[52]
Self-powered origami paper-based electrochemical biosensors						
Single folding	Adenosine	Biotin-labeled aptamers immobilized on streptavidin-functionalized MBs	Electrochemical readout with a digital multimeter	Standard solutions	Up tp 250 μM 11.8 μM	[53]
Multiple folding	Adenosine triphosphate	ssDNA immobilized on a chemiluminescence-photoelectrochemical system composed of ABEI−AuNPs, p-iodophenol, and thioglycolic acid-capped CdS NPs	Electrochemical readout with a digital multimeter	Human serum	1–1000 pM 0.2 pM	[54]

Table 1. *Cont.*

Type of Origami	Analyte	Sensing (Bio)Component	Electrochemical Technique	Matrix	Linear Range LOD	Ref.
Pop-up	Adenosine triphosphate	Aptamer hybridized with GOx-labelled ssDNA and ssDNA immobilized on AuNPs	Electrochemical readout with a digital multimeter and Differential pulse voltammetry (DPV)	Standard solutions	10–5000 nM 3 nM	[55]

AgNPs, silver nanoparticles; MBs, magnetic beads; MWCNTs, multiple wallet carbon nanotubes; THI, thionine; AuNPs, gold nanoparticles; Ag@BSA, bovine serum protein-stabilized silver nanoparticles; spAb, monoclonal mouse anti-human TFF3 solid-phase Ab; mpAb, monoclonal rabbit anti-human TFF3 mobile-phase antibody; 2° mpAb, Biotinylated goat anti-rabbit secondary mobile-phase antibody; dsDNA, double strand DNA; GO, graphene oxide; IL, ionic liquid; GN, graphene; ABEI, N-(aminobutyl)-N-(ethylisoluminol); ssDNA, single strand DNA sequence; CdS NPs, cadmium sulfide nanoparticles; GOx, glucose oxidase.

2. Origami Paper-Based Electrochemical Enzymatic Biosensors

In the overall scenario of biosensors, the electrochemical enzymatic biosensor is the type of biosensor widely investigated and developed, taking into account that the most famous electrochemical biosensor is the one based on the glucose oxidase enzyme. In the case of glucose oxidase biosensor, the target analyte is glucose, namely the enzymatic substrate, and thus the electrochemical output is proportional to the amount of the target analyte.

Within the paper-based device for glucose detection, an easy configuration has been reported by Liang et al. [29], where the capability to move the pad near the classical printed electrochemical sensor was used to eliminate the electrochemical interferences in the case of glucose detection. In detail, before folding the enzymatic pad (thus when the enzyme is not present on the working electrode surface), the interferents such as ascorbate, urate, and paracetamol were completely consumed by a simple electrolysis step. Then, the enzymatic pad, with the enzyme loaded by drop casting, was put into contact with the working electrode allowing for the coulometric detection of glucose in the range of 0 to 24 mM, covering the diabetic range with recovery comprised in the range of 98–102%.

Then, the connection of the enzymatic pad to the working electrode allowed for the coulometric detection of glucose in the range from 0 to 24 mM, covering the diabetic range with recovery comprised in the range of 98–102%.

The configuration reported by Li et al. [30] combined two layers, encompassing one layer for the electrochemical cell, which was drawn with a pencil, and one layer for the enzymatic pad, containing glucose oxidase. In this case, the porosity of the paper was used to load both the glucose oxidase on the enzymatic pad and the electrochemical mediator, namely ferrocenecarboxylic acid, on the electrochemical cell layer. To make the analysis, the end-user has to add only the sample. This biosensor is characterized by a linear range comprised between 1 and 12 mM and a detection limit of 0.05 mM.

The porosity of the paper and the vertical microfluidics have been also exploited for the easy and accurate sampling of sweat to deliver a pump-free wearable device. Li et al. [31] designed an origami double enzymatic biosensor with a multi-layer structure to boost the sweat diffusion, the entrapment of the enzymes, and the accessibility of electrolytes for the detection of both glucose and lactate using glucose oxidase and lactate oxidase, respectively. Indeed, as designed, they vertically increased the area of the hydrophilic region for fostering the diffusion of sweat along the vertical direction into the paper substrate, through capillary-driven force (Figure 1A). This configuration allowed for avoiding fluid accumulation, ensuring that the biosensor was capable of detecting the biomarkers in the instant sweat rather than the accumulated sweat. In that way, the accuracy of instant detection of sweat composition is largely improved, furnishing the punctual concentration value, instead of a mean of values. The developed paper-based origami device demonstrated a dynamic range comprised between 0.08 and1.25 mM with a detection limit of 17.05 µM in the case of glucose biosensor, while in the range of 0.3–20.3 mM with a detection limit of 3.73 µM in the case of lactate biosensor, values useful to detect the physiological level of these biomarkers in sweat.

By exploiting the foldability of the paper, Wang et al. [32] designed a "pop-up" electrochemical paper-based analytical device inspired by pop-up greeting cards and children's books for measuring the beta-hydroxybutyrate by simply re-adapting a glucometer and a commercial beta-hydroxybutyrate kit (Figure 1B). The pop-up structure enables the easy change of the fluidic path and the control of the timing for implementing the easiness of the analysis, displaying a good linear fit in the clinically relevant range of 0.1 to 6.0 mM (R^2 = 0.96). The authors highlighted that the limit of detection of 0.3 mM is comparable with the limit of detection of commercial test strips, but the standard deviation is smaller in the latter case.

Contrariwise to biosensors able to detect the enzymatic substrate, the inhibitive biosensors have as the target analyte the enzyme inhibitors, and their electrochemical output is inversely proportional to the amount of the target analyte [56]. Furthermore, for the quantification of the inhibitor, the addition of the substrate for the measurement of the enzymatic activity in the absence and in the presence of the inhibitor is mandatory. The origami paper-based electrochemical biosensors allow for several smart configurations able to detect the inhibitors of enzymes by improving the simplicity of the measure.

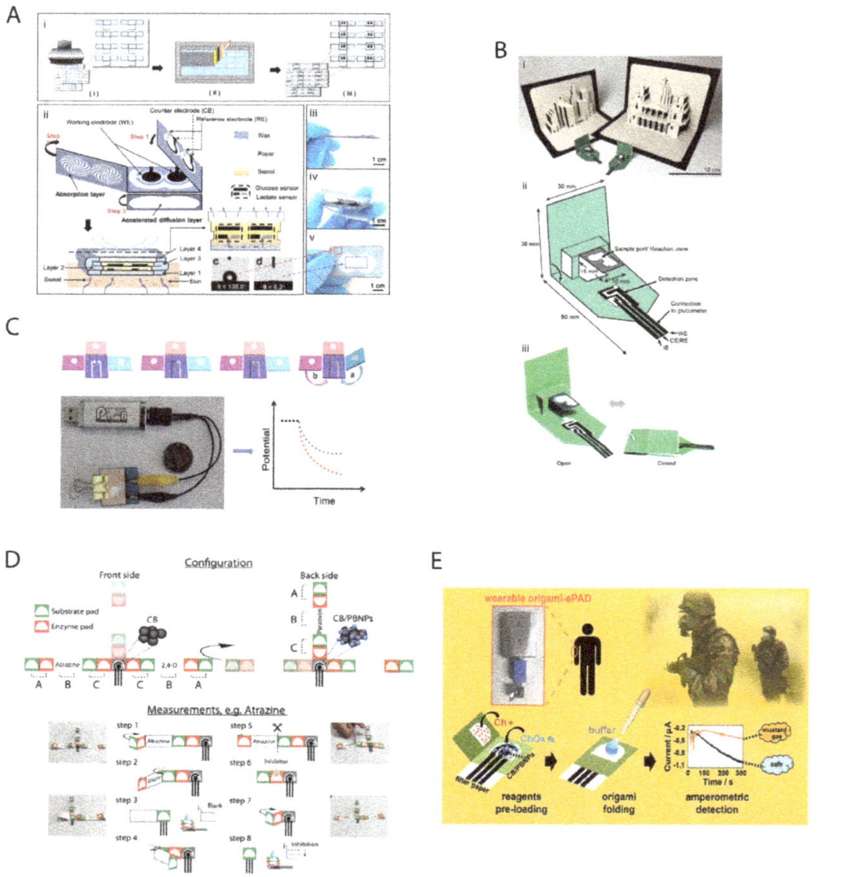

Figure 1. (**A**) A highly integrated sensing paper for wearable electrochemical sweat analysis. Reprinted with permission from [31], 2020 Elsevier; (**B**) a paper-based "pop-up" electrochemical device for analysis of Beta-Hydroxybutyrate. Reprinted with permission from [32], 2016 American Chemical Society; (**C**) a three-dimensional origami paper-based device for potentiometric biosensing. Reprinted with permission from [33], 2016 WILEY-VCH Verlag GmbH & Co; (**D**) origami multiple paper-based electrochemical biosensors for pesticide detection. Reprinted with permission from [10], 2018 Elsevier; (**E**) a wearable origami-like paper-based electrochemical biosensor for sulfur mustard detection. Reprinted with permission from [12], 2019 Elsevier.

Ding et al. reported [33] the first potentiometric origami paper-based biosensor using butyrylcholinesterase as enzyme and organophosphorus pesticides as target analytes, these pesticides being able to irreversibly inhibit this enzyme. In detail, the authors used as sensing layer a polymeric membrane designed for high sensitivity and selectivity toward butyrylcholine using heptakis(2,3,6-tri-o-methyl)-b-cyclodextrin as an ionophore. In the presence of enzymatic by-product choline, the sensor showed lower potential response than the butyrylcholine cation, and this ion-selective electrode membrane was selected for the detection of butyrylcholinesterase activity. Regarding the 3D structure, the origami biosensor was conceived with a test pad surrounded by three folding pads for loading the enzymatic substrate (substrate pad), the enzyme (enzymatic pad), and the sample (sample pad) (Figure 1C). The substrate pad was folded above the test pad and some μL of substrate solution was added onto the pad. Successively, the enzyme pad was folded and some μL of the enzymatic solution was spotted onto the enzyme zone. The sample pad was then folded on the top, and some μL of sample solution were added to the sample zone. Since it is an irreversible inhibition, for which an incubation time (reaction between the enzyme and inhibitor) increases the sensitivity, the sample pad and the enzyme pad were clamped by two binder clips to ensure a contact for 5 min. After that, the sample pad was unfolded, the inhibited enzyme pad was folded onto the substrate pad, and thus the enzymatic activity was measured by adding the phosphate buffer solution. In this proof of concept, the authors observed a dynamic range comprised between 0.1 and 1.0 nM and a detection limit of 0.06 nM using methyl parathion as target organophosphate. In this configuration, the presence of several pads allowed to avoid the different reservoirs for enzymatic reaction with the substrate and incubation time, making the whole process measurement suitable to be carried out in a miniaturized system.

The foldability of the paper and the easiness to be embedded as well as to cut the pads have been exploited by our group to design origami paper-based devices able to detect different classes of pesticides by using different types of enzymes able to be differently inhibited [10]. Furthermore, we exploited the porosity of the paper to load both the enzyme and the enzymatic substrate, delivering a reagent-free device, which requires the end- users to only add the sample or distilled water. This origami paper-based device was designed by integrating two different office paper screen-printed electrodes (placed back-to-back in the front and the backside of the origami system) with foldable filter paper strips, pre-loading different enzymes and enzymatic substrates on the latter. In detail, the analysis of different classes of pesticides, namely organophosphorus insecticides, phenoxy-acid herbicides, and triazine herbicide, was achieved using butyrylcholinesterase, alkaline phosphatase, and tyrosinase enzymes for the detection, respectively. For each class of pesticide, a filter paper strip was configured as follows: two pads were pre-loaded with the enzyme (red pads), and the other two pads were pre-loaded with the substrate (green pads) by simple drop-casting of the reagent solutions (A,C in Figure 1D). This configuration allows for the measurement of the enzymatic activity both without inhibitors and in the presence of inhibitors. In detail, when a distal couple of pads is folded to contact the electrochemical cell printed on the office paper, the non-inhibited enzymatic activity can be measured by the drop-casting distilled water. The choice of distilled water relies on the fact that the pads already contain the enzyme, its substrate and buffer salts, and the distilled water can dissolve them. For the sample measurement in the presence of inhibitors, some μL of water sample are loaded only onto the red pad containing the enzyme for the incubation time (5 min), then the pads are folded to contact the electrochemical cell printed on the office paper and the measure is recorded. The degree of inhibition was evaluated by chronoamperometric mode and the suitability with real samples was assessed by adding the river water samples without any treatment, thanks to the capability of paper to block the particulate, disclosing the feature of the paper pad of enabling the sample treatment. The paper-based device demonstrated the capability to quantify paraoxon, 2,4-dichlorophenoxyacetic acid, and atrazine at ppb level in both standard solutions and river water sample with good accuracy, demonstrating that exploiting the features of paper is possible to treat the sample, store the reagents, and make the measurement using a foldable paper-based device.

The porosity of the paper in the origami configuration was also exploited for the detection of the target analyte in the gas phase without any external sampling system, as demonstrated by our group [12] in the case of choline oxidase biosensors for chemical warfare agents, i.e., mustard agents, knowing that these compounds can reversibly inhibit the choline oxidase enzyme. In detail, an origami device constituted of two layers has been fabricated, one layer in which the electrochemical cell is printed, and the enzyme and phosphate salts are pre-loaded, and the second layer in which the substrate, i.e., choline, is pre-loaded, obtaining a biosensor ready-to-use without the necessity of additional reagents for the analysis (Figure 1E). The measurement was carried out by overlapping the two origami layers and exposing the biosensor to the mustard agents-containing aerosol. The porosity of paper allowed the layers of the origami to be impregnated by the aerosol and thus to dissolve the reagents, enabling the inhibition against the enzymatic reaction to be measured. This paper-based device was first tested with a toxic mustard agent simulant, (bis-(2-chloroethyl) amine), and subsequently with a real mustard agent in compliance with the appropriate security measurements at the Bundeswehr Institute of Pharmacology and Toxicology (Munich, Germany) in both solution and aerosol phase, reaching a detection limit equal to 1 mM and 0.019 g/m^3 in the liquid and aerosol phase, respectively. This work highlighted that the paper-based device is able to overcome the constrain of the polyester printed sensors, that in the case of aerosol require an external engineering sampling system [57], resulting in the need for a more complex analytical device.

3. Origami Paper-Based Electrochemical DNA Sensors

Selective electrochemical biosensors are obtained by employing nucleic acid sequences as specific recognition elements. Biosensors based on nucleic acids have been widely explored over the last years due to their manifold physical, chemical, and biological properties [58]. Among them, aptasensors have been widely explored [59], basing on the use of aptamers, namely small artificial sequences of single-stranded DNA or RNA (ssDNA or ssRNA) capable of selectively recognize the specific target, including proteins, peptides, and various small molecules of analytical interest [60]. The key properties of nucleic acids rely on both the high specificity of the hybridization between complementary strands of DNA/RNA and the possibility to be conjugated to a sensing element (e.g., electroactive species, recognition elements), which confers to them an extreme versatility. During the last decades, the advances achieved by synthetic biology have converged the knowledge of genetic, chemistry, engineering, and biology to develop the technology for re-designing genes or proteins, taking inspiration from nature and going beyond it. The realization of synthetic antibodies, as well as synthetic nucleic acid molecules, have opened the way toward a wider horizon in the biosensor field. Indeed, nowadays nanotechnology allows for a fine design of synthetic nucleic acid molecules, providing tailor-made sensing materials for a variety of applications by using simple procedures and instrumentation. These advantageous features have significantly boosted the development of cost-effective nucleic acid-based biosensors, also reaching the field of paper-based sensors during the last years [61]. Several strategies have been explored for integrating nucleic acid molecules as sensing elements in the cellulosic network of paper, showing that the paper matrix is a suitable platform for realizing nucleic acid-based biosensors [62–64].

In the framework of origami sensors, Henry's group contributed with their experience in paper-based sensors to realize a nucleic acid-based biosensor designed with an origami configuration [34]. The origami sensor was developed by covalently immobilizing a pyrrolidinyl peptide nucleic acid (acpcPNA) onto partially oxidized cellulose paper. In detail, the procedure employed consisted of a mild oxidative treatment, based on LiCl in $NaIO_4$, on the cellulose matrix to generate aldehyde groups functional to the immobilization of the acpcPNA in the presence of $NaBH_3CN$. The authors took advantage of the covalently binding of the acpcPNA to the cellulose network to build up a device that can be multiply regenerated by simply replacing the overlaying pad, avoiding any leaching of the acpcPNA. The nucleic acid probe was immobilized on the screen-printed working electrode, while the reference and counter electrodes were screen-printed on a second pad, to be overlapped for the measurement (Figure 2A).

The device was tested for the detection of mycobacterium tuberculosis as an example application. In detail, a synthetic 15-base oligonucleotide of mycobacterium tuberculosis was chosen as target ssDNA able to hybridize with the immobilized acpcPNA, obtaining an electrochemical DNA sensor with a label-free approach. The complex formation resulted in the variation of the charge transfer resistance, monitored by electrochemical impedance spectroscopy using $[Fe(CN)_6]^{3-/4-}$ as the electrochemical probe. The analytical performance of the developed origami was proved for the detection in real clinical samples confirmed by the comparison with PCR reference method, allowing for a linear response in the range from 2 nM to 200 nM of mycobacterium tuberculosis and limits of detection and quantification in the nanomolar level.

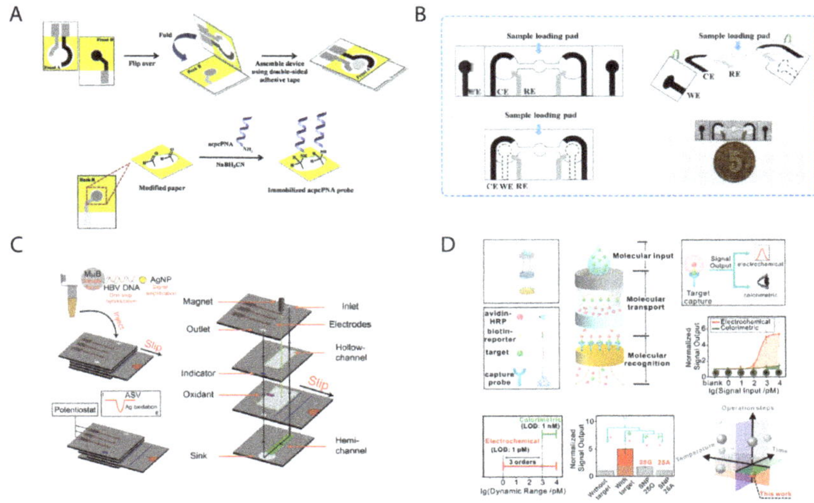

Figure 2. (**A**) Electrochemical impedance-based DNA sensor using pyrrolidinyl peptide nucleic acids for tuberculosis detection. Reprinted with permission from [34], 2018 Elsevier; (**B**) microfluidic origami nano-aptasensor for peanut allergen Ara h1 detection. Reprinted with permission from [35], 2021 Elsevier; (**C**) detection of Hepatitis B Virus DNA with a paper electrochemical sensor. Reprinted with permission from [36], 2015 American Chemical Society; (**D**) molecular threading-dependent mass transport in paper origami for single-step electrochemical DNA sensors. Reprinted with permission from [37], 2018 American Chemical Society.

The origami folding was recently combined with the horizontal transport by capillary forces by Jiang et al. [35]. The authors have developed a microfluidic aptasensor on a single piece of chromatography paper patterned by polydimethylsiloxane, with microchannels and screen-printed electrodes (Figure 2B). The working electrodes were obtained separately from the counter and reference electrodes. The origami was designed with a symmetrical geometry, having two fold lines to be bent for completing two analogous electrochemical cells. This configuration allowed for a dual-target detection or screening at a time. The working electrode was electrodeposited with black phosphorus nanosheets decorated with an aptamer as the molecular recognition probe, being specific for the recognition of the peanut allergen Ara h1 without the need for labeling. The detection was carried out by using a ferro/ferricyanide redox probe in differential pulse voltammetry, achieving a good linear response in the range of 50–1000 ng/mL and a detection limit of 21.6 ng/mL.

A more sophisticated origami architecture was conceived by Crooks' research group, who realized an origami that incorporates slip layers (oSlip) for the detection of DNA from hepatitis B virus (HBV) [36]. In detail, the "slip pad" consisted of a moveable layer of paper that can be used to initiate on-chip chemical reactions at the desired time. This component was provided with hollow channels for realizing a one-step device assembled in a sandwich-like configuration (Figure 2C). The layering of this device was exploited for selecting the paper layers when needed to load the reagents or to control the flow through the sandwich device.

The device exploits both silver nanoparticle and magnetic microbeads for obtaining an amplification of the response equal to a factor of 250,000 and ~25-fold, respectively. The magnetic microbeads are used as mobile solid phase to support the capture probe, which was designed to bind on separated portions of the sequence both the ssDNA from HBV and the ssDNA labeled to the AgNPs. Paper allows for the concentration of these particles at the electrode, enabling signal amplification. In detail, the sample containing the magnetic microbeads-DNA-AgNPs, obtained after one-step hybridization, was injected in the inlet hollow and allowed to flow through the layers; as the solution reaches the bottom layer, it flows upward through the outlet hollow and solves a colorimetric indicator, revealing that the device is ready for the electrochemical assay. The reaction is then activated when the slip layer is moved to its functional position, enabling the dissolution of $KMnO_4$ to oxide Ag contained in AgNPs to Ag^+, which is detected in voltammetric mode with a detection limit of 85 pM. The presence of ~250,000 Ag atoms in each AgNP explains the amplification of the signal resulting from this configuration.

To verify the performance of this origami, the voltammetric response was compared with a conventional electrochemical cell, reproducing the HBV-DNA sandwich assay. Interestingly, the response of the multilayer paper device was significantly higher than the response resulting from the conventional electrochemical cell. The authors attributed this improvement of the signal to the geometric confinement of Ag^+ near the working electrode, ensured by the wax hydrophobic barrier that limits the possibility of diffusion of the oxidized Ag^+. This effect is combined with the direct application of a magnet at the working electrode, which ensures the close proximity of the magnetic microbeads to the electrode.

A noteworthy study of the unique properties of origami-like devices was conducted by Ye et al. in 2018 [37], focusing on the advantages that the 3D vertical flow configuration of paper-based origami devices can provide. Indeed, they designed an origami geometry to realize a molecular threading-dependent transport system for the controlled and directional transport of biomolecules. The vertical configuration was conceived to enable molecular recognition and enzymatic reaction with a programmed order. Importantly, the authors drew attention to the hierarchical structure of the paper material, which is highlighted as a versatile and multitasking material among many examples in nature, having the mass storage capability and the capillary properties as key features.

The advantage of a 3D vertical configuration relies on the possibility to overcome the Lucas–Washburn law, which limits the mass transport distance to about 3–4 cm in 2D lateral flow devices. This was shown by comparing the transport of methylene blue between a 2D paper device, on which the diffusion was driven only by capillary forces, and a 3D paper architecture (obtained by folding the paper strip), observing a sharply greater coloring for the latter (Figure 2D).

After demonstrating the concept by visual mode, the authors have applied the origami device for multiple components storage and transport, namely DNA and proteins. In detail, biotin-DNA and avidin-HRP were dropped on dedicated areas of paper (preserving chips) and transported through the 3D configuration from a reaction area to a detection area (reaction and detection chips, respectively). The transport was carried out under controlled conditions, by programing the 3D folding of the paper device. A cascade reaction was triggered as soon as the biomolecule reach the reaction area, modified with AuNPs, where a capture probe was immobilized (ssDNA and tetrahedral DNA nanostructures). The foldability of paper was herein skillfully exploited to drive the cascade reactions also exploiting the porosity of paper to transport the regents to the desired site or, on the other hand, to store the immobilized receptors. Interestingly, such a complex design allowed for a rapid and single-step measurement of a target DNA molecule with picomolar detection limit and the capability of distinguishing a single base mismatch.

4. Origami Paper-Based Electrochemical Immunosensors

Immunosensors are analytical devices used to detect the binding event between an antibody (Ab) and an antigen (Ag) with the formation of a stable complex [65]. For immunosensor development, either Ab or Ag can be immobilized on the surface of different transducers, producing several immunosensor configurations with high sensitivity and selectivity thanks to the high specificity of antigen–antibody interactions [66–68].

Recently, paper-based immunosensors have been used as for the development of point-of-care testing kits exploiting the lateral flow assay (LFA) technology [69]. The LFA-based paper-based sensors generally consist of a sample pad, a conjugate pad, a nitrocellulose membrane, and an absorption pad. In these sensors, test analytes pass horizontally from the sample pad to the test section and the absorption pad by capillary force, allowing the binding between antigens and labelled antibodies on the conjugation pad for a sensitive detection. Although LFA-based electrochemical immunosensors have provided relatively short assay times (approximately 10–20 min), low-cost analysis, simple handling, and ease of mass production [70], it has some limitations such as relatively low sensitivity, limited sample volume, and difficulties in making multiple measurements [71].

In the last decade, three-dimensional microfluidic paper devices have been developed to improve the lateral flow-based sensors and to overcome their limitations leading to the development of paper-based vertical-flow immunosensor, which allows rapid vertical flow assay systems with controlled vertical flow and a separate measuring area [72]. For instance, Bhardwaj et al. [38] developed a vertical flow-based paper immunosensor using a different pore size sample pad for the electrochemical detection of the influenza virus H1N1 in both standard buffer solution and saliva samples. This lateral flow-based paper immunosensor consists of a different pore size sample pad (a pad characterized by pores with different sizes i.e., larger pores with a 11 μm diameter and smaller pores with a 0.45 μm diameter), a conjugate pad, a nitrocellulose membrane strip, and an absorption pad, all of which are vertically stacked one upon the other onto a polyester backing film. The three-electrode area was defined on a nitrocellulose membrane strip by wax printing (Figure 3A). The different pore size pad allowed them to increase the binding efficiency of antigen-horseradish peroxidase-tagged antibodies on the conjugate pad and concentrate the antigen–antibody complexes by providing the optimal residual time, fast detection (~6 min), and high sensitivity, with a limit of detection lower than 5 PFU mL^{-1} for saliva samples. Moreover, the porosity of sample pads acted not only as flow/volume control components, but also as a filter that facilitates small-sized biological particles such as viruses to pass through, while retaining larger particles, which is useful for the detection of complex fluids.

A paper-based electrochemical vertical flow paper-based device was also developed by Wang et al. [39], for the detection of 17β-E2 using multi-walled carbon nanotubes/thionine/gold nanoparticles composites synthesized and coated directly on the screen-printed working electrode for the immobilization of anti-E2. The device, which consisted of four layers, was made on four pieces of cellulose filter papers whose size was 10.5 mm × 35.0 mm. The samples were injected from the sample inlet, flowed through the microchannel and entered the filter hole. After filtering, they finally arrived at the reaction site. This vertical flow-based structure offers different advantages, including small volume of required samples, a simple procedure of sample handling, and an increased sensitivity, automated flow-injection and samples filtration.

As already discussed, the possibility to fold the paper to obtain different origami geometries allows for the fabrication of 3D structure without the need for other materials, e.g., double side tape, to integrate the different layers together. This strategy revealed to be advantageous also for the development of origami-based immunoassays. For instance, Li et al. [40] designed an immunosensing device for the detection of prostate antigen using two waxed pads, one for the screen-printed counter and reference electrodes and the other one for the working electrode, which were folded to create the 3D structure (Figure 3B). The detection of prostate antigen was carried out using an electrochemical enzymatic redox cycling constituted by glucose oxidase as an enzyme label, 3,3′,5,5′-tetramethylbenzidine as a redox electrochemical mediator, and glucose as the enzymatic substrate. The porosity of the paper was also exploited to grow gold nanoparticles directly on the surfaces of cellulose fibres in the working electrode where subsequently, manganese oxide nanowires were electrodeposited to form a network with large surface areas. The proposed method successfully fulfilled the highly sensitive detection of prostate antigen with a linear range of 0.005 ng/mL–100 ng/mL with a detection limit of 0.0012 ng/mL.

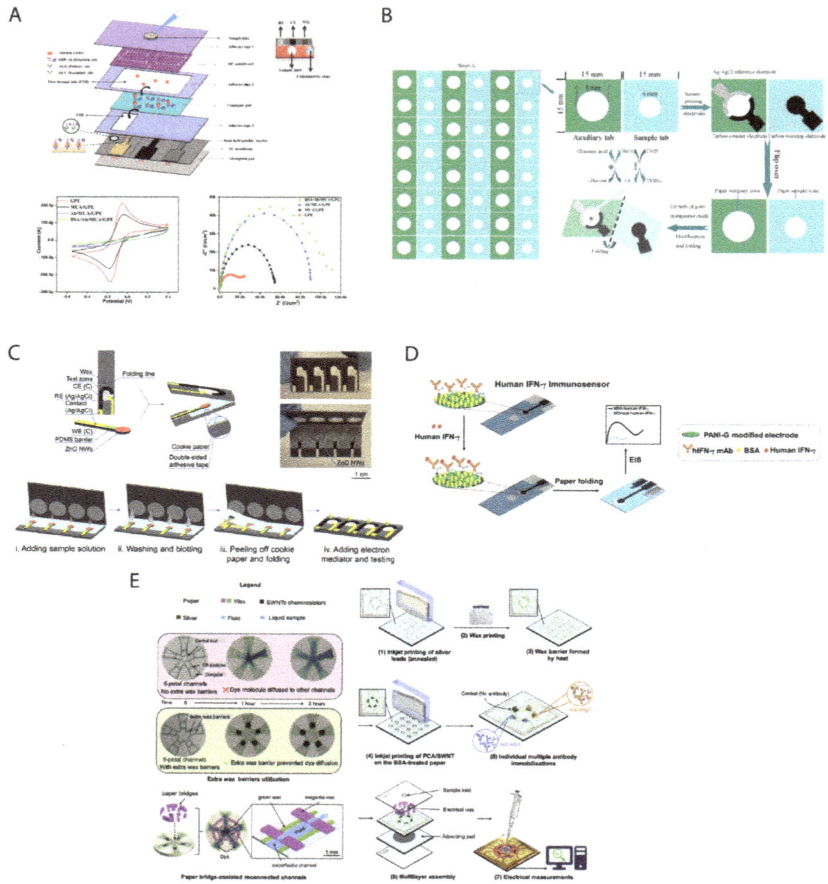

Figure 3. (**A**) Vertical flow-based paper immunosensor for rapid electrochemical and colorimetric detection of influenza virus using a different pore size sample pad. Reprinted with permission from [38], 2018 Elsevier; (**B**) growth of gold-manganese oxide nanostructures on a 3D origami device for glucose-oxidase label based electrochemical immunosensor. Reprinted with permission from [40], 2014 Elsevier; (**C**) a microfluidic paper-based origami nanobiosensor for label-free, ultrasensitive immunoassays. Reprinted with permission from [41], 2016 WILEY-VCH Verlag GmbH & Co; (**D**) label-free paper-based electrochemical impedance immunosensor for human interferon gamma detection. Reprinted with permission from [42], 2018 Elsevier; (**E**) an origami immunosensor for multiplexed analyte detection in body fluids. Reprinted with permission from [44], 2020 Elsevier.

Another example of origami configuration was reported by Li et al. [41], who exploited the paper folding to create a biosensor able to detect HIV p24 antigen in human serum with a low detection limit of 300 fg/mL (>33 times lower than that of a commercial p24 antigen test kit), integrating hydrothermally synthesized zinc oxide nanowires and electrochemical impedance spectroscopic technique. In this case, paper was exploited for the in situ growth of zinc oxide nanowires directly on a carbon working electrode and then the zinc oxide nanowires were functionalized with p24 antibodies. The device consisted of two pieces of cellulose paper: (i) one piece of paper containing a hydrophilic paper test zone patterned via solid wax printing, and the carbon counter and the silver/silver chloride reference screen-printed electrodes, and (ii) another piece of paper including the carbon working electrode, on which zinc oxide nanowires are directly synthesized in situ (Figure 3C). The hydrothermal growth of ZnO NWs on the paper substrate included two steps: first, an uniform coating on the paper with a seeding layer of ZnO nanoparticles, which provides the starting points of the ZnO-NW growth; then, the directional nucleation of ZnO NWs from the seeding layer. In this second step the ZnO-NP-coated paper was immersed in an aqueous solution of zinc salt and other chemicals at an elevated temperature for the growth of ZnO NWs. The authors chose the origami structure and arranged the working electrode on a piece of paper separated from the origami paper layer for two main reasons. First, the zinc oxide nanowires on the working electrode have intimate contact with the solution contained in the test zone, yielding enhanced electrochemical performance; Moreover, the hydrothermal growth and surface biofunctionalization of zinc oxide nanowires involve heating and immersion in solutions which could compromise the hydrophobicity of wax barriers if the working electrode was printed directly on the origami paper layer.

The same approach was used by Reucha et al. [42] for the development of an origami electrochemical platform for sensitive detection of human IFN-γ. Indeed, the origami consisted of two wax patterned separated pads (Figure 3D). The working electrode was designed separately from the counter and reference electrodes to reduce the consumption of the reagents and the sample volume, as well as to prevent reference and counter electrodes contamination with proteins during the preparation of the immunosensor. For the detection of human IFN-γ, a monoclonal human IFN-γ antibody was immobilized on the polyaniline modified graphene screen-printed paper electrode and electrochemical impedance spectroscopy was used for detection of human IFN-γ in a range of 5–1000 pg/mL with a detection limit of 3.4 pg/mL.

The possibility of designing immunosensors with more complex configurations has been achieved by folding different layers of paper to realize one origami structure for the simultaneously and selectively detection of different analytes. Indeed, Sun et al. [43] proposed an origami multiplexed enzyme-free electrochemical immunosensor for the detection of human chorionic gonadotropin, prostate-specific antigen, and carcinoembryonic antigen. The electrochemical immunosensor is characterized by zinc oxide nanorods which provide high number of sites for conjugating the capture antibodies and reduced graphene oxide which improves the electronic transmission rate. The current signal is generated from the reduction of H_2O_2 and further amplified by a subsequent signal labels-promoted deposition of silver. The origami device was comprised of an auxiliary pad surrounded by three sample pads of the same size (20 mm $\times$ 20 mm). The electrode array consisted of a screen-printed Ag/AgCl reference electrode and a carbon counter electrode on the auxiliary zone and three screen-printed carbon working electrodes on the three paper sample zones, respectively, which are functionalized with three different capture probes for their specific analytes. Between each pad and auxiliary pad, an unprinted line (1 mm in width) was defined as a fold line which ensures that the paper sample zones on the three pads were properly and exactly aligned to the auxiliary pad after folding. Under optimal conditions, the proposed immunosensor exhibit excellent precision, high sensitivity, and a detection limit of 0.0007 mIU/mL for human chorionic gonadotropin, 0.35 pg/mL for prostate-specific antigen, and 0.33 pg/mL for carcinoembryonic antigen.

Another example of multiplexed 3D electrochemical immunosensor is proposed by Shen et al. [44] with a label-free field-effect transistor/chemiresistor-based immunosensor. This sensor consisted of pyrene carboxylic acid-modified single-walled carbon nanotubes deposited by quantitative inkjet printing with an optimal three-dimensional semiconductor density on the paper substrate. Monoclonal anti-human antibodies were individually immobilized onto the SWCNTs surface to achieve a highly sensitive and specific detection of human serum albumin and human immunoglobulin G with detection limit of 1.5 pM. The origami biosensor composed of five-petal shaped hydrophilic channels on paper was designed to equally split one sample into five aliquots for individual sensing channels by the capillary force (Figure 3E). In details, the origami was composed of three layers: (1) top layer for sample-splitting and paper-bridging; (2) middle layer with chemiresistor biosensor arrays for multiplexed detections; and (3) bottom layer with sufficient absorbing capability. Unlike the manually assembled ones, origami devices benefited from the precise trimming by CO_2 laser cutting that helped facile alignment of the three layers thereby reducing the human labor and error.

In alternative to the foldable configurations, Crooks' research group applied the approach of the slipping pads, already discussed for the DNA-based origami biosensors [36], also in the application field of origami-inspired immunosensors, aiming to further improve the possibilities of designing reconfigurable structures of origami devices. They proposed two different origami slip pads (*o*SLIP) where the device is fabricated by paper folding and operated by paper slipping. The first paper-based immunosensor is based on quantitative detection of silver nanoparticle labels linked to a magnetic microbead support via a ricin immunosandwich with a detection limit of 34 pM [45]. The sensor platform comprised four wax-patterned paper layers. The three carbon electrodes were stencil-printed on the lower layer 1, which displays the inlet and the outlet reservoirs. Layer 2 contained a hollow channel and a paper reservoir loaded with a blue dye, the latter indicating that the device was ready for measurement as soon as the colouring occurred (due to the sample flow through the default path within the microfluidic channels). Layer 3, namely the slip layer, contained both a hollow channel and a paper pad for dried oxidant storage. Finally, layer 4 consisted of a hydrophilic layer (hemichannel) and a sink pad that drove a continuous flow of fluid through the device until its capacity is filled. The oSlip is firstly assembled by folding the paper. The assay began by injecting the pre-formed ricin immunocomposite into the oSlip inlet. Then, the ricin immunocomposite is concentrated under the first carbon electrode by the magnetic field. When the blue colour appeared due to the flow reaching the dye at the outlet, the pre-dried chemical oxidant was slipped (by pulling Layer 3 until a green indicator line became visible) reaching a direct contact with the ricin immunocomposite. Finally, the dissolved Ag^+ ions are electrodeposited on the electrode as metallic Ag for 200 s and then stripped off. This work has demonstrated how the combination of folding components and layers that can be slipped into and out of the origami can provide sophisticated configurations that allow for fine control of the different steps for the device application.

The second origami slip immunosensor structure was used to detect the kidney disease marker Trefoil Factor 3 (TFF3) in human urine [46]. The sensor is based on a quantitative metallo-immunoassay able to determine TFF3 concentrations via electrochemical detection of environmentally stable silver nanoparticle labels attached to magnetic microbeads via a TFF3 immunosensor. For the electrochemical detection, a one-step assay was performed characterized by incubating TFF3, the AgNP/2°mpAb/mpAb conjugate, and the microbead/pAb conjugate simultaneously. After this single step, the immunocomplex was washed three times and then it was injected into the inlet of the oSlip for the measurement. Similarly, to the work previously described [64], this device design is characterized by a hollow channel and a hemichannel, which enable the microbeads to flow rapidly through the origami, and a slip-layer switch, which allowed for time-controlling of the reagent delivery. Moreover, the *o*Slip configuration allows for the preconcentration of all necessary reagents making easier the work of end-user and allowing for the detection of TFF3 in human urine in the concentration range comprised between 0.03 and 7.0 μg/mL.

5. Origami Paper-Based Electrochemical MIP Sensors

Molecular imprinting is a powerful technique based on a mimic approach, inspired by natural receptors to achieve molecular recognition [73]. This strategy aims to reproduce antibody-like binding properties or enzyme-like catalytic activities by using a polymeric template markable with the target analyte, in order to leave an imprint that can ensure a selective and specific recognition. Usually, MIPs are obtained by the cross-linking of functional monomers, which are polymerized under specific conditions in the presence of template molecules by covalent, non-covalent, or hydrogen interactions [74]. After the polymerization, the template molecules are removed from the polymer, thus leaving cavities that are able to act with a key-lock mechanism when the template molecules are detected in a working sample [75]. This approach allows for exploiting the molecular memory imprinted in the polymer by the target molecule itself, which is first used as a template and then ensures the high affinity in shape and size at the measurement stage.

MIPs have been used for a variety of applications ranging from chromatographic separation to molecular sensing [76,77], but also serving as a tool for drug delivery [78]. In the field of electrochemical sensors, they are known as smart modification materials characterized by short synthesis time, largely improved storage stability, and cost-effectiveness. One of the main challenges for this technology is the imprinting of biomacromolecules, such as proteins, because of their complex structures and large sizes. Moreover, the use of MIPs for electrochemical devices is typically affected by low sensitivity due to the poor conductivity and electrocatalytic activity of most of the common polymers employed. In order to overcome these limitations, various approaches have been explored, including strategies to improve the surface imprinting efficiency or the use of conductive nanomaterials.

Over recent years, the paper was showed to be a suitable platform for the in-situ synthesis of MIPs, and its properties have contributed to the realization of MIPs, such as the porosity and the adsorptive cellulosic network [79,80]. Also in this field, the versatility of paper has disclosed new possibilities for expanding the applicability of MIP technology.

Very recently, the MIP technology was integrated with the benefits carried by origami structured paper devices. Amatatongchai et al. [47] applied a simple origami configuration to develop a paper-based biosensor for serotonin detection. The biosensor was based on a graphite-paste electrode modified with a MIP composed of Fe_3O_4@Au nanoparticles encapsulated with imprinted silica. Circular hydrophobic areas were realized on filter paper by alkyl ketene dimer inkjet printing. The three-electrode cell was screen-printed using a graphite paste comprising graphite powder, carbon nanotubes, and mineral oil in correspondence with the hydrophilic areas. After printing the conductive paste on a single piece of paper, the Fe_3O_4@Au@SiO_2-MIP nanocomposite was drop-cast onto the working electrode. Thus, the sensor was ready to be folded, overlapping the two circular hydrophilic areas, and to be used for the measurement (Figure 4A).

The configuration of this origami device was further studied by varying the number of layers overlapped and by evaluating the effect of increasing volumes of the sample. As introduced above, each paper layer presented a circular hydrophilic region delimited by the alkyl ketene dimer hydrophobic barrier. After vertically aligning the hydrophilic areas, a serotonin sample was drop-cast on the hydrophilic region on the top of the origami sensor. An increment of the voltammetric response was observed when a second paper layer was overlapped on the underlying screen-printed paper layer, while the addition of a third paper layer did not change the sensor response significantly. Although this evidence would be worthy of further investigation, the authors did not provide a critical explanation and continued the study by choosing the double-layer configuration. Regarding the effect of the sample volume, the increment from 10 μL to 20 μL resulted in increasing the signal recorded. The authors ascribed this behavior to the facilitated diffusion of the sample through the layers when using a larger volume. However, they highlighted that a further increase of the volume can be responsible for an overload of the sample and the eventual occurrence of noise due to wetting of the electrical contact. Importantly, the authors pointed out how the use of paper layering to design a folded origami configuration was capable of avoiding the direct contact of the sample with the electrode surface, especially important for the real samples having complex matrices to reduce the possible interfering effects on the sensor performance. Moreover, the high surface-to-volume ratio offered by the paper-fiber matrix allows for a high impregnation of the sample within the sensor material, favoring the detection. The origami sensor modified with the Fe_3O_4@Au@SiO_2-MIP nanocomposite was proved to be suitable for the detection of serotonin in pharmaceutical and urine samples, with good analytical performances in terms of precision and high tolerance toward interfering chemicals.

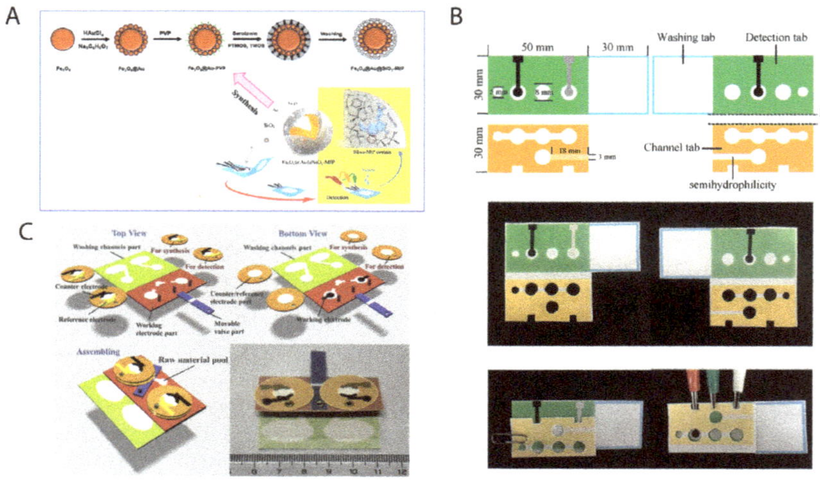

Figure 4. (**A**) Highly sensitive and selective electrochemical paper-based device using a graphite screen-printed electrode modified with molecularly imprinted polymers coated Fe_3O_4@Au@SiO_2 for serotonin determination. Reprinted with permission from [47], 2019 Elsevier; (**B**) ultrasensitive microfluidic paper-based electrochemical/visual biosensor based on spherical-like cerium dioxide catalyst for miR-21 detection. Reprinted with permission from [48], 2019 American Chemical Society; (**C**) the strategy of antibody-free biomarker analysis by in-situ synthesized molecularly imprinted polymers on movable valve paper-based device. Reprinted with permission from [49], 2019 Elsevier.

Further advances in the application of MIP to paper origami biosensors were achieved by Yu's group [48], who proposed an ultrasensitive sensing platform for the rapid and accurate detection of glycoproteins. The origami device was composed of three regions: a detection pad, a channel pad, and a washing pad. The electrodes were screen-printed on the detection pad, with the working one printed on the reverse side. A circular area (8 mm in diameter) was placed in correspondence with the working electrode but on the obverse side. A smaller circular area (5 mm in diameter) was placed laterally and used as an inlet zone for the sample. The channel pad was overlapped on the detection pad by alternatively folding along two fold lines to switch the sensor from the washing mode (fold line 1) to the detection mode (fold line 2). Using fold line 2, the channel pad allows the sample to flow from the inlet zone to the electrodes, thanks to the capillary forces through the main hydrophilic channel, thus enabling the electrochemical measurements. For the washing step, a buffer was added in the 8-mm circular pool to wash the detecting area. A semi-hydrophilic channel was designed on the channel pad to draw the excess of washing solution by capillary forces, driving it toward the washing pad. The semi-hydrophilic channel was conceived to further protect the detection area from being contaminated by the washing liquid (Figure 4B).

A glycoprotein-based MIP was synthesized on the working electrode after the in-situ growth of Au nanorod in conjunction with 4-mercaptophenylboronic acid as a recognition element, able to form a covalent cyclic ester with the target glycoprotein. The detecting principle relied on AuNPs immobilized on the surface of SiO_2 and functionalized with 4-mercaptophenylboronic acid nanoparticles able to link DNA molecules, obtaining a SiO_2@Au/dsDNA/CeO_2 nanocomposite. Double strands of DNA were allowed to form on the surface of SiO_2@Au through a hybridization chain reaction in the presence of two hairpin DNAs. Thus, CeO_2 nanoparticles were bound to the DNA probes through an amidation reaction. The target glycoprotein, namely ovalbumin (OVA), was recognized by the boronate affinity-based MIP, enabling the sandwich interaction with the SiO_2@Au/dsDNA/CeO_2 nanocomposite. The detection was hence obtained by the reduction of Ce^{4+} to Ce^{3+} upon the addition of 1-naphthol reagent. In this configuration, the SiO_2@Au/dsDNA/CeO_2 nanocomposite served for obtaining higher electron transfer efficiency and larger surface area for the DNA immobilization at the same time.

The morphological and electrochemical suitability of this origami device was analyzed in detail. Micrographs of the chromatography paper revealed a framework of cellulosic fibers that was responsible to provide a biocompatible, incompact microenvironment for the in situ growing of Au nanorods. The successful formation of the MIP was verified as well. The comprehensive characterization of the electrochemical features of the Au-μPADs/MIPs as well as the promising analytical performance achieved with standard samples enabled the authors to apply the origami device to real samples, consisting of 100-fold diluted egg white samples, resulting in high analytical accuracy.

A sophisticated example of the integration of microfluidic properties of paper-based analytical devices with the surface bio-molecularly-imprinted technique was reported by Qi et al. [49] for the selective and sensitive clinical detection of carcino-embryonic antigen. Interestingly, a movable valve and circular rotating pieces of paper were used to allow for a modulable control of the device. This bio-molecularly-imprinted device was designed to carry out antibody-free biomarker analysis by in-situ synthesized MIP. This strategy allows for the direct detection of antigens avoiding the issue of the antibody preservation typical for enzyme-linked immunosorbent assay.

The origami configuration was conceived for applying a controlled overlapping of the sample areas and the working electrode area upon folding the origami. The use of a movable valve was chosen to realize a configuration suitable for a multi-step electropolymerization process, achieving satisfactory results also over long times of polymerization (~1 h). Moreover, the possibility to rotate the circular parts, on which the counter and reference electrodes were printed, was used to allow the electrochemical cell to be completed both before and after the origami folding, depending on the need.

In detail, four different paper components were designed to fabricate the origami device: (i) the working electrode part hosting two working electrodes; (ii) the counter/reference electrode part, composed of four independent rotating circular pads; (iii) the washing part, with two washing channels; and (iv) the movable valve. The latter was connected to the working electrode part by means of rivets and was placed in its functional position when necessary. The working electrode part and the washing part were obtained on the same piece of paper and designed to be overlapped by folding the origami. The washing part presented hydrophobic channels and two waste pools for transporting waste solution (Figure 4C).

The origami biosensors were prepared by surface imprinting technology, exploiting the sophisticated architecture of the origami device. Two identical sets of reaction units, each one configured as described above, were available on the same device in a symmetrical geometry, one of which was served as a blank baseline. A sequential modification of the working electrode with graphene oxide, chitosan and glutaraldehyde was carried out with the origami unfolded. When carcino-embryonic antigen was added as a molecular template, the counter/reference electrode parts were rotated to overlap with the working electrode areas, allowing for completing the electrochemical cell. Dopamine was added into the synthetic material pool, close to the working electrodes but separated by hydrophobic regions; thus, the movable valve was moved to connect the pool to the working electrode and let dopamine flow to reach the cell for the electropolymerization process by cyclic voltammetry. Hence, the origami was folded, to overlap the working electrode areas with the pools present on the washing area, and the dropping of an eluent allowed to remove the excess of template carcino-embryonic antigen.

The rotating parts of this origami were exploited also for the measurement of standard and real samples. Firstly, keeping the origami unfolded and the circular parts and the valve away from the working electrode area, the samples were loaded on the working electrode parts for a resting period, during which the analyte diffused inside the MIP structure. Thus, the origami was folded in order to overlap the washing part with the working electrode areas and let the excess sample solution being removed by a washing step. Finally, the circular part was rotated to overlap the counter and reference electrodes on the working electrode, and the voltammetric detection was performed.

The performances of such complex origami were carefully interrogated by electrochemical studies. Importantly, the suitability of the movable valve to allow the dopamine to flow to the working electrode area was successfully verified by comparison with direct drop-casting on the working electrode. After a comprehensive characterization of the surface molecularly imprinted process, the analytical performance of the Bio-MIP-ePADs was determined, obtaining satisfactory analytical features in terms of sensitivity, selectivity, and accuracy.

6. Origami Paper-Based Electrochemical Cell-Based Biosensors

The use of cells as biosensing elements for electrochemical biosensors turns to be necessary to obtain direct information about the cytotoxicity of chemicals as well as to monitor the inter-related effects of cytotoxicity on cell physiology. As is well known, the toxic effects occurring into living cells can trigger a cascade of reactions typically involving several components of the cell, thus requiring an in-situ and/or in vivo monitoring for understanding the overall consequences.

Various types of cyto-sensors, including chemiluminescent, fluorescent, electrochemical, electrochemiluminescent, and surface-enhanced Raman scattering cyto-sensors have been developed. Especially in the biomedical field, the sensitive and selective detection of cancer cells is fundamental to provide an early diagnosis and to plan a proper therapy.

Recently, the use of paper has been introduced as a novel platform for cytological or histological studies. Paper-based cyto-devices have begun to stand out in the field of sensors offering the attractive potential of transporting the cytological/histological research and applications on versatile, simple, miniaturized, and low-cost devices with an in-situ and rapid read-out. Firstly, the analytical methods applied for the study of paper-based cyto-devices exploited the colorimetric approach and techniques of fluorescence imaging. The development of this sector also in the field of the electrochemical sensor will provide an important contribution to multiple areas, including basic scientific advancement, clinical diagnostics, therapeutics, and energy storage/production.

Few studies have been reported about electrochemical cyto-devices with an origami principle. A first example was conceived by Su et al. [50] for obtaining an in vivo-like cell culture for human acute promyelocytic leukemia cells (HL-60), used as a proof-of-concept. Cell-targeting aptamers, namely KH1C12, were chosen as molecular probes for specific recognition of HL-60 cell.

The device was fabricated on a single sheet of flat paper, divided into two main squared areas: a paper cell pad (red square) as the working electrode area, and a paper auxiliary pad (blue square) with the counter and reference electrode (Figure 5A). The former was composed of four circular paper cell zones defined by the wax pattern, while the latter presented a single circular hydrophilic zone. The two areas were supposed to be perfectly overlapped upon bending on the folding line, thus completing the paper electrochemical cell.

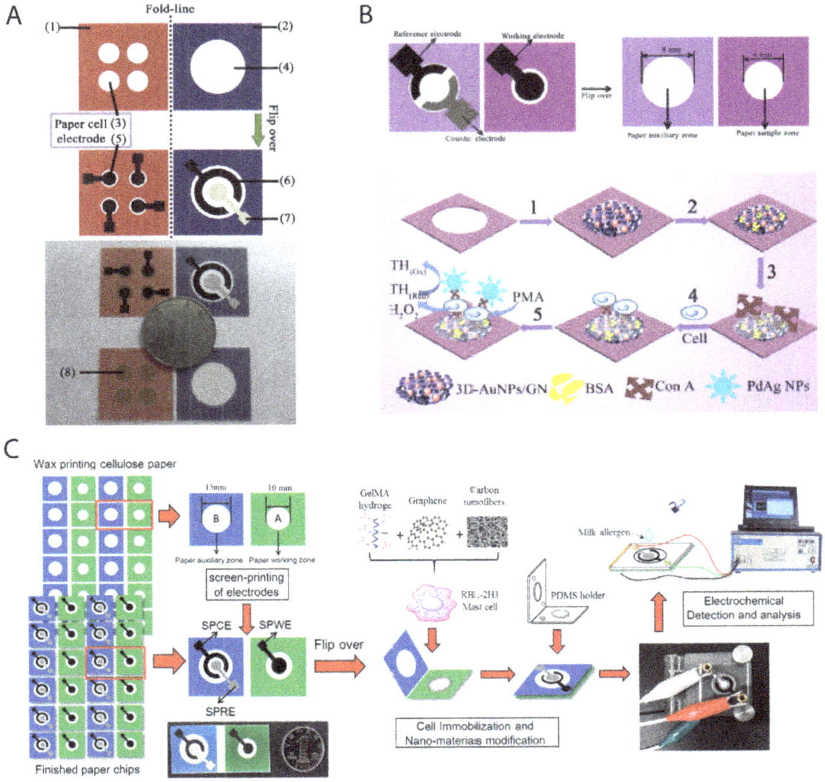

Figure 5. (**A**) Paper-based electrochemical cyto-device for sensitive detection of cancer cells and in situ anticancer drug screening. Reprinted with permission from [50], 2014 Elsevier; (**B**) electrochemical K-562 cells sensor based on origami paper device for point-of-care testing. Reprinted with permission from [51], 2015 Elsevier; (**C**) a novel electrochemical mast cell-based paper biosensor for the rapid detection of milk allergen casein. Reprinted with permission from [52], 2019 Elsevier.

The preparation of the cyto-origami device was carried out through the immobilization of the biocomponents on the paper cell area, previously functionalized with AuNPs to improve the conductivity and ensure the immobilization process. In detail, the conductive paper cell zones were modified with KH1C12 aptamers, capable of immobilizing a homogeneous HL-60 cell suspension. For the measurement, the paper auxiliary pad is folded on the paper cell pad and clamped within two homemade circuit boards. The detection process starts with the loading of the horseradish peroxidase-labeled folic acid, used as electrochemical bio-probes into each paper cell zone, followed by the addition of o-phenylenediamine and H_2O_2 through the paper auxiliary pad. The folate receptor is specifically recognized by the previously immobilized HL-60 cell and enables the amplification of the electrochemical signal thanks to the labeled horseradish peroxidase enzyme, which catalyzes the oxidation of o-phenylenediamine by H_2O_2. The enzymatic activity is finally recorded produce by differential pulse voltammetry.

This cyto-origami device was tested for anticancer drug screening, by monitoring the apoptosis induced by the action of the anticancer drug. In detail, cell apoptosis is associated with the translocation of the membrane phosphatidylserine from the inner to the outer side of the plasma membrane. In this case, horseradish peroxidase-labeled annexin-V was used because it was able to interact with the phosphatidylserine, available only when cell apoptosis occurs. This study showed further advantages of exploiting the 3D culture of cancer cells in the paper cell areas.

Shortly after, Ge et al. [51] continued the study on leukemia cell lines by developing an origami paper-based device for the detection of K-562 (Figure 5B). The origami structure exploited in this case was similar in principle to the one above discussed. The electroanalytical performance of the sensor was improved by using a nanocomposite based on AuNPs, graphene, and an ionic liquid, while the cells were captured by specific binding with concanavalin A (Con A) and cell surface mannose. Phorbol 12-myristate-13-acetate was used to stimulate the endogenous generation of H_2O_2 in the cells, to be detected by voltammetric detection. The authors reported a promising detection limit for cell concentration, equal to 200 cells/mL.

Another example of a cell-based origami biosensor with an analogous configuration (Figure 5C) was recently developed by Jiang et al. [52]. The authors wanted to mimic the physiological conditions directly on the paper environment to realize a biosensor for the recognition of food allergens using mast cells. In detail, casein was chosen as the major allergen in cow's milk. A composite based on graphene and a hydrogel was used to enhance the conductivity of the paper working electrode, where the Rat basophilic mast cells were immobilized by simple drop-casting. The detection of casein was performed by differential pulse voltammetry allowing the authors to measure this allergen with a limit of detection of 32 ng/mL.

7. Self-Powered Origami Paper-Based Electrochemical Biosensors

Among the emerging challenges in the sensor field and beyond, there is a need to develop self-powered devices able to operate independently, wirelessly, and sustainably. A substantial breakthrough in the sensor power supply concern occurred when the concept of self-powered sensors was first proposed in 2006 [81]. Such systems consist of sensors devised to generate an electric signal when mechanically or chemically activated, without the need for an external power source [82]. Self-powered devices have been proposed as clean energy harvesters, with the ambitious purpose to create autonomous tools that can harvest the energy from the environment to power themself, rather than from conventional power supplies [83].

This achievement would allow revolutionizing the fields of sensing, data transmitting, data processing, energy harvesting, and energy storage. In the framework of electrochemical sensors for analytical applications, self-powered wearable electronics would provide on-site and continuous monitoring of biomedical parameters, representing a significant advance for the monitoring during therapies and early diagnosis.

The researchers' efforts are now dedicated to investigating flexible, miniaturizable, and versatile substrates for the development of efficient self-powered devices. Paper has been again highlighted as a promising platform material for providing significant advances in this field [84].

One of the very first examples of self-powered paper-based biosensors devised with origami characteristics was reported by the research group of Crooks [53]. The authors developed a self-powered biosensor able to generate a current without the need for an external power source, realizing a platform in principle adaptable to a range of target molecules, including antibodies, DNAzymes, or aptazymes.

The whole sensor was printed with a symmetric geometry on a single layer of paper (Figure 6A). To avoid the proximity of the printed conductive inks to the microfluidic pattern, the paper was divided into two sides supposed to be folded for the measurement. On one side, an inlet area was fabricated by wax printing, from which a pair of identical channels was originated. The channels recombined at the end of their pathways in a region destined to form two half electrochemical cells. On the second side, two electrodes were obtained by screen-printing of conductive carbon ink. When the paper is folded at the predefined fold line, the electrodes were overlapped on the region where the channels recombine, completing the two half electrochemical cells. After reagents preloading and drying, the device is folded and sealed by lamination, which serves also for connecting the device to copper wires for electrochemical connection. This strategy allows the device to retain the sample avoiding losses due to evaporation or other alterations, without the need for adhesives that can cause contamination or nonspecific adsorption.

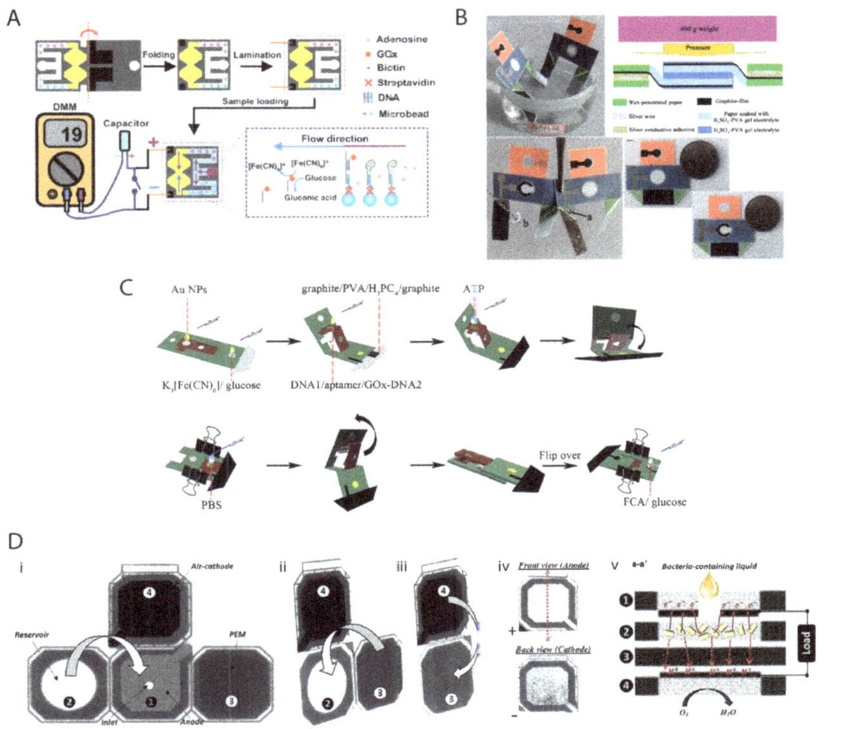

Figure 6. (**A**) Aptamer-based origami paper analytical device for electrochemical detection of adenosine. Reprinted with permission from [53], 2012 WILEY-VCH Verlag GmbH & Co; (**B**) photoelectrochemical lab-on-paper device based on an integrated paper supercapacitor and internal light source. Reprinted with permission from [54], 2013 American Chemical Society; (**C**) a self-powered origami paper analytical device with a pop-up structure for dual-mode electrochemical sensing of ATP assisted by glucose oxidase-triggered reaction. Reprinted with permission from [55], 2019 Elsevier; (**D**) power-on-paper: origami-inspired fabrication of 3D microbial fuel cells. Reprinted with permission from [85], 2017 Elsevier.

During the preparation of the biosensor, biotin-labeled aptamers immobilized on streptavidin-functionalized microbeads were entrapped into the initial part of the channels. The detection principle relied on the immobilized aptamer as a recognition element for adenosine, chosen as the target analyte for a proof-of-concept study. The adenosine binding caused the release of a GOx-labelled ssDNA, which was able to catalyze the oxidation of glucose, finally resulting in the conversion of redox probe $[Fe(CN)_6]^{3-}$ to $[Fe(CN)_6]^{4-}$. The difference in concentrations of the redox probe between the sensing half-cell and control half-cell results in a voltage that is accumulated. In this way, the sensor behaves similarly to a battery, able to charge a capacitor. By connecting the system to a digital multimeter, the accumulated capacitor is immediately discharged giving a current readout. The level of accumulated voltage was proved to increase linearly with an increasing concentration of glucose oxidase. The origami biosensor provided detection of adenosine up to 250 μM, with a detection limit of 11.8 μM. The authors highlighted that the integration of the capacitor for obtaining the accumulation of the concentration-dependent voltage allowed for increasing the sensitivity of the assay 15.5-fold in comparison with the direct measurement of the current.

Later on, Ge et al. realized an origami device that combined a dual chemiluminescence-photoelectrochemical detection principle with a paper supercapacitor, resulting in the collection of the generated photocurrent [54]. The sensor was obtained on wax-patterned paper shaped with a particular geometry. A reaction area (reported in red in Figure 6B) was bordered with a collection area (reported in blue in Figure 6B), from which two rectangular "legs" was originated (reported in green in Figure 6B). Hydrophilic circular pools were present on the reaction and collection areas, with a screen-printed working electrode and counter and reference electrodes, respectively, conceived for being overlapped and allowing for the formation of an electrochemical cell. The wax-patterned hydrophobic legs were designed to be diagonally folded and overlapped, providing the paper supercapacitor. This component was electrically connected to the reference electrode, being able to be charged after the chemiluminescence-photoelectrochemical reactions occurred. The supercapacitor was provided of drawn thin-film graphite electrodes that were soaked with H_2SO_4-PVA gel electrolyte and left to solidify.

The origami biosensor was adapted for the detection of adenosine triphosphate in human serum samples as a model example, using a specific binding aptamer for molecular reorganization. The principle of the assay relied on the use of N-(aminobutyl)-N-(ethylisoluminol)-functionalized gold nanoparticles ABEI−AuNPs−H_2O_2 as a chemiluminescence-inducing system and p-iodophenol as a chemiluminescence enhancer. Thioglycolic acid-capped water-soluble cadmium sulfide nanoparticles were employed as the active species to generate photocurrents, being capable of absorbing the chemiluminescence emitted by the ABEI−AuNPs− H_2O_2 system thus generating the photocurrent. The system is triggered when ATP is present as a target molecule, able to form a complex with two ssDNAs separately immobilized on both the AuNPs and the CdS NPs, which draws the AuNPs sufficiently close to the cadmium sulfide nanoparticles enabling the chemiluminescence-photocurrent system. H_2O_2 was used as a sacrificial electron donor to provide the electrons for the valence-band holes and complete the photocurrent generation cycle. The paper supercapacitor allowed for the storage of the photocurrent produced (for a period of 180 s), which was amplified and recorded as soon as the device was connected to a digital multimeter.

The porosity and large surface area of the cellulose fibers were exploited to in-situ modified the working electrode by assembling poly(dimethyldiallylammonium chloride) and carbon nanotubes, which enhanced the conductivity of the paper and provided fast electron-transfer to ensure the efficiency of the photoelectrochemical signal through the chemiluminescence-inducing system. Indeed, the comparison of the developed configuration with a device obtained on ITO support revealed that the role of the modified paper was determinant for the resulting performance of the sensor, enhancing the response of about 3-fold in intensity.

An interesting example was given by Yu's research group, who demonstrated that the self-powered technology on paper can be successfully integrated with origami devices by designing a pop-up mechanism [55]. The resulting biosensor provided dual-mode read-out, exploiting both direct voltametric measurement and the supercapacitor mode. The two read-out modes were designed for ATP detection as a model analyte. The pop-up structure was a key feature to control the microfluidic pathways of the device as well as the incubation times of the assay. The device was composed of an electrode area, the detection zone, the reaction zone, the hollow zone, and the supercapacitor system. The reaction zone and detection areas were spatially separated by a hydrophobic pattern, allowing the operator to connect them at the desired moment. Indeed, the detection area could be overlapped on the reaction area by applying a mechanical pressure in the clockwise direction, so that the working electrode and the carbon wire were put into contact.

During the preparation of the assay, the origami was kept unfolded. The hydrophilic paper in correspondence of the working electrode (detection zone) was modified with AuNPs to improve the conductivity as well as to allow the specific ssDNA and the aptamer to be sequentially immobilized by the affinity of thiol groups with gold. The aptamer was designed to undergo hybridization with the first ssDNA immobilized, and on the other hand to hybridize with a second ssDNA molecule labeled with glucose oxidase enzyme. Finally, ATP was incubated in the detection area. In the meanwhile, potassium ferricyanide and glucose were loaded on the reaction zone; moreover, the paper supercapacitor was obtained in the unprinted area of the origami by the deposition of graphite and PVA/H_3PO_4 electrolyte.

At the moment of the detection, the detection zone was overlapped with the reaction zone by exploiting the pop-up structure. The GOx-labelled ssDNA was released owing to the capture of ATP by aptamer and transferred to the reaction zone by vertical microfluidic, thus catalyzing the oxidation of glucose. The presence of pre-loaded ferricyanide acted as an electrochemical probe to acquire a DPV signal. Importantly, the concentration of species in the reaction zone resulted in the accumulation of a voltage that charged the paper supercapacitor. The connection of the system to a digital multimeter allowed for the acquisition of the resulting self-generated current. The origami biosensor was able to detect ATP concentration ranging from 10 to 5000 nM both in the direct voltametric measurement and the supercapacitor modes, with a limit of detection in the nanomolar level.

Another worthy of note approach is focused on the exploitation of microbial cells that can be exploited to produce green electric power. Indeed, the microorganisms' capacity to catalyze the degradation of the organic compounds, such as from organic waste material, represents a sustainable approach for generating clean energy, particularly promising for those environments having limited energy sources. In detail, the physiological metabolism of microbial cells can be converted into reusable energy by transferring the resulting electron exchanges to an external electrode. In light of these attractive features, the integration of microbial fuel cells with microfluidic paper-based devices represents an ambitious purpose that could bridge the gap between this fascinating green approach and its practical realization.

A first attempt in this direction was made by Mohammadifar et al. [85], who built up a foldable origami device based on microbial fuel cells with the interesting capability of rapidly generating power with a small amount of bacteria-containing liquid, thanks to the rapid adsorption provided by the paper platform. An initial 2D geometry was obtained on paper, divided into four functional foldable areas, namely the anode pad, the reservoir pad, the cation exchange membrane, and the cathode pad (Figure 6D). The paper was patterned with wax to delimit hydrophilic areas. The cation exchange membrane was characterized by hydrophobic properties and proton/cation conductance. An anodic material was applied on the middle pad (i.e., the anode pad), which presented an inlet hole for injection of the bacterial liquid. The left tab was used as a paper reservoir to store the bacterial cells and the endogenous organic fuels. The origami was obtained by folding the tabs along the folding lines and the layers were kept together with an adhesive spray. After the bacterial cells were inoculated through the inlet hole, the protons/cations produced during the microbial metabolism diffused through the hydrophobic cation exchange membrane toward the cathode, reacting with oxygen. The authors investigated several composites with anodic properties to optimize both the bacterial attachment and the electron transfer, drawing attention to the dependence of the output voltages on the choice of the modifying material applied on paper. This study provided new insights into the construction of self-powered paper-based biosensors and further disclosed the perspectives for paper origami devices. Importantly, it shows how paper can be designed to create cathodic and anodic regions by exploiting different materials and a programmed folding. Moreover, the use of microbial cells and power generation on paper can play a significant role in future advances towards self-generated sustainable devices.

8. Conclusions and Perspective

The development of electrochemical paper-based biosensors has been started ca. 10 years ago by replacing the plastic-based support with the paper used as a substrate to print the electrochemical cell as well as microfluidic patterns with the overriding goal to deliver a pump-free microfluidic device characterized by an easy and cost-effective fabrication. The several features of the paper in the time have been exploited step by step, by fabricating analytical devices with additional features. For example, the porosity of the paper has been successively exploited to store the reagents for delivering a reagent-free biosensor or to treat the sample by simply adding the sample at the opposite of the layer in which the electrochemical cell is printed. The foldability of the paper has been recently used to design a 3D structure system to avoid the operator in performing multistep analysis, to control the timing, to reduce the sample manipulation, and to implement the multi-analysis. For instance, in the case of the origami paper-based devices for multiclass pesticide detection, the use of multiple pads combined with office-paper printed electrochemical sensors allowed to detect of three different classes of pesticides by only folding the pads, adding the distilled water, and cut the used pads used after the measure. In the case of the complex sample as whole blood, the use of several layers for vertical microfluidics combined with printed electrochemical sensors delivered an easy-to-use analytical tool for the application of precision medicine. Furthermore, the possibility to use the origami system with different layers loaded with different reagents can create an embedded device ready for the detection of the target analyte in the gas phase. Another additional value is related to the low cost of the paper rendering the device suitable for a single use, and thus overcoming the fouling problem. At the state of the art, the paper-based devices need to be industrialized by manufacturing them at the large scale. In addition, to fill in the gap between the bench and the market, a robust folding and unfolding procedure needs to be addressed. In the next future, we are confident that the recent efforts in printed electronics will pave the way for a new direction in origami paper-based electrochemical devices, with the final aims to fabricate in the next future integrated fully printed analytical tools, able, with the only addition of the sample, to automatically treat the sample, to contain the reagents, and to detect the signal.

Author Contributions: F.A., N.C., V.C. and D.M. writing, original draft preparation, review and editing. All authors have read and agreed to the published version of the manuscript.

Funding: This research was funded by Lazio Innova, Regione Lazio, grant number: A0375-2020-3656, Project E-Crome.

Institutional Review Board Statement: Not applicable.

Informed Consent Statement: Not applicable.

Data Availability Statement: The data presented in this study are available on request from the corresponding author.

Acknowledgments: F.A. thanks the Project E-Crome, A0375-2020-3656, Lazio Innova, Regione Lazio.

Conflicts of Interest: The authors declare no conflict of interest.

References

1. Anastas, P.T.; Warner, J.C. Principles of Green Chemistry. *Green Chem. Theory Pract.* **1998**, *29*, 1–23.
2. Nowak, P.M.; Wietecha-Posłuszny, R.; Pawliszyn, J. White Analytical Chemistry: An Approach to Reconcile the Principles of Green Analytical Chemistry and Functionality. *TrAC Trends Anal. Chem.* **2021**, *138*, 116223. [CrossRef]
3. Cinti, S.; Fiore, L.; Massoud, R.; Cortese, C.; Moscone, D.; Palleschi, G.; Arduini, F. Low-Cost and Reagent-Free Paper-Based Device to Detect Chloride Ions in Serum and Sweat. *Talanta* **2018**, *179*, 186–192. [CrossRef] [PubMed]
4. Cinti, S.; Moscone, D.; Arduini, F. Preparation of Paper-Based Devices for Reagentless Electrochemical (Bio)Sensor Strips. *Nat. Protoc.* **2019**, *14*, 2437–2451. [CrossRef] [PubMed]
5. Caratelli, V.; Ciampaglia, A.; Guiducci, J.; Sancesario, G.; Moscone, D.; Arduini, F. Precision Medicine in Alzheimer's Disease: An Origami Paper-Based Electrochemical Device for Cholinesterase Inhibitors. *Biosens. Bioelectron.* **2020**, *165*, 112411. [CrossRef] [PubMed]

6. Bagheri, N.; Mazzaracchio, V.; Cinti, S.; Colozza, N.; Di Natale, C.; Netti, P.A.; Saraji, M.; Roggero, S.; Moscone, D.; Arduini, F. Electroanalytical Sensor Based on Gold-Nanoparticle-Decorated Paper for Sensitive Detection of Copper Ions in Sweat and Serum. *Anal. Chem.* **2021**, *93*, 5225–5233. [CrossRef]
7. Cinti, S.; Cusenza, R.; Moscone, D.; Arduini, F. Paper-Based Synthesis of Prussian Blue Nanoparticles for the Development of Whole Blood Glucose Electrochemical Biosensor. *Talanta* **2018**, *187*, 59–64. [CrossRef]
8. Scordo, G.; Moscone, D.; Palleschi, G.; Arduini, F. A Reagent-Free Paper-Based Sensor Embedded in a 3D Printing Device for Cholinesterase Activity Measurement in Serum. *Sens. Actuators B Chem.* **2018**, *258*, 1015–1021. [CrossRef]
9. Jemmeli, D.; Marcoccio, E.; Moscone, D.; Dridi, C.; Arduini, F. Highly Sensitive Paper-Based Electrochemical Sensor for Reagent Free Detection of Bisphenol A. *Talanta* **2020**, *216*, 120924. [CrossRef]
10. Arduini, F.; Cinti, S.; Caratelli, V.; Amendola, L.; Palleschi, G.; Moscone, D. Origami Multiple Paper-Based Electrochemical Biosensors for Pesticide Detection. *Biosens. Bioelectron.* **2019**, *126*, 346–354. [CrossRef]
11. Bagheri, N.; Cinti, S.; Nobile, E.; Moscone, D.; Arduini, F. Multi-Array Wax Paper-Based Platform for the Pre-Concentration and Determination of Silver Ions in Drinking Water. *Talanta* **2021**, *232*, 122474. [CrossRef] [PubMed]
12. Colozza, N.; Kehe, K.; Dionisi, G.; Popp, T.; Tsoutsoulopoulos, A.; Steinritz, D.; Moscone, D.; Arduini, F. A Wearable Origami-like Paper-Based Electrochemical Biosensor for Sulfur Mustard Detection. *Biosens. Bioelectron.* **2019**, *129*, 15–23. [CrossRef] [PubMed]
13. Colozza, N.; Sassolini, A.; Agosta, L.; Bonfanti, A.; Hermansson, K.; Arduini, F. A Paper-Based Potentiometric Sensor for Solid Samples: Corrosion Evaluation of Reinforcements Embedded in Concrete Structures as a Case Study. *ChemElectroChem* **2020**, *7*, 2274–2282. [CrossRef]
14. Colozza, N.; Tazzioli, S.; Sassolini, A.; Agosta, L.; di Monte, M.G.; Hermansson, K.; Arduini, F. Multiparametric Analysis by Paper-Assisted Potentiometric Sensors for Diagnostic and Monitoring of Reinforced Concrete Structures. *Sens. Actuators B Chem.* **2021**, *345*, 130352. [CrossRef]
15. Martinez, A.W.; Phillips, S.T.; Carrilho, E.; Thomas, S.W.; Sindi, H.; Whitesides, G.M. Simple Telemedicine for Developing Regions: Camera Phones and Paper-Based Microfluidic Devices for Real-Time, Off-Site Diagnosis. *Anal. Chem.* **2008**, *80*, 3699–3707. [CrossRef] [PubMed]
16. Martinez, A.W.; Phillips, S.T.; Whitesides, G.M. Three-Dimensional Microfluidic Devices Fabricated in Layered Paper and Tape. *Proc. Natl. Acad. Sci. USA* **2008**, *105*, 19606–19611. [CrossRef]
17. Martinez, A.W.; Phillips, S.T.; Whitesides, G.M.; Carrilho, E. Diagnostics for the Developing World: Microfluidic Paper-Based Analytical Devices. *Anal. Chem.* **2010**, *82*, 3–10. [CrossRef] [PubMed]
18. Dungchai, W.; Chailapakul, O.; Henry, C.S. Electrochemical Detection for Paper-Based Microfluidics. *Anal. Chem.* **2009**, *81*, 5821–5826. [CrossRef]
19. Nery, E.W.; Kubota, L.T. Sensing Approaches on Paper-Based Devices: A Review. *Anal. Bioanal. Chem.* **2013**, *405*, 7573–7595. [CrossRef]
20. Adkins, J.; Boehle, K.; Henry, C. Electrochemical Paper-Based Microfluidic Devices: Microfluidics and Miniaturization. *Electrophoresis* **2015**, *36*, 1811–1824. [CrossRef]
21. Mettakoonpitak, J.; Boehle, K.; Nantaphol, S.; Teengam, P.; Adkins, J.A.; Srisa-Art, M.; Henry, C.S. Electrochemistry on Paper-Based Analytical Devices: A Review. *Electroanalysis* **2016**, *28*, 1420–1436. [CrossRef]
22. Zhu, G.; Yin, X.; Jin, D.; Zhang, B.; Gu, Y.; An, Y. Paper-Based Immunosensors: Current Trends in the Types and Applied Detection Techniques. *TrAC Trends Anal. Chem.* **2019**, *111*, 100–117. [CrossRef]
23. Noviana, E.; McCord, C.P.; Clark, K.M.; Jang, I.; Henry, C.S. Electrochemical Paper-Based Devices: Sensing Approaches and Progress toward Practical Applications. *Lab Chip* **2020**, *20*, 9–34. [CrossRef] [PubMed]
24. Paschoalino, W.J.; Kogikoski, S.; Barragan, J.T.C.; Giarola, J.F.; Cantelli, L.; Rabelo, T.M.; Pessanha, T.M.; Kubota, L.T. Emerging Considerations for the Future Development of Electrochemical Paper-Based Analytical Devices. *ChemElectroChem* **2019**, *6*, 10–30. [CrossRef]
25. Ozer, T.; McMahon, C.; Henry, C.S. Advances in Paper-Based Analytical Devices. *Annu. Rev. Anal. Chem.* **2020**, *13*, 85–109. [CrossRef] [PubMed]
26. Noviana, E.; Carrão, D.B.; Pratiwi, R.; Henry, C.S. Emerging Applications of Paper-Based Analytical Devices for Drug Analysis: A Review. *Anal. Chim. Acta* **2020**, *1116*, 70–90. [CrossRef] [PubMed]
27. Mazurkiewicz, W.; Podrażka, M.; Jarosińska, E.; Kappalakandy Valapil, K.; Wiloch, M.; Jönsson-Niedziółka, M.; Witkowska Nery, E. Paper-Based Electrochemical Sensors and How to Make Them (Work). *ChemElectroChem* **2020**, *7*, 2939–2956. [CrossRef]
28. Krikstolaityte, V.; Ding, R.; Hui, E.C.; Lisak, G. Paper as Sampling Substrates and All-Integrating Platforms in Potentiometric Ion Determination. *TrAC Trends Anal. Chem.* **2020**, *133*, 116070. [CrossRef]
29. Liang, B.; Zhu, Q.; Fang, L.; Cao, Q.; Liang, X.; Ye, X. An Origami Paper Device for Complete Elimination of Interferents in Enzymatic Electrochemical Biosensors. *Electrochem. Commun.* **2017**, *82*, 43–46. [CrossRef]
30. Li, W.; Qian, D.; Wang, Q.; Li, Y.; Bao, N.; Gu, H.; Yu, C. Fully-Drawn Origami Paper Analytical Device for Electrochemical Detection of Glucose. *Sens. Actuators B Chem.* **2016**, *231*, 230–238. [CrossRef]
31. Li, M.; Wang, L.; Liu, R.; Li, J.; Zhang, Q.; Shi, G.; Li, Y.; Hou, C.; Wang, H. A Highly Integrated Sensing Paper for Wearable Electrochemical Sweat Analysis. *Biosens. Bioelectron.* **2021**, *174*, 112828. [CrossRef]
32. Wang, C.-C.; Hennek, J.W.; Ainla, A.; Kumar, A.A.; Lan, W.-J.; Im, J.; Smith, B.S.; Zhao, M.; Whitesides, G.M. A Paper-Based "Pop-up" Electrochemical Device for Analysis of Beta-Hydroxybutyrate. *Anal. Chem.* **2016**, *88*, 6326–6333. [CrossRef] [PubMed]

33. Ding, J.; Li, B.; Chen, L.; Qin, W. A Three-Dimensional Origami Paper-Based Device for Potentiometric Biosensing. *Angew. Chem. Int. Ed.* **2016**, *55*, 13033–13037. [CrossRef] [PubMed]
34. Teengam, P.; Siangproh, W.; Tuantranont, A.; Vilaivan, T.; Chailapakul, O.; Henry, C.S. Electrochemical Impedance-Based DNA Sensor Using Pyrrolidinyl Peptide Nucleic Acids for Tuberculosis Detection. *Anal. Chim. Acta* **2018**, *1044*, 102–109. [CrossRef]
35. Jiang, H.; Guo, Q.; Zhang, C.; Sun, Z.; Weng, X. Microfluidic Origami Nano-Aptasensor for Peanut Allergen Ara H1 Detection. *Food Chem.* **2021**, *365*, 130511. [CrossRef]
36. Li, X.; Scida, K.; Crooks, R.M. Detection of Hepatitis B Virus DNA with a Paper Electrochemical Sensor. *Anal. Chem.* **2015**, *87*, 9009–9015. [CrossRef]
37. Ye, D.; Li, L.; Li, Z.; Zhang, Y.; Li, M.; Shi, J.; Wang, L.; Fan, C.; Yu, J.; Zuo, X. Molecular Threading-Dependent Mass Transport in Paper Origami for Single-Step Electrochemical DNA Sensors. *Nano Lett.* **2019**, *19*, 369–374. [CrossRef]
38. Bhardwaj, J.; Sharma, A.; Jang, J. Vertical Flow-Based Paper Immunosensor for Rapid Electrochemical and Colorimetric Detection of Influenza Virus Using a Different Pore Size Sample Pad. *Biosens. Bioelectron.* **2019**, *126*, 36–43. [CrossRef]
39. Wang, Y.; Luo, J.; Liu, J.; Li, X.; Kong, Z.; Jin, H.; Cai, X. Electrochemical Integrated Paper-Based Immunosensor Modified with Multi-Walled Carbon Nanotubes Nanocomposites for Point-of-Care Testing of 17β-Estradiol. *Biosens. Bioelectron.* **2018**, *107*, 47–53. [CrossRef]
40. Li, L.; Xu, J.; Zheng, X.; Ma, C.; Song, X.; Ge, S.; Yu, J.; Yan, M. Growth of Gold-Manganese Oxide Nanostructures on a 3D Origami Device for Glucose-Oxidase Label Based Electrochemical Immunosensor. *Biosens. Bioelectron.* **2014**, *61*, 76–82. [CrossRef]
41. Li, X.; Liu, X. A Microfluidic Paper-Based Origami Nanobiosensor for Label-Free, Ultrasensitive Immunoassays. *Adv. Healthc. Mater.* **2016**, *5*, 1326–1335. [CrossRef]
42. Ruecha, N.; Shin, K.; Chailapakul, O.; Rodthongkum, N. Label-Free Paper-Based Electrochemical Impedance Immunosensor for Human Interferon Gamma Detection. *Sens. Actuators B Chem.* **2019**, *279*, 298–304. [CrossRef]
43. Sun, G.; Zhang, L.; Zhang, Y.; Yang, H.; Ma, C.; Ge, S.; Yan, M.; Yu, J.; Song, X. Multiplexed Enzyme-Free Electrochemical Immunosensor Based on ZnO Nanorods Modified Reduced Graphene Oxide-Paper Electrode and Silver Deposition-Induced Signal Amplification Strategy. *Biosens. Bioelectron.* **2015**, *71*, 30–36. [CrossRef]
44. Shen, Y.; Modha, S.; Tsutsui, H.; Mulchandani, A. An origami electrical biosensor for multiplexed analyte detection in body fluids. *Biosens. Bioelectron.* **2021**, *171*, 112721. [CrossRef]
45. Cunningham, J.C.; Scida, K.; Kogan, M.R.; Wang, B.; Ellington, A.D.; Crooks, R.M. Paper Diagnostic Device for Quantitative Electrochemical Detection of Ricin at Picomolar Levels. *Lab Chip* **2015**, *15*, 3707–3715. [CrossRef]
46. DeGregory, P.R.; Tsai, Y.-J.; Scida, K.; Richards, I.; Crooks, R.M. Quantitative Electrochemical Metalloimmunoassay for TFF3 in Urine Using a Paper Analytical Device. *Analyst* **2016**, *141*, 1734–1744. [CrossRef] [PubMed]
47. Amatatongchai, M.; Sitanurak, J.; Sroysee, W.; Sodanat, S.; Chairam, S.; Jarujamrus, P.; Nacapricha, D.; Lieberzeit, P.A. Highly Sensitive and Selective Electrochemical Paper-Based Device Using a Graphite Screen-Printed Electrode Modified with Molecularly Imprinted Polymers Coated Fe3O4@Au@SiO2 for Serotonin Determination. *Anal. Chim. Acta* **2019**, *1077*, 255–265. [CrossRef]
48. Sun, X.; Jian, Y.; Wang, H.; Ge, S.; Yan, M.; Yu, J. Ultrasensitive Microfluidic Paper-Based Electrochemical Biosensor Based on Molecularly Imprinted Film and Boronate Affinity Sandwich Assay for Glycoprotein Detection. *ACS Appl. Mater. Interfaces* **2019**, *11*, 16198–16206. [CrossRef]
49. Qi, J.; Li, B.; Zhou, N.; Wang, X.; Deng, D.; Luo, L.; Chen, L. The Strategy of Antibody-Free Biomarker Analysis by in-Situ Synthesized Molecularly Imprinted Polymers on Movable Valve Paper-Based Device. *Biosens. Bioelectron.* **2019**, *142*, 111533. [CrossRef]
50. Su, M.; Ge, L.; Ge, S.; Li, N.; Yu, J.; Yan, M.; Huang, J. Paper-Based Electrochemical Cyto-Device for Sensitive Detection of Cancer Cells and in Situ Anticancer Drug Screening. *Anal. Chim. Acta* **2014**, *847*, 1–9. [CrossRef]
51. Ge, S.; Zhang, L.; Zhang, Y.; Liu, H.; Huang, J.; Yan, M.; Yu, J. Electrochemical K-562 Cells Sensor Based on Origami Paper Device for Point-of-Care Testing. *Talanta* **2015**, *145*, 12–19. [CrossRef]
52. Jiang, D.; Ge, P.; Wang, L.; Jiang, H.; Yang, M.; Yuan, L.; Ge, Q.; Fang, W.; Ju, X. A Novel Electrochemical Mast Cell-Based Paper Biosensor for the Rapid Detection of Milk Allergen Casein. *Biosens. Bioelectron.* **2019**, *130*, 299–306. [CrossRef]
53. Liu, H.; Xiang, Y.; Lu, Y.; Crooks, R.M. Aptamer-Based Origami Paper Analytical Device for Electrochemical Detection of Adenosine. *Angew. Chem. Int. Ed.* **2012**, *51*, 6925–6928. [CrossRef]
54. Ge, L.; Wang, P.; Ge, S.; Li, N.; Yu, J.; Yan, M.; Huang, J. Photoelectrochemical Lab-on-Paper Device Based on an Integrated Paper Supercapacitor and Internal Light Source. *Anal. Chem.* **2013**, *85*, 3961–3970. [CrossRef] [PubMed]
55. Liu, Y.; Cui, K.; Kong, Q.; Zhang, L.; Ge, S.; Yu, J. A Self-Powered Origami Paper Analytical Device with a Pop-up Structure for Dual-Mode Electrochemical Sensing of ATP Assisted by Glucose Oxidase-Triggered Reaction. *Biosens. Bioelectron.* **2020**, *148*, 111839. [CrossRef] [PubMed]
56. Amine, A.; Arduini, F.; Moscone, D.; Palleschi, G. Recent Advances in Biosensors Based on Enzyme Inhibition. *Biosens. Bioelectron.* **2016**, *76*, 180–194. [CrossRef]
57. Arduini, F.; Neagu, D.; Dall'Oglio, S.; Moscone, D.; Palleschi, G. Towards a Portable Prototype Based on Electrochemical Cholinesterase Biosensor to Be Assembled to Soldier Overall for Nerve Agent Detection. *Electroanalysis* **2012**, *24*, 581–590. [CrossRef]
58. Kavita, V. DNA Biosensors-A Review. *J. Bioeng. Biomed. Sci.* **2017**, *7*, 222. [CrossRef]

59. Akki, S.U.; Werth, C.J. Critical Review: DNA Aptasensors, Are They Ready for Monitoring Organic Pollutants in Natural and Treated Water Sources? *Environ. Sci. Technol.* **2018**, *52*, 8989–9007. [CrossRef]
60. Dhiman, A.; Kalra, P.; Bansal, V.; Bruno, J.G.; Sharma, T.K. Aptamer-Based Point-of-Care Diagnostic Platforms. *Sens. Actuators B Chem.* **2017**, *246*, 535–553. [CrossRef]
61. Scognamiglio, V.; Antonacci, A.; Lambreva, M.D.; Litescu, S.C.; Rea, G. Synthetic Biology and Biomimetic Chemistry as Converging Technologies Fostering a New Generation of Smart Biosensors. *Biosens. Bioelectron.* **2015**, *74*, 1076–1086. [CrossRef]
62. Srisomwat, C.; Yakoh, A.; Chuaypen, N.; Tangkijvanich, P.; Vilaivan, T.; Chailapakul, O. Amplification-Free DNA Sensor for the One-Step Detection of the Hepatitis B Virus Using an Automated Paper-Based Lateral Flow Electrochemical Device. *Anal. Chem.* **2021**, *93*, 2879–2887. [CrossRef] [PubMed]
63. Teengam, P.; Siangproh, W.; Tuantranont, A.; Henry, C.S.; Vilaivan, T.; Chailapakul, O. Electrochemical Paper-Based Peptide Nucleic Acid Biosensor for Detecting Human Papillomavirus. *Anal. Chim. Acta* **2017**, *952*, 32–40. [CrossRef]
64. Cinti, S.; Cinotti, G.; Farolo, C.; Nguyen, E.P.; Caratelli, V.; Moscone, D.; Arduini, F.; Merkoci, A. Experimental Comparison in Sensing Breast Cancer Mutations by Signal ON and Signal OFF Paper-Based Electroanalytical Strips. *Anal. Chem.* **2020**, *92*, 1674–1679. [CrossRef] [PubMed]
65. Lin, J.; Ju, H. Electrochemical and Chemiluminescent Immunosensors for Tumor Markers. *Biosens. Bioelectron.* **2005**, *20*, 1461–1470. [CrossRef] [PubMed]
66. Holford, T.R.J.; Davis, F.; Higson, S.P.J. Recent Trends in Antibody Based Sensors. *Biosens. Bioelectron.* **2012**, *34*, 12–24. [CrossRef] [PubMed]
67. Takahashi, Y.; Shevchuk, A.I.; Novak, P.; Murakami, Y.; Shiku, H.; Korchev, Y.E.; Matsue, T. Simultaneous Noncontact Topography and Electrochemical Imaging by SECM/SICM Featuring Ion Current Feedback Regulation. *J. Am. Chem. Soc.* **2010**, *132*, 10118–10126. [CrossRef]
68. Lee, H.J.; Lee, S.H.; Yasukawa, T.; Ramón-Azcón, J.; Mizutani, F.; Ino, K.; Shiku, H.; Matsue, T. Rapid and Simple Immunosensing System for Simultaneous Detection of Tumor Markers Based on Negative-Dielectrophoretic Manipulation of Microparticles. *Talanta* **2010**, *81*, 657–663. [CrossRef]
69. Jauset-Rubio, M.; Svobodová, M.; Mairal, T.; McNeil, C.; Keegan, N.; Saeed, A.; Abbas, M.N.; El-Shahawi, M.S.; Bashammakh, A.S.; Alyoubi, A.O.; et al. Ultrasensitive, Rapid and Inexpensive Detection of DNA Using Paper Based Lateral Flow Assay. *Sci. Rep.* **2016**, *6*, 37732. [CrossRef]
70. Zhu, X.; Shah, P.; Stoff, S.; Liu, H.; Li, C. A Paper Electrode Integrated Lateral Flow Immunosensor for Quantitative Analysis of Oxidative Stress Induced DNA Damage. *Analyst* **2014**, *139*, 2850–2857. [CrossRef]
71. Sajid, M.; Kawde, A.-N.; Daud, M. Designs, Formats and Applications of Lateral Flow Assay: A Literature Review. *J. Saudi Chem. Soc.* **2015**, *19*, 689–705. [CrossRef]
72. Oh, Y.K.; Joung, H.-A.; Kim, S.; Kim, M.-G. Vertical Flow Immunoassay (VFA) Biosensor for a Rapid One-Step Immunoassay. *Lab Chip* **2013**, *13*, 768. [CrossRef] [PubMed]
73. Tarannum, N.; Khatoon, S.; Dzantiev, B.B. Perspective and Application of Molecular Imprinting Approach for Antibiotic Detection in Food and Environmental Samples: A Critical Review. *Food Control.* **2020**, *118*, 107381. [CrossRef]
74. Blanco-López, M.C.; Lobo-Castañón, M.J.; Miranda-Ordieres, A.J.; Tuñón-Blanco, P. Voltammetric Sensor for Vanillylmandelic Acid Based on Molecularly Imprinted Polymer-Modified Electrodes. *Biosens. Bioelectron.* **2003**, *18*, 353–362. [CrossRef]
75. Dai, C.; Liu, C.; Wei, J.; Hong, H.; Zhao, Q. Molecular Imprinted Macroporous Chitosan Coated Mesoporous Silica Xerogels for Hemorrhage Control. *Biomaterials* **2010**, *31*, 7620–7630. [CrossRef] [PubMed]
76. Hasanzadeh, M.; Shadjou, N.; de la Guardia, M. Cytosensing of Cancer Cells Using Antibody-Based Molecular Imprinting: A Short-Review. *TrAC Trends Anal. Chem.* **2018**, *99*, 129–134. [CrossRef]
77. Arabi, M.; Ostovan, A.; Bagheri, A.R.; Guo, X.; Wang, L.; Li, J.; Wang, X.; Li, B.; Chen, L. Strategies of Molecular Imprinting-Based Solid-Phase Extraction Prior to Chromatographic Analysis. *TrAC Trends Anal. Chem.* **2020**, *128*, 115923. [CrossRef]
78. Zaidi, S.A. Molecular Imprinting: A Useful Approach for Drug Delivery. *Mater. Sci. Energy Technol.* **2020**, *3*, 72–77. [CrossRef]
79. Li, W.; Zhang, X.; Li, T.; Ji, Y.; Li, R. Molecularly Imprinted Polymer-Enhanced Biomimetic Paper-Based Analytical Devices: A Review. *Anal. Chim. Acta* **2021**, *1148*, 238196. [CrossRef]
80. Díaz-Liñán, M.C.; López-Lorente, A.I.; Cárdenas, S.; Lucena, R. Molecularly Imprinted Paper-Based Analytical Device Obtained by a Polymerization-Free Synthesis. *Sens. Actuators B Chem.* **2019**, *287*, 138–146. [CrossRef]
81. Wang, Z.L. Piezoelectric Nanogenerators Based on Zinc Oxide Nanowire Arrays. *Science* **2006**, *312*, 242–246. [CrossRef]
82. Cho, E.; Mohammadifar, M.; Choi, S. A Single-Use, Self-Powered, Paper-Based Sensor Patch for Detection of Exercise-Induced Hypoglycemia. *Micromachines* **2017**, *8*, 265. [CrossRef] [PubMed]
83. Xia, K.; Du, C.; Zhu, Z.; Wang, R.; Zhang, H.; Xu, Z. Sliding-Mode Triboelectric Nanogenerator Based on Paper and as a Self-Powered Velocity and Force Sensor. *Appl. Mater. Today* **2018**, *13*, 190–197. [CrossRef]
84. Yuan, L.; Xiao, X.; Ding, T.; Zhong, J.; Zhang, X.; Shen, Y.; Hu, B.; Huang, Y.; Zhou, J.; Wang, Z.L. Paper-Based Supercapacitors for Self-Powered Nanosystems. *Angew. Chem. Int. Ed.* **2012**, *51*, 4934–4938. [CrossRef] [PubMed]
85. Mohammadifar, M.; Zhang, J.; Yazgan, I.; Sadik, O.; Choi, S. Power-on-Paper: Origami-Inspired Fabrication of 3D Microbial Fuel Cells. *Renew. Energy* **2018**, *118*, 695–700. [CrossRef]

biosensors MDPI

Review

Hybrid Technologies Combining Solid-State Sensors and Paper/Fabric Fluidics for Wearable Analytical Devices

Meritxell Rovira [1], César Fernández-Sánchez [1,2] and Cecilia Jiménez-Jorquera [1,*]

[1] Instituto de Microelectrónica de Barcelona (IMB-CNM), CSIC, Campus UAB, Bellaterra, 08193 Barcelona, Spain; meritxell.rovira@csic.es (M.R.); cesar.fernandez@csic.es (C.F.-S.)
[2] CIBER de Bioingeniería, Biomateriales y Nanomedicina (CIBER-BBN), Jordi Girona 18-26, 08034 Barcelona, Spain
* Correspondence: cecilia.jimenez@csic.es

Abstract: The development of diagnostic tools for measuring a wide spectrum of target analytes, from biomarkers to other biochemical parameters in biological fluids, has experienced a significant growth in the last decades, with a good number of such tools entering the market. Recently, a clear focus has been put on miniaturized wearable devices, which offer powerful capabilities for real-time and continuous analysis of biofluids, mainly sweat, and can be used in athletics, consumer wellness, military, and healthcare applications. Sweat is an attractive biofluid in which different biomarkers could be noninvasively measured to provide rapid information about the physical state of an individual. Wearable devices reported so far often provide discrete (single) measurements of the target analytes, most of them in the form of a yes/no qualitative response. However, quantitative biomarker analysis over certain periods of time is highly demanded for many applications such as the practice of sports or the precise control of the patient status in hospital settings. For this, a feasible combination of fluidic elements and sensor architectures has been sought. In this regard, this paper shows a concise overview of analytical tools based on the use of capillary-driven fluidics taking place on paper or fabric devices integrated with solid-state sensors fabricated by thick film technologies. The main advantages and limitations of the current technologies are pointed out together with the progress towards the development of functional devices. Those approaches reported in the last decade are examined in detail.

Keywords: wearables; paper microfluidics; fabric microfluidics; solid-state sensors; electrochemical (bio)sensor; clinical analysis

Citation: Rovira, M.; Fernández-Sánchez, C.; Jiménez-Jorquera, C. Hybrid Technologies Combining Solid-State Sensors and Paper/Fabric Fluidics for Wearable Analytical Devices. *Biosensors* **2021**, *11*, 303. https://doi.org/10.3390/bios11090303

Received: 5 August 2021
Accepted: 26 August 2021
Published: 28 August 2021

1. Introduction

In recent years, efficient, simple, and low-cost diagnostic tools for decentralized rapid analysis of biomarkers at home, the point-of-care (POC), the point-of-need, or over-the-counter (OTC) have been highly demanded. Reports show that the diagnostic market size will be doubled within the next five years, this being recently boosted by the COVID-19 pandemic. The pace at which these devices are developed has been possible with the advancements achieved in materials, electronics, and manufacturing techniques. Two main groups can be identified regarding the device ability to provide a qualitative (yes/no) or quantitative result. While the former have shown widespread application mainly due to the implementation of easy and very cost-effective instrument-free lateral-flow strip devices [1], the development of quantitative approaches has been more challenging considering the strict requirements in sensitivity, specificity, and precision that should be fulfilled.

Quantitative POC devices are ideally compact and comprise miniaturized sensor components integrated in a device that allows for the sample uptake, processing, and analysis in an automatic or semi-automatic fashion. This can be done with fluidic approaches and electronic or photonic components for sensor interrogation and signal recording and interpretation.

The different elements of these POC devices are detailed and examined in the following sections.

1.1. Microfluidic Elements

Microfluidic elements are required in POC devices considering that the sample volumes usually available in clinical diagnosis are in the scale of microliters or below. Microchannels showing different geometries are included together with flow cells for sensor implementation, as well as reagent reservoirs and actuating components such as pumps and valves [2] for driving the sample and required reagents towards the device sensing area. The integration of pumps and valves is a key issue in microfluidics, and many devices reported so far rely on external components that make them of limited use for decentralized analyses. Few of them have gone into the market being the iSTAT system commercialized by Abbot the most representative one [3].

Among the different pumping strategies that have been assessed in the field of microfluidics, those that rely on flow by capillary action can suit a great variety of applications that make use of small liquid volumes, and the overall sample analysis requires very few steps. The inherent porosity of paper and other related materials enables the liquid to flow through them by capillary action without the need for an external pumping source [4]. In this context, the concept of microfluidic paper-based analysis devices (μPADs) has been created in recent years and provides a feasible low-cost alternative to traditional laboratory tests for the diagnosis of many common diseases and disorders [5]. Emanuel Elizalde et al. investigated the capillary inhibition in paper-like substrates to better understand fluid transport in terms of the macroscopic geometry of the flow domain [6]. This study concluded that capillary driven flow is determined by two main factors, that is the physicochemical properties of the interfaces involved and the geometry of the flow. Besides, paper has advantages over other materials, including low cost, simple production, good chemical stability, flexibility, biocompatibility, lightweight, and hydrophilicity. In addition, its surface can be easily modified, cut, folded, and/or stacked. This material has also been the basis of the well-known lateral flow devices with the most popular example being the glucose meter [7] and the very recently developed COVID-19 POC devices [8].

Regarding textiles, gaps between fibers in the fabric provide capillary channels for liquids to wick along threads without the need of an external pump. Their capillarity properties can be modified with standardized methods commonly used in paper microfluidics such as wax printing, photolithography, or cutting [9,10]. Besides, to get smart clothes, new composite and nanomaterials have been found to be promising candidates to obtain improved performance of textile based wearable devices [11].

1.2. Sensor Types

The integration of the sensing part on POC devices is diverse and has generated a huge number of reviews. Detection is an important challenge in the context of POCs, considering that the sensing equipment may be miniaturized, portable, easy to use, and non-expensive. Although a large variety of methods have been used for the detection of analytes, colorimetric sensing is surely the most prevalent [12,13]. Colorimetric chemicals can be easily integrated in paper and textile substrates and results can be visually evaluated under the naked eye (yes/no answer) and may also provide semiquantitative results by using a calibration chart, a handheld reader, or the camera of a smartphone. An interesting paper describing the measurement of pH and lactate on a textile substrate shows the simplicity of this approach for semiquantitative analysis [14]. However, they cannot provide real-time, continuous, and quantitative information of the target analytes being analyzed.

For this, it has been shown that chemical sensors mostly based on electrochemical transduction modes enable analyte continuous monitoring, providing rapid and reliable quantitative results, and can be adapted for different healthcare applications. However, integration of both paper and fabric-based microfluidics and chemical sensor devices has been challenging, and different innovative designs have been reported so far, as will be explained in the following sections. Electrochemical techniques offer high sensitivity, low detection limits, possibility of enhancing selectivity by applying different potential of modifying the electrodes, and inexpensive. In order to facilitate the integration of electrochemical sensors in POC structures, miniaturized solid-state sensors should be used. Their size can be scaled to micro- and nano-level, they can be easily implemented in fluidic systems, they require low amounts of sample, electrodes can be shaped as required and they can be integrated into almost any type of platform or substrate. The technologies used for solid state electrochemical sensors are generally thin film deposition (e.g., photolithography, evaporation and chemical vapor deposition of metal layers) and thick film deposition (e.g., inkjet printing and screen printing). Most of the works described in this review use the later technique. Screen printing (SP) is a popular electrode fabrication technique as it uses inexpensive equipment, can be automatized, can be scaled up for mass production, and it is user-friendly [15]. It uses screens or meshes to pattern printed electrodes on a wide range of substrate materials. Conductive ink is spread onto previously designed screens or meshes with the desired electrode geometries placed on the substrate and they are thermally cured at 60–90 °C for several minutes. Screen-printed electrodes (SPEs) are not only easy to fabricate but also can be simply modified by adding reagents and particles to the inks. However, the electrochemical properties of SPEs are unfortunately not as good as traditional metallic electrodes in terms of electron transfer kinetics and electrode resistance. Inkjet printing is now becoming a great alternative to SP due to the possibility to fabricate more miniaturized patterns and integrate printed circuits in the same substrate [16]. As an example, an organic electrochemical transistor (OET) inkjet printed on a flexible polyimide foil (PI) and applied to sweat measurement is described in reference [16,17].

All these techniques use substrates for wearable devices that are compatible with the sensor fabrication technique requirements (i.e., temperature, mechanical robustness, etc.). The most usual materials are polymethyl methacrylate (PMMA), with polyimide (PI) belonging to the brand Kapton being one of the most popular, and polyethylene terephthalate (PET).

Regarding thin film deposition techniques for sensor fabrication, the most representative are those for fabrication of ISFETs with silicon substrates and flexible substrates [18–20].

Among electrochemical detection, potentiometric ion selective electrodes are usually employed in POCs to quantify analytes that are not redox active (i.e., ions and pH) [21]. The change in voltage signal between the working and the reference electrode at close to zero-current conditions when the ionophore binds to the analyte is proportional to the activity of the analyte in the sample. This concentration dependence is described by the Nernst equation and confers to potentiometry the capability to quantify. In Figure 1, the schematic of a potentiometric sensor is shown.

In order to determine species that are redox active, amperometric and voltamperometric techniques can be used [21]. These are the most usual techniques for the detection of biomarkers like glucose and lactate as the corresponding enzymes catalyze redox reactions. These amperometric sensors consist of a biocatalyzer, an electronic mediator, and an electrochemical transductor. The intensity recorded when the redox active specie is oxidized/reduced could be related to the amount of analyte by Faraday's law and more specifically to the Cottrell equation when the chronoamperometry technique is used—a potential is fixed between the working and reference electrode—and the reaction is controlled by diffusion (Figure 1).

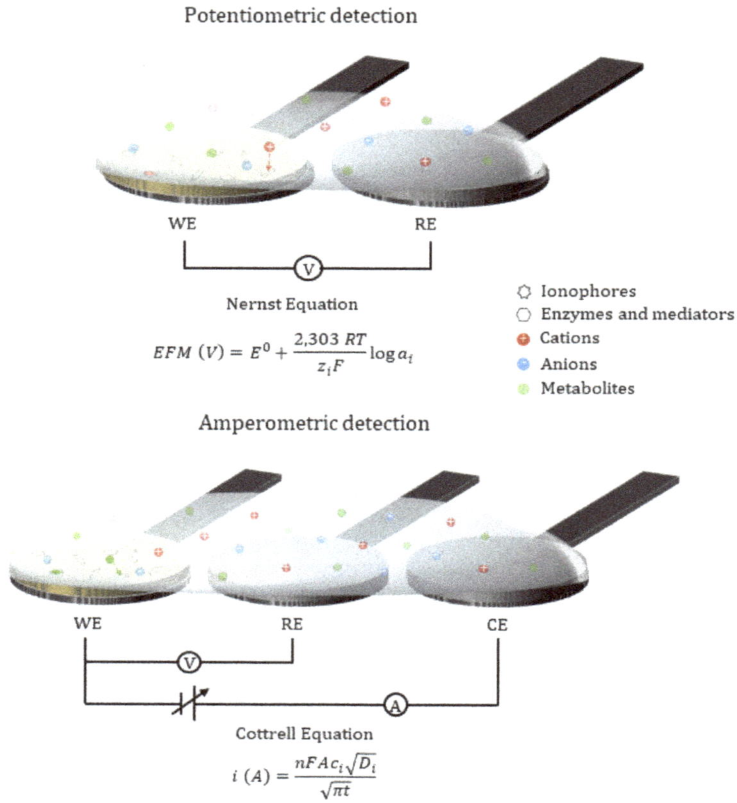

Figure 1. Potentiometric and amperometric sensors schemes and mechanisms of detection.

1.3. Applications in Healthcare

Many resources have been recently invested in order to improve people's healthcare by prevention and continuous diagnosis. Pharmaceutical companies such as Roche or Janssen are developing their own wearable devices for diagnosis and better understanding of diseases such as Parkinson's, Huntington's, or heart-related injuries. Furthermore, an increase in the number of diabetes cases has pushed medical tech companies into wearables business and one of the most popular diabetes monitoring devices is the FreeStyle Libre by Abbott [22]. It is a 14-day system that invasively and continuously monitors glucose levels. As shown in Figure 2a, it is worn on the back of the upper arm and results are transmitted to a reader or smartphone.

One of the last revolutions in the area of health monitoring and POC devices is in sports. Companies such as Fitbit, Garmin, and Apple have developed new platforms that enable the monitoring of consumer's own basic health (e.g., physical activity, heart rate, and sleep stages) in their daily lives [23]. However, reliable monitoring of (bio)chemical parameters has not been achieved yet and important investments have been made in this direction.

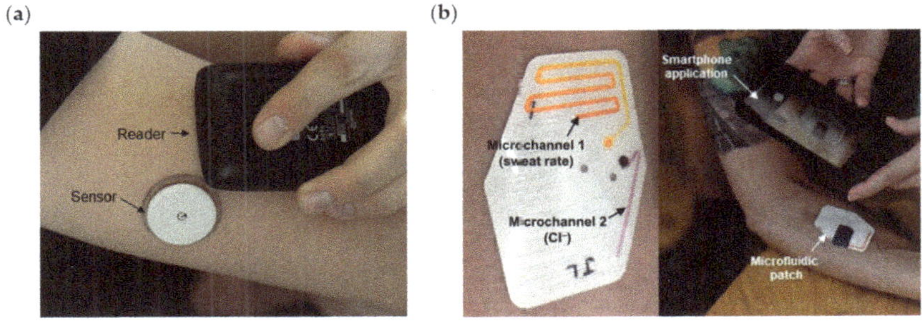

Figure 2. (**a**) Picture of FreeStyle Libre system from Abbot, consisting of a sensor worn on the arm and a handheld reader. (**b**) Picture of Gx Sweat Patch from Epicore Biosystems worn on the arm and read with a smartphone. Reprinted with permission from [25] for panel (**a**). Reprinted with permission from [24], Copyright 2020 American Association for the Advancement of Science for panel (**b**).

One of the major challenges of this decade is the continuous, noninvasive monitoring of biomarkers in sweat by means of wearable devices. The analysis of several biomarkers provides useful information about the physiological state of a person. Sweat is composed mainly of water (99%), but it also contains electrolytes such as sodium, chloride, and potassium, and metabolites like urea, pyruvate, and lactate. Sweating leads to the loss of water and electrolytes due to thermoregulatory action of sweat during exercise. The measurement of these electrolytes could provide information of the individual health status. Regarding to commercial wearable devices that measure sweat biomarkers, only the Gx Sweat Patch from Epicore Biosystems in partnership with Gatorade and Pepsico is capable of continuously measure biochemical markers in sweat. The device components are shown in Figure 2b. It provides a colorimetric feedback measurement of sweat rate, fluid loss, and sodium loss in real time [24].

The development of wearable devices based on electrochemical sensors for in situ sweat biomarker detection was first published in 2010 by the group of Diamond from Dublin University that developed a Na ion selective electrode (ISE) wearable device and was applied to compare the change in concentration of this target species during exercise done by patients suffering from Cystic Fibrosis (CF) and healthy patients [26]. Since then, several groups have been developing different wearable devices for monitoring biomarkers (pH, sodium, chloride, potassium, and lactate) in sweat. In most of them, no microfluidic components are applied and just absorbable surface in direct contact with the sensing element is used for collecting sweat and then being measured.

Some of the wearable devices described in the literature integrate the sensor in the same paper or fabric. This can be modified with conducting organic (polyaniline, polypyrrole, polytiophene, etc.) or inorganic (metal, metal oxide, etc.) materials to be conductive enough to perform the transduction of the chemical signal to the electronic signal. Techniques such as dip coating, printing, or sputtering are used to modify paper and fabric fibers [27,28]. For instance, a paper-based glucose sensor integrated into a standard Band-Aid adhesive patch is described in [29]. It used a 3D paper-based fuel cell configuration that enabled glucose monitoring without external power supply. On the other hand, it has been recently reported a wearable origami-like paper-based platform for sulfur mustard detection [30]. A wax pattern was defined onto a filter paper sheet and the three-electrode system was printed with conductive inks. They used different layers of paper to pre-load reagents needed for the enzymatic bioassay separately. In a similar approach, a paper-based wearable platform for continuous and real-time analysis of lactate and glucose in sweat was lately developed [31]. At first, fluidic channels and reservoirs were defined with hydrophobic wax in a paper sheet. Later, electrodes were screen-printed and, at last, paper was folded into a multi-layer structure that enabled a vertical flow of sweat.

The group of Wang was early in describing a SP biosensor demonstrating the potentialities of sensor integration in clothing for sport and defense applications [32]. A more simple approach to build electrochemical sensors for pH, K, and NH_4 using cotton yarns is described by Andrade group in [33]. Yarns were turned into electrical conductors through a carbon nanotube ink and covered with a polymeric membrane to obtain ion-selective electrodes. Through the placement of these sensors in a band-aid, this approach could be easily implemented in a wearable device. Another interesting approach made use of a single cotton yarn as gate of an organic electrochemical transistor (OET) to detect adrenaline for hydration and stress human monitoring [34].

However, to date, these devices integrating the sensing element in paper and fabrics are far to provide reliable quantitative results. In addition, they usually lose microfluidic capabilities of the substrate.

Generally, when microfluidic components are considered, PDMS has been the material of choice for most of the microfluidic devices developed in research laboratories. PDMS has been undoubtedly key for the development of microfluidics due to the ease of producing structures like microchannels of different geometries as well as other fluidic components and of integrating sensor devices allowing rapid prototyping and proof-of-concept demonstrations [35]. However, this material can show drawbacks depending on the channel size like the formation of bubbles (mild wettability), long time of response, and poor chemical compatibility. PMMA has also been used in microfluidic structures. An example is that described by the group of Diamond for colorimetric pH detection in sweat [36].

Other materials like PET, SU-8, and COC/COP have also been considered to different extent [17,20,37]. As already pointed out, paper is an alternative material for microfluidics in point-of-care devices in general and wearable devices in particular due to its capacity of driving the sweat to the sensor areas while working, but also to collect the sample and accumulate it after carrying out measurements in continuous applications. The use of paper channels ensures obtaining fresh sweat from the skin surface that is continuously driven through the microfluidic channel by capillary action towards the sensors. Besides, fabrics are materials equivalent to paper as they also use capillary forces to drive the samples over the sensors and they show the particularity of producing channels themselves without the need of patterning [9,10,38–41]. Devices that integrate paper or fabric materials meet the "ASSURED" criteria (affordable, sensitive, specific, user-friendly, rapid and robust, equipment-free, and deliverable to users) set by the World Health Organization (WHO) for diagnostic tests [42].

This review describes all reported works focused on paper- and fabric-based microfluidic devices that integrate solid-state sensors applied to wearable systems for sweat analysis. The hybridization of the sensor component layer with the microfluidic component enables the continuous monitoring of the target analytes, using low-powered simple instrumentation that can be supplied wirelessly. However, the integration of both parts is one of the major difficulties in wearable devices. Its potential has not been exploited yet because of the technological challenges still to be worked out. The potential of such type of devices is foreseen and that is why it deserves being summarized in this short review.

2. Paper-Based Microfluidics for Wearable Devices

Among the many strategies toward capillary-driven microfluidics, paper is considered the most attractive and promising substrate material. It has unique advantages such as its extremely low cost and its great mechanical properties (flexible, light, and thin). Its porous structure provides capillary transport, capacity of storage of reagents, air permeability useful to avoid air bubble formation, filtration, and high surface to volume ratio that results in short time for the analysis [43]. Due to its inherent biocompatibility, biodegradability, sustainability, and disposability, as well as chemical and biological inertness, it is particularly important for the immobilization of bioreagents, and it also fits into the so-called circular economy model.

Among the different paper-like materials available in the market, filter papers of different grades show special features, in terms of physicochemical properties in terms porosity or overall structure, which make them especially suited for microfluidic devices. Traditional paper is made of cellulose but porous flexible substrates made of nitrocellulose (NC), bio-source photonic crystals, and flexible softshell particles have also been considered as paper materials [44].

The first example of a paper microfluidic structure was described in 1937 by Yagoda Herman [45], who patterned hydrophilic spot tests on a paper substrate using paraffin. Using the same technique, in 1949, Müller et al. reported a paraffin-patterned paper microfluidic channel [46] for improving the performance of a paper chromatographic device. However, it was not until the first decade of the 21st century that paper microfluidics drew the attention of the scientific community [47]. In 2007, Whitesides' group at Harvard University introduced the first paper-based device for the detection of glucose, fabricated by a lithographic method and pioneered the term "paper microfluidics" [48]. Since then, a large variety of fabrication techniques for producing paper-based microfluidic devices have been developed.

Paper is patterned to construct the required microfluidic architecture and two main approaches have been applied in this regard: (1) two-dimensional shaping/cutting of paper and (2) patterning hydrophilic/hydrophobic areas on paper. The former mainly comprises cutting the paper with a cutter plotter capable of shaping paper into previously customized designed or with a CO_2 laser cutting apparatus. The latter is based on creating a hydrophilic/hydrophobic contrast by one of the following approaches: (a) Physical blocking of the paper pore structure. (b) Physical deposition of a hydrophobic agent on the paper surfaces. (c) Chemical modification of the paper surface.

Regarding the implementation of a detection approach into a wearable device, some features should be considered. This should be intuitive to operate but also portable and inexpensive, fulfilling the requirements for in field analysis. We can distinguish three different approaches for paper-based analysis: the naked eye, digital colorimetry, by using a smartphone for example that can read the color and intensity of a spot and instrumental analysis in which the signal is recorded by an instrument [49].

Colorimetric detection is the most popular one in paper-microfluidic devices, as paper offers a bright, high-contrast, and colorless background for color change readings. Numerous paper-based assays, such as the new COVID-19 antigen test and the pregnancy test, are simply interpreted under the naked eye and provide with a yes/no answer for qualitative analysis. Digital colorimetry based on the measurement of color intensity or pixel counting has also been of widespread use mainly because the only instrument that is required is a digital camera, which we find in every smartphone or portable device. However, semiquantitative analysis can be carried out using these approaches.

Instrumental analysis is necessary for quantitative biomarker analysis. Among the different options available, electrochemical detection has been the most widely used. Electrochemical paper-based analytical devices combine the inherent advantages of electrochemical detection, such as high sensitivity and low detection limits (LODs), affordability and the possibility of enhancing selectivity and sensitivity by applying different electrochemical techniques or using electrochemical cells made of different materials and showing specific configurations, with those of paper, already mentioned above [50].

Table 1 summarizes the main characteristics of some wearable devices that combine electrochemical analysis and paper-based microfluidics.

The first work showing the feasible integration of a microfluidic structure together with a solid-state sensor was reported by Heikenfeld's group [51]. He developed a flexible patch containing a Na ISE and a paper microfluidic structure for continuous measuring this ion in sweat during exercise practice. One of the most relevant features of this patch was being battery-free and communicating wirelessly to an android smartphone. It used a radio-frequency inductively powered (RFID) chip in order to obtain the power for sensors reading. Unfortunately, due to the limitations of the analogue digital converter on the RFID, the Na sensor response was lower than the expected one. It also exhibited a slight drift (3–5 mV) due to the instability of the Ag/AgCl reference electrode. Its simple microfluidics design enabled low response times (30 s), but it was unable to be used for long-time measurements during exercise performance.

Following this work, a flexible microfluidic platform for simultaneously and selectively measuring lactate, pH, and sodium in sweat by integrating amperometric and potentiometric detection approaches (Figure 3a) [52] was reported. Sensors consist of arrays of flexible 50 µm diameter flexible microneedles implemented in microfluidic paper channels. The sensor analytical performance showed that they stabilized very rapidly, within 10 s of starting the measurements, this being a relevant feature for continuous monitoring of the target biomarkers. Long-term in vitro studies carried out with this device showed stability and repeatability for over 8, 5, and 4 months for the pH, sodium, and lactate sensors, respectively. Sample flow control in the device is achieved by using papers of different grades. On-body tests were carried out on six healthy males during cycle ergometry and treadmill running, obtaining reliable results for all the sensors that were comparable to those recorded in previously reported works.

More recently, a wearable patch for lactate monitoring in sweat that make use of an osmotic-capillary microfluidic pumping approach for controlling the flow [53], was developed. This was possible by incorporating an hydrogel film at the sweat collector area of the device, which, due to the high content of ions, had the effect of pumping by osmotic force and enabled continuous extraction of fluid from the skin surface [54] while the paper-based channel drove the fluid to the sensor by capillary wicking. This paper only reported in vitro tests, where the lactate sensor showed a linear response in the relevant physiological range of 5 mM to 20 mM.

In a more complex approach, a three-dimensional paper-based microfluidic structure was used in the production of a patch for glucose detection in sweat [55]. A patterned paper was folded into an origami structure with five equal layers in order to form the 3D flow channel. This microfluidics design included five layers: sweat collector, vertical channel, transverse channel, electrode layer, and sweat evaporator (Figure 3b). This configuration enabled the separation of the electrodes from skin, fast response, and continuous evaporation leading to continuous renewal of the sample over the electrodes. The 3D device was tested using a flow injection system that simulate the flow of sweat. The results displayed a good repeatability and linearity in glucose concentration ranging from 0 to 1.9 mM covering the range of glucose physiological concentrations in human sweat (range: 0.25–1.5 mM). However, the sensitivity of the sensor was lower than in the test without microfluidics due to the absence of stirring. On-body tests were performed placing the system on the forearm of three subjects. The results showed that glucose concentration decreased gradually during exercise (from 1.5 mM to 0.4 mM) and that there was a significant variability between the three subjects. A second generation of this device was developed that integrated a potassium electrochemical sensor. The new version of the device was tested for real-time monitoring of this target species [56]. It showed a small size, and the results reported a quasi-Nernstian response, 5 s response time, and a stable performance within 1 week. An interesting feature that was studied in this work is the patterning of the sweat collection area. Three designs were tested: circular, radial, and snowflake-like structures. The circular pattern covered the maximum of grids among the three of them, but it required the largest sweat volume (20 µL) to fill it and so a long time for sweat to reach the electrodes zone was necessary before a signal could be measured. In the radial design the area in contact with the skin was reduced, and so the volume required for filling it. The specific structure made the collector area to be more efficient for sweat collection and transport. The snowflake-like design showed more branches of smaller width, reducing the hydrophilic area contacting the skin even further. Therefore, the snowflake-like pattern showed the most efficient sweat collection and the fastest response. Real-time analysis was performed on volunteers during cycling with the device attached to the forearm. Results were consistent with the previous literature reporting potassium measurements in sweat. However, the mechanism of sweat potassium loss is still unclear.

The hybridization of paper microfluidics and sensors defined on a different substrate can also be used for non-chemical based systems. An example of this is reported in [57] where a sensor for measuring sweat rate is described, which consisted of printed interdigitated electrodes and a serpentine shaped paper-based microfluidic channel. By calculating the fluid volume per serpentine line, the time values at which the fluid crosses the fingers of the interdigitated electrodes, and by measuring the admittance between the two interdigitated electrodes, sweat rate could be monitored. The device could hold sweat volumes up to 82 μL in 30 min filling time and could be placed on various locations of the body for extended periods. This device was validated in vitro.

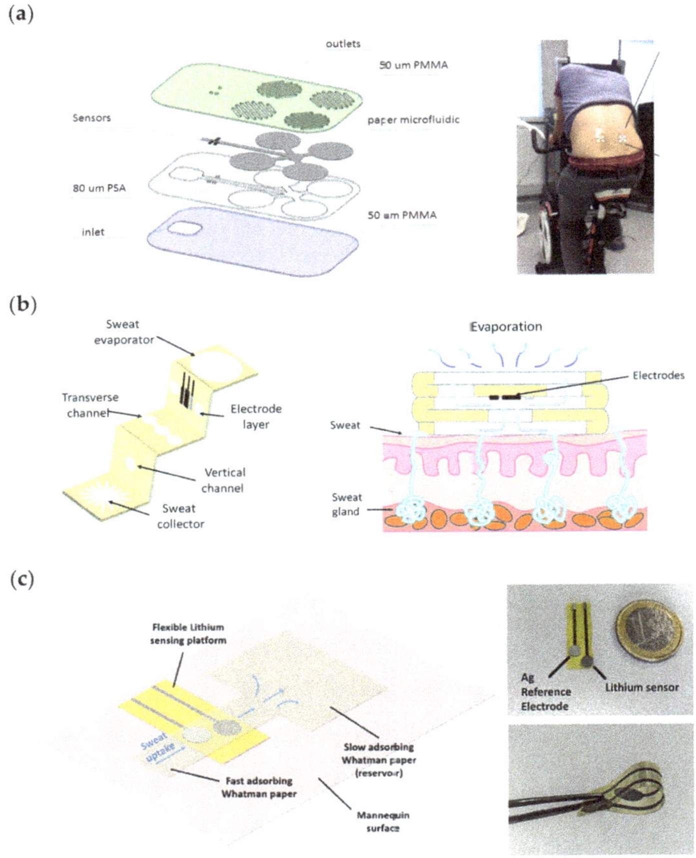

Figure 3. (**a**) Assembly of the different layers of the multi-sensing patch for sodium, pH, and lactate monitoring and picture of the device on-body. (**b**) Schematic of the three-dimensional paper-based device for glucose monitoring, including the different parts and its application on the human skin. (**c**) Photographs of the lithium-sensing layer and schematics of the wearable system including paper-microfluidics. Reprinted with permission from [52] and Copyright 2016 MDPI for panel (**a**). Reprinted with permission from [55] and Copyright 2019 The Royal Society for panel (**b**). Reprinted with permission from [58] and Copyright 2021 IEEE for panel (**c**).

In a recent paper, a wearable electrochemical sensing platform was developed to monitor lithium in sweat for therapeutic drug monitoring of people suffering from bipolar disorder [58]. A potentiometric lithium ion-selective electrode (ISE) was integrated on a flexible polyimide (PI) substrate. The microfluidics design was simple but very effective. A fast-absorbing paper was implemented on the sample collection area and sensing regions to ensure the effective renewal of sweat on the electrode area. Quasi-Nernstian behavior of the sensor was proved with artificial sweat samples, providing good analytical capability for the target application. Neither on-body tests nor tests simulating typical sweat flow rate (1.5 μL cm^{-2} min^{-1})) were reported, so the response time of the sensor and the maximum operating time of the platform are unknown. This device has lately completed with more sensors and cotton has been used instead paper as microfluidic element [59].

3. Fabric-Based Microfluidics for Wearable Devices

As a result of the continuous efforts put to find new substrate materials for microfluidic platforms, simple and inexpensive fabric-based microfluidics have emerged. This research arrived soon after the paper-based microfluidic analytical technology in 2010 [9,10]. As well as in paper microfluidics, gaps between fibers in the fabric provide capillary channels for liquids to wick along threads without the need of an external pump. Their capillarity properties can be modified with standardized methods commonly used in paper-microfluidics such as wax printing, photolithography, or cutting.

Fabric-based microfluidic technologies are affordable, flexible, robust, user-friendly, quick, and scalable. They can be fabricated in 2D or 3D configurations to transport fluids depending on the application [60]. Both paper and fabric can be combined to take advantage of the special features of both materials. Thread can transport liquids with no need of patterned barriers, while paper can facilitate the immobilization and in turn integration of enzymes and reagents [9,61]. In 2013, Pan and coworkers proposed a novel interfacial microfluidic transport principle using micropatterned superhydrophobic textiles [62]. It consisted of stitching patterns of hydrophilic cotton yarns on a superhydrophobic cotton textile treated with fluoropolymer microparticles. The system was developed aiming at achieving a continuous and controlled sample flow but lacked a biochemical testing application.

Just like in paper microfluidic devices, thread is also suitable for fabricating microfluidic electroanalytical systems. There are many different types of fibers that can be used for the construction of fluidic channels, which can be combined with electrodes for quantitative analyses. Among them, rayon, cotton, nylon, polyester, wool, or silk threads could be used that differ on the wicking rate. If the application requires fast-absorption areas, fibers with high wicking rates may be used. They can also be treated with plasma to increase their wicking rate [63].

Few reported applications are focused on the hybridization of textile-microfluidics and solid-state sensors [64–66]. C. W. Bae et al. developed a fully stretchable sweat sensing patch for the detection of glucose with a non-enzymatic nanoporous gold (NPG)-based sensor and a textile microfluidic component (Figure 4a) [64]. Tests with different dyed water solutions showed filling times of ~16 min and good renewal of the solution. Sensitivities of 57 μA cm^{-2} mM $^{-1}$ were achieved without any hysteresis effect. A drawback of the system was that when sample volume was not enough, the sensor generated a false signal, thus being a serious obstacle continuous monitoring. To tackle it, the number of inlets in the sample collector area and the overall collector size were adapted considering the absorption rate of the fabric and the channel width.

Table 1. Main characteristics of the reported devices integrating paper and fabric microfluidics and electrochemical sensors on flexible substrates.

Integration of Substrate, Sensors and Microfluidics	Detection Technique	Sensor Technique/Type	Marker	Response Time *	Sensitivity (Linear Range)	Ref.
Paper microfluidics						
Paper microfluidics + ISE and RE on flexible Kapton substrate	Pot	Pd and Ag electrodeposited on patterned Cu electrodes for WE and RE	Sodium	30 s	0.3 mV/mM (10–90 mM)	[51]
PMMA layers and paper microfluidics + 6 electrodes placed inside a paper channel	Pot/Amp	Pt and Ag flexible microneedles for 3 WE, 2 RE and CE; pH IrOx membrane; Na WE coated with PEDOT; lactate LOx in BSA/PU + SPEES/PES	pH/ Sodium/ Lactate	10 s	pH: 71.9 mV/dec Na: 56 mV/dec	[52]
A hydrogel and paper microfluidics + WE, RE and CE on PI substrate	Amp	SP WE with Prussian Blue/carbon ink, RE with Ag/AgCl ink and CE with carbon ink; LOx and Nafion drop-casted in WE	Lactate	16–20 min	0.03 µA/(mM·mm^2) (5–20 mM) LOD: 6 mM	[53]
3D wax-printed paper microfluidics + WE, RE and CE on PET substrate	Amp	SP WE and CE with Prussian Blue/graphite ink and RE with Ag/AgCl ink; GOx drop-casted in WE	Glucose	-	35.7 µA/(mM·cm^2) (0–1.9 mM) LOD: 5 µM	[55]
3D wax-printed paper microfluidics + WE and RE on PET substrate	Pot	SP WE and RE with Ag/AgCl and carbon inks; WE coated with PEDOT:PSS	Potassium	5 s	61.8 mV/dec (1–32 mM)	[56]
Paper microfluidics + electrode on PI film	Imp	SP electrode with carbon ink	Sweat rate	30 min	-	[57]
Paper microfluidics + ISE and RE on a flexible PI substrate	Pot	WE (ISE) and RE by photolithography patterning	Lithium	-	56.8 mV/dec (2 mM–1 M) LOD: 1.7 mM	[58]
Fabric microfluidics						
Cotton fabric and PU nanofiber cover + WE, RE and CE on flexible PDMS substrate	Amp	Nanoporous Au WE and CE vacuum-deposited and Ag/AgCl RE	Glucose	16.15 min	57.6 µA/(mM·cm^2)	[61]
Thread + ISE and RE on PET film	Pot	SP ISE and RE with graphite and Ag/AgCl inks resp.; ISE and RE coated with PEDOT and PVB resp.	Sodium	8–10 min	56.7 mV/dec	[65]
3D printed platform containing thread microfluidics and ISEs on PET substrate	Pot	Pt CE, Ag pseudoRE and ISEs SP with carbon ink; ISEs and REs coated with PEDOT and POT	Sodium/ Potassium	8 min	Na: 52.4 and 56.4 mV/dec for PEDOT and POT resp. K: 45.7 and 54.3 mV/dec for PEDOT and POT resp.	[66]

* Time to respond to changes in concentration. Pot: Potentiometric; Amp: Amperometric; Imp: Impedimetric; WE: Working electrode; RE: Reference electrode; CE: Counter electrode; SP: Screen-printed; PI: Polyimide; PET: Polyethylene terephthalate; PDMS: Polydimethylsiloxane; LOD: Limit of detection; GOx: Glucose Oxidase; IrOx: iridium oxide; PU: polyurethane; BSA: bovine serum albumin; SPEES/PES: sulphonated polyesther ether sulphone—polyether sulphone; LOx: Lactate Oxidase; PEDOT:PSS: poly(3,4-ethylenedioxythiophene) polystyrene sulfonate; PEDOT: poly(3,4-ethylenedioxythiophene); PVB: Polyvinyl butyral; POT: poly(3-octylthiophene-2,5-diyl).

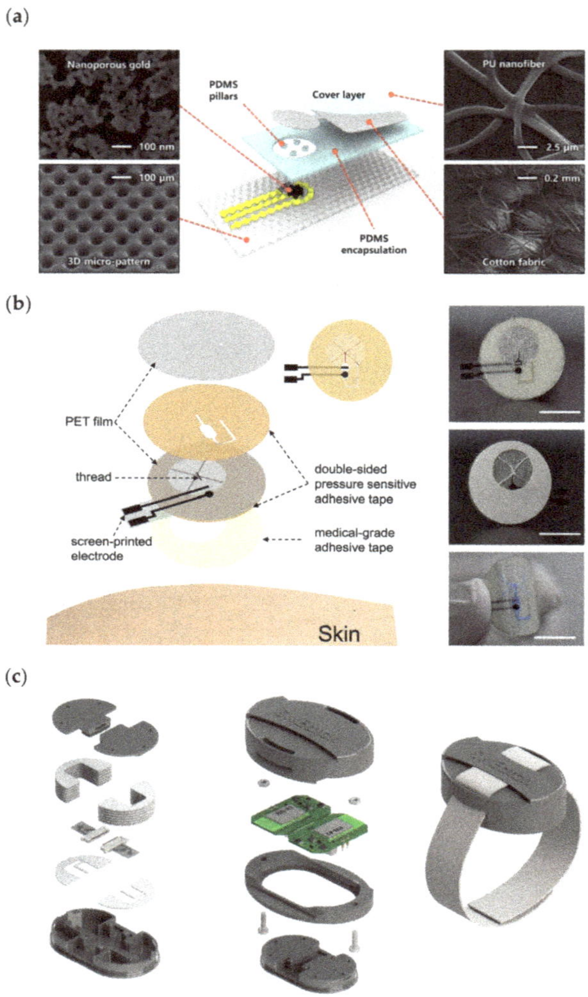

Figure 4. (**a**) Pictures of the glucose sensing patch on the forearm, demonstrating its conformality under compression and stretching, and schematic of the different layers conforming the device showing their SEM images. (**b**) Schematic showing the different components of the patch for sodium monitoring, and photographs of the device on the arm and forehead of an individual (threads colored with blue dye). (**c**) Schematics of the different layers of the microfluidic unit, of the platform body and view of the fully enclosed 3D printed SwEatch platform. Reprinted with permission from [64] and Copyright 2019 American Chemical Society for panel (**a**). Adapted with permission from [65] and Copyright 2020 Elsevier for panel (**b**). Adapted with permission from [66] and Copyright 2019 Elsevier for panel (**c**).

Recently, a wearable device for the detection of sodium in sweat was reported [65] that consisted of fabric microfluidics and screen-printed electrodes implemented on PET films (Figure 4b). A patterned double-sided PSA was sandwiched between two PET films, forming a channel of 280 μm height in which a thread was placed. It required 8–11 min to fill the detection reservoir with just one inlet, this being a rather long time for continuous monitoring. Long-term stability tests resulted in a low drift (below 1 mV after 1000 s). On body Na monitoring provided concentrations in the normal physiological range.

Diamond's group recently developed a watch type platform for monitoring Na and K in sweat [66]. They had previously reported the system called "SwEatch" only for monitoring sodium ions [67]. The microfluidics was based on two threads that draw the sweat from the skin to both Na and K sensors (Figure 4c). On-body trials showed sharp increases in the signal for both Na and K taking place when sweat reached the electrodes after 8 min. The device detected a lower Na concentration than expected, probably due to sweating rate, skin temperature, or electrolyte reabsorption.

The main characteristics of the previously described are summarized in Table 1.

4. Conclusions and Outlook

The research status and characteristics of paper- and fabric-based fluidics integrated with solid-state sensors was reviewed. All the works reported so far included sensors mainly fabricated by screen-printing techniques. One of the main drawbacks of this type of sensors is that they require large sample volumes, and the miniaturization capacity of microfluidics cannot be fully exploited. However, it was shown that all the reported approaches show great versatility and conformability (2D and 3D structures), their properties can be modified to control the sample flow, they are cost-effective and flexible to be incorporated into wearables, they do not require pumps or valves, and they allow continuous measurement. Having the sensing layer and the microfluidics in different substrates enables the integration of compact and simple electronics.

Few works test their devices on-body, which shows the difficulty of the final integration and stability of the measurements in real conditions. The ones assessed in a real scenario show promising results that successfully compared with previously reported information, rather than contrasting them with standard methods. Further work should be done in order to improve the sampling of biofluids and to be able to fully validate the devices.

Comparing the microfluidic characteristics of paper and fabric, it is clear that fabric-microfluidic devices have longer time responses than paper-based ones due to more limited wettability of the applied fabric. Fabric also displays a wide variety of interfiber gap sizes, resulting in a worse reproducibility than paper. On the other hand, thread has great tensile strength, flexibility, ability to easily form 3D structures and simplicity of being combined into wearable materials. Therefore, the optimal selection of an adequate material platform relies on the requirements of the application.

The current proof-of-concept demonstrations show a great opportunity for simplifying basic health monitoring, enabling self-diagnosis without needing complex instrumentation. However, the embedding of such devices with wearable electronics for direct sample collection and continuous monitoring deserves further investigation. Implementing microfabricated sensors in these wearable systems would offer clear advantages in terms of miniaturization, fabrication, reproducibility, degree of integration of not only sensor and microfluidic components, but also all the required electronics. It is still in its early stages, as it requires progress in innovation, knowledge and skills in fabrication techniques and detection methods to enable these platforms to reach maturity.

Author Contributions: Writing—original draft preparation, M.R.; writing—review and editing C.F.-S. and C.J.-J.; Supervision, C.J.-J. All authors have read and agreed to the published version of the manuscript.

Funding: This research received funding from SNF-SINERGIA (N°CRSII5_177255), Switzerland.

Institutional Review Board Statement: Not applicable.

Informed Consent Statement: Not applicable

Acknowledgments: The authors gratefully acknowledge the funding from SNF-SINERGIA under the project WeCare.

Conflicts of Interest: The authors declare no conflict of interest.

References

1. Eltzov, E.; Guttel, S.; Kei, A.L.Y.; Sinawang, P.D.; Ionescu, R.E.; Marks, R.S. Lateral Flow Immunoassays – from Paper Strip to Smartphone Technology. *Electroanalysis* **2015**, *27*, 2116–2130. [CrossRef]
2. Weisgrab, G.; Ovsianikov, A.; Costa, P.F. Functional 3D Printing for Microfluidic Chips. *Adv. Mater. Technol.* **2019**, *4*, 1900275. [CrossRef]
3. Abbot An Easy, Integrated Point-of-Care Testing (POCT) Solution for a Broad Range of Clinical Settings. Available online: https://www.pointofcare.abbott/int/en/offerings/istat (accessed on 30 July 2021).
4. Gutiérrez-Capitán, M.; Baldi, A.; Fernández-Sánchez, C. Electrochemical paper-based biosensor devices for rapid detection of biomarkers. *Sensors* **2020**, *20*, 967. [CrossRef]
5. Tseng, C.-C.; Kung, C.-T.; Chen, R.-F.; Tsai, M.-H.; Chao, H.-R.; Wang, Y.-N.; Fu, L.-M. Recent advances in microfluidic paper-based assay devices for diagnosis of human diseases using saliva, tears and sweat samples. *Sens. Actuators B Chem.* **2021**, *342*. [CrossRef]
6. Elizalde, E.; Urteaga, R.; Berli, C.L.A. Rational design of capillary-driven flows for paper-based microfluidics. *Lab Chip* **2015**, *15*, 2173–2180. [CrossRef]
7. Dimeski, G.; Jones, B.W.; Tilley, V.; Greenslade, M.N.; Russell, A.W. Glucose meters: Evaluation of the new formulation measuring strips from Roche (Accu-Chek) and Abbott (MediSense). *Ann. Clin. Biochem.* **2010**, *47*, 358–365. [CrossRef]
8. Wang, C.; Shi, D.; Wan, N.; Yang, X.; Liu, H.; Gao, H.; Zhang, M.; Bai, Z.; Li, D.; Dai, E.; et al. Development of spike protein-based fluorescence lateral flow assay for the simultaneous detection of SARS-CoV-2 specific IgM and IgG. *Analyst* **2021**, *146*, 3908–3917. [CrossRef] [PubMed]
9. Li, X.; Tian, J.; Shen, W. Thread as a versatile material for low-cost microfluidic diagnostics. *ACS Appl. Mater. Interfaces* **2010**, *2*, 1–6. [CrossRef] [PubMed]
10. Bhandari, P.; Narahari, T.; Dendukuri, D. Fab-Chips: A versatile, fabric-based platform for low-cost, rapid and multiplexed diagnostics. *Lab Chip* **2011**, *11*, 2493–2499. [CrossRef] [PubMed]
11. Hatamie, A.; Angizi, S.; Kumar, S.; Pandey, C.M.; Simchi, A.; Willander, M.; Malhotra, B.D. Review - Textile Based Chemical and Physical Sensors for Healthcare Monitoring. *J. Electrochem. Soc.* **2020**, *167*. [CrossRef]
12. Cate, D.M.; Adkins, J.A.; Mettakoonpitak, J.; Henry, C.S. Recent Developments in Paper-Based Microfluidic Devices. *Anal. Chem.* **2014**, *87*, 19–41. [CrossRef]
13. Nery, E.W.; Kubota, L.T. Sensing approaches on paper-based devices: A review. *Anal. Bioanal. Chem.* **2013**, *405*, 7573–7595. [CrossRef] [PubMed]
14. Promphet, N.; Rattanawaleedirojn, P.; Siralertmukul, K.; Soatthiyanon, N.; Potiyaraj, P.; Thanawattano, C.; Hinestroza, J.P.; Rodthongkum, N. Non-invasive textile based colorimetric sensor for the simultaneous detection of sweat pH and lactate. *Talanta* **2019**, *192*, 424–430. [CrossRef]
15. Noviana, E.; McCord, C.P.; Clark, K.M.; Jang, I.; Henry, C.S. Electrochemical paper-based devices: Sensing approaches and progress toward practical applications. *Lab Chip* **2019**, *20*, 9–34. [CrossRef]
16. Gao, M.; Li, L.; Song, Y. Inkjet printing wearable electronic devices. *J. Mater. Chem. C* **2017**, *5*, 2971–2993. [CrossRef]
17. Demuru, S.; Kunnel, B.P.; Briand, D. Real-Time Multi-Ion Detection in the Sweat Concentration Range Enabled by Flexible, Printed, and Microfluidics-Integrated Organic Transistor Arrays. *Adv. Mater. Technol.* **2020**, *5*, 1–9. [CrossRef]
18. Nakata, S.; Arie, T.; Akita, S.; Takei, K. Wearable, Flexible, and Multifunctional Healthcare Device with an ISFET Chemical Sensor for Simultaneous Sweat pH and Skin Temperature Monitoring. *ACS Sens.* **2017**, *2*, 443–448. [CrossRef]
19. Zhang, J.; Rupakula, M.; Bellando, F.; Garcia Cordero, E.; Longo, J.; Wildhaber, F.; Herment, G.; Guérin, H.; Ionescu, A.M. Sweat Biomarker Sensor Incorporating Picowatt, Three-Dimensional Extended Metal Gate Ion Sensitive Field Effect Transistors. *ACS Sens.* **2019**, *4*, 2039–2047. [CrossRef] [PubMed]
20. Garcia-Cordero, E.; Bellando, F.; Zhang, J.; Wildhaber, F.; Longo, J.; Guérin, H.; Ionescu, A.M. Three-Dimensional Integrated Ultra-Low-Volume Passive Microfluidics with Ion-Sensitive Field-Effect Transistors for Multiparameter Wearable Sweat Analyzers. *ACS Nano* **2018**, *12*, 12646–12656. [CrossRef]
21. Diamond, D. Wiley: Principles of Chemical and Biological Sensors. 1998. Available online: https://www.wiley.com/en-us/Principles+of+Chemical+and+Biological+Sensors-p-9780471546191 (accessed on 25 August 2021).
22. Abbott FreeStyle Libre. Available online: https://www.freestyle.abbott/us-en/home.html (accessed on 23 August 2021).
23. Bunn, J.A.; Navalta, J.W.; Fountaine, C.J.; Reece, J.D. Current State of Commercial Wearable Technology in Physical Activity Monitoring 2015–2017. *Int. J. Exerc. Sci.* **2018**, *11*, 503. [PubMed]
24. Baker, L.B.; Model, J.B.; Barnes, K.A.; Anderson, M.L.; Lee, S.P.; Lee, K.A.; Brown, S.D.; Reimel, A.J.; Roberts, T.J.; Nuccio, R.P.; et al. Skin-interfaced microfluidic system with personalized sweating rate and sweat chloride analytics for sports science applications. *Sci. Adv.* **2020**, *6*, eabe3929. [CrossRef]
25. Thirunavukkarasye-Raveendran Blutzuckermessgerät FreeStyle Libre Mit Auslesegerät, Am Oberarm Befestigt. Available online: https://commons.wikimedia.org/wiki/File:FreeStyle_libre_am_Oberarm_und_Auslesegerät-5.JPG#globalusage (accessed on 26 August 2021).
26. Schazmann, B.; Morris, D.; Slater, C.; Beirne, S.; Fay, C.; Reuveny, R.; Moyna, N.; Diamond, D. A wearable electrochemical sensor for the real-time measurement of sweat sodium concentration. *Anal. Methods* **2010**, *2*, 342–348. [CrossRef]

27. Kumar, S.; Pandey, C.M.; Hatamie, A.; Simchi, A.; Willander, M.; Malhotra, B.D. Nanomaterial-Modified Conducting Paper: Fabrication, Properties, and Emerging Biomedical Applications. *Glob. Chall.* **2019**, *3*, 1900041. [CrossRef]
28. Ghahremani Honarvar, M.; Latifi, M. Overview of wearable electronics and smart textiles. *J. Text. Inst.* **2017**, *108*, 631–652. [CrossRef]
29. Cho, E.; Mohammadifar, M.; Choi, S. A single-use, self-powered, paper-based sensor patch for detection of exercise-induced hypoglycemia. *Micromachines* **2017**, *8*, 265. [CrossRef]
30. Colozza, N.; Kehe, K.; Dionisi, G.; Popp, T.; Tsoutsoulopoulos, A.; Steinritz, D.; Moscone, D.; Arduini, F. A wearable origami-like paper-based electrochemical biosensor for sulfur mustard detection. *Biosens. Bioelectron.* **2019**, *129*, 15–23. [CrossRef] [PubMed]
31. Li, M.; Wang, L.; Liu, R.; Li, J.; Zhang, Q.; Shi, G.; Li, Y.; Hou, C.; Wang, H. A highly integrated sensing paper for wearable electrochemical sweat analysis. *Biosens. Bioelectron.* **2021**, *174*, 112828. [CrossRef] [PubMed]
32. Yang, Y.L.; Chuang, M.C.; Lou, S.L.; Wang, J. Thick-film textile-based amperometric sensors and biosensors. *Analyst* **2010**, *135*, 1230–1234. [CrossRef] [PubMed]
33. Guinovart, T.; Parrilla, M.; Crespo, G.A.; Rius, F.X.; Andrade, F.J. Potentiometric sensors using cotton yarns, carbon nanotubes and polymeric membranes. *Analyst* **2013**, *138*, 5208–5215. [CrossRef] [PubMed]
34. Coppedè, N.; Tarabella, G.; Villani, M.; Calestani, D.; Iannotta, S.; Zappettini, A. Human stress monitoring through an organic cotton-fiber biosensor. *J. Mater. Chem. B* **2014**, *2*, 5620–5626. [CrossRef]
35. Zhang, S.; Zahed, M.A.; Sharifuzzaman, M.; Yoon, S.; Hui, X.; Barman, S.C.; Sharma, S.; Yoon, H.S.; Park, C.; Park, J.Y. A wearable battery-free wireless and skin-interfaced microfluidics integrated electrochemical sensing patch for on-site biomarkers monitoring in human perspiration. *Biosens. Bioelectron.* **2021**, *175*, 112844. [CrossRef] [PubMed]
36. Curto, V.F.; Fay, C.; Coyle, S.; Byrne, R.; O'Toole, C.; Barry, C.; Hughes, S.; Moyna, N.; Diamond, D.; Benito-Lopez, F. Real-time sweat pH monitoring based on a wearable chemical barcode micro-fluidic platform incorporating ionic liquids. *Sensors Actuators B Chem.* **2012**, *171–172*, 1327–1334. [CrossRef]
37. Li, D.; Yu, H.; Pu, Z.; Lai, X.; Sun, C.; Wu, H.; Zhang, X. Flexible Microfluidics for Wearable Electronics. *Flex. Wearable Electron. Smart Cloth.* **2020**, 213–235. [CrossRef]
38. Safavieh, R.; Zhou, G.Z.; Juncker, D. Microfluidics made of yarns and knots: From fundamental properties to simple networks and operations. *Lab Chip* **2011**, *11*, 2618–2624. [CrossRef]
39. Nilghaz, A.; Wicaksono, D.H.B.; Gustiono, D.; Majid, F.A.A.; Supriyanto, E.; Kadir, M.R.A. Flexible microfluidic cloth-based analytical devices using a low-cost wax patterning technique. *Lab Chip* **2012**, *12*, 209–218. [CrossRef] [PubMed]
40. Ballerini, D.R.; Li, X.; Shen, W. Patterned paper and alternative materials as substrates for low-cost microfluidic diagnostics. *Microfluid. Nanofluidics* **2012**, *13*, 769–787. [CrossRef]
41. Nilghaz, A.; Ballerini, D.R.; Shen, W. Exploration of microfluidic devices based on multi-filament threads and textiles: A review. *Biomicrofluidics* **2013**, *7*, 1–16. [CrossRef]
42. Mabey, D.; Peeling, R.W.; Ustianowski, A.; Perkins, M.D. Diagnostics for the developing world. *Nat. Rev. Microbiol.* **2004**, *2*, 231–240. [CrossRef]
43. Akyazi, T.; Basabe-Desmonts, L.; Benito-Lopez, F. Review on microfluidic paper-based analytical devices towards commercialisation. *Anal. Chim. Acta* **2018**, *1001*, 1–17. [CrossRef] [PubMed]
44. Chu, T.; Chu, J.; Gao, B.; He, B. Modern evolution of paper-based analytical devices for wearable use: From disorder to order. *Analyst* **2020**, *145*, 5388–5399. [CrossRef]
45. Herman, Y. Test Paper. US Patent US2129754, 13 September 1938.
46. Muller, R.H.; Clegg, D.L. Automatic Paper Chromatography. *Anal. Chem.* **1949**, *21*, 1123–1125. [CrossRef]
47. Fernández-Sánchez, C.; McNeil, C.J.; Rawson, K.; Nilsson, O. Disposable noncompetitive immunosensor for free and total prostate-specific antigen based on capacitance measurement. *Anal. Chem.* **2004**, *76*, 5649–5656. [CrossRef] [PubMed]
48. Martinez, A.W.; Phillips, S.T.; Butte, M.J.; Whitesides, G.M. Patterned Paper as a Platform for Inexpensive, Low-Volume, Portable Bioassays. *Angew. Chem.* **2007**, *119*, 1340–1342. [CrossRef]
49. Salentijn, G.I.J.; Grajewski, M.; Verpoorte, E. Reinventing (Bio)chemical Analysis with Paper. *Anal. Chem.* **2018**, *90*, 13815–13825. [CrossRef]
50. Yáñez-Sedeño, P.; Campuzano, S.; Pingarrón, J.M. Screen-printed electrodes: Promising paper and wearable transducers for (bio)sensing. *Biosensors* **2020**, *10*, 76. [CrossRef]
51. Rose, D.P.; Ratterman, M.E.; Griffin, D.K.; Hou, L.; Kelley-Loughnane, N.; Naik, R.R.; Hagen, J.A.; Papautsky, I.; Heikenfeld, J.C. Adhesive RFID sensor patch for monitoring of sweat electrolytes. *IEEE Trans. Biomed. Eng.* **2015**, *62*, 1457–1465. [CrossRef] [PubMed]
52. Anastasova, S.; Crewther, B.; Bembnowicz, P.; Curto, V.; Ip, H.M.; Rosa, B.; Yang, G.Z. A wearable multisensing patch for continuous sweat monitoring. *Biosens. Bioelectron.* **2017**, *93*, 139–145. [CrossRef] [PubMed]
53. Yokus, M.A.; Saha, T.; Fang, J.; Dickey, M.D.; Velev, O.D.; Daniele, M.A. Towards Wearable Electrochemical Lactate Sensing using Osmotic-Capillary Microfluidic Pumping. *Proc. IEEE Sensors* **2019**, 1–4. [CrossRef]
54. Shay, T.; Dickey, M.D.; Velev, O.D. Hydrogel-enabled osmotic pumping for microfluidics: Towards wearable human-device interfaces. *Lab Chip* **2017**, *17*, 710–716. [CrossRef] [PubMed]
55. Cao, Q.; Liang, B.; Tu, T.; Wei, J.; Fang, L.; Ye, X. Three-dimensional paper-based microfluidic electrochemical integrated devices (3D-PMED) for wearable electrochemical glucose detection. *RSC Adv.* **2019**, *9*, 5674–5681. [CrossRef]

56. Liang, B.; Cao, Q.; Mao, X.; Pan, W.; Tu, T.; Fang, L.; Ye, X. An Integrated Paper-Based Microfluidic Device for Real-Time Sweat Potassium Monitoring. *IEEE Sens. J.* **2021**, *21*, 9642–9648. [CrossRef]
57. Yokus, M.A.; Agcayazi, T.; Traenkle, M.; Bozkurt, A.; Daniele, M.A. Wearable Sweat Rate Sensors. *Proc. IEEE Sens.* **2020**, 2–5. [CrossRef]
58. Criscuolo, F.; Cantu, F.; Taurino, I.; Carrara, S.; De Micheli, G. A Wearable Electrochemical Sensing System for Non-Invasive Monitoring of Lithium Drug in Bipolar Disorder. *IEEE Sens. J.* **2021**, *21*, 9649–9656. [CrossRef]
59. Criscuolo, F.; Hanitra, I.N.; Aiassa, S.; Taurino, I.; Oliva, N.; Carrara, S.; De Micheli, G. Wearable multifunctional sweat-sensing system for efficient healthcare monitoring. *Sens. Actuators B Chem.* **2021**, *328*. [CrossRef]
60. Zhang, C.; Su, Y.; Liang, Y.; Lai, W. Microfluidic Cloth-Based Analytical Devices: Emerging Technologies and Applications. *Biosens. Bioelectron.* **2020**, *168*, 112391. [CrossRef]
61. Xiao, G.; He, J.; Qiao, Y.; Wang, F.; Xia, Q.; Wang, X.; Yu, L.; Lu, Z.; Li, C.-M. Facile and Low-Cost Fabrication of a Thread/Paper-Based Wearable System for Simultaneous Detection of Lactate and pH in Human Sweat. *Adv. Fiber Mater.* **2020**, *2*, 265–278. [CrossRef]
62. Xing, S.; Jiang, J.; Pan, T. Interfacial microfluidic transport on micropatterned superhydrophobic textile. *Lab Chip* **2013**, *13*, 1937–1947. [CrossRef] [PubMed]
63. Tan, W.; Powles, E.; Zhang, L.; Shen, W. Go with the capillary flow. Simple thread-based microfluidics. *Sens. Actuators B Chem.* **2021**, *334*, 129670. [CrossRef]
64. Bae, C.W.; Toi, P.T.; Kim, B.Y.; Lee, W.I.; Lee, H.B.; Hanif, A.; Lee, E.H.; Lee, N.E. Fully Stretchable Capillary Microfluidics-Integrated Nanoporous Gold Electrochemical Sensor for Wearable Continuous Glucose Monitoring. *ACS Appl. Mater. Interfaces* **2019**, *11*, 14567–14575. [CrossRef] [PubMed]
65. Ma, B.; Chi, J.; Xu, C.; Ni, Y.; Zhao, C.; Liu, H. Wearable capillary microfluidics for continuous perspiration sensing. *Talanta* **2020**, *212*, 120786. [CrossRef] [PubMed]
66. Pirovano, P.; Dorrian, M.; Shinde, A.; Donohoe, A.; Brady, A.J.; Moyna, N.M.; Wallace, G.; Diamond, D.; McCaul, M. A wearable sensor for the detection of sodium and potassium in human sweat during exercise. *Talanta* **2020**, *219*, 121145. [CrossRef] [PubMed]
67. Glennon, T.; O'Quigley, C.; McCaul, M.; Matzeu, G.; Beirne, S.; Wallace, G.G.; Stroiescu, F.; O'Mahoney, N.; White, P.; Diamond, D. 'SWEATCH': A Wearable Platform for Harvesting and Analysing Sweat Sodium Content. *Electroanalysis* **2016**, *28*, 1283–1289. [CrossRef]

biosensors

MDPI

Article

Carbon-Coated Superparamagnetic Nanoflowers for Biosensors Based on Lateral Flow Immunoassays

Amanda Moyano [1], Esther Serrano-Pertierra [1], María Salvador [2], José Carlos Martínez-García [2], Yolanda Piñeiro [3], Susana Yañez-Vilar [3], Manuel Gónzalez-Gómez [3], José Rivas [3], Montserrat Rivas [2] and M. Carmen Blanco-López [1,*]

[1] Department of Physical and Analytical Chemistry & Institute of Biotechnology of Asturias, University of Oviedo, c/Julián Clavería 8, 33006 Oviedo, Spain; moyanoamanda@uniovi.es (A.M.); serranoesther@uniovi.es (E.S.-P.)

[2] Department of Physics & IUTA, University of Oviedo, Campus de Viesques, 33204 Gijón, Spain; salvadormaria@uniovi.es (M.S.); jcmg@uniovi.es (J.C.M.-G.); rivas@uniovi.es (M.R.)

[3] Department of Applied Physics, University of Santiago de Compostela, Campus Vida, 15782 Santiago de Compostela, Spain; yolanda.fayoly@gmail.com (Y.P.); susana.yanez@usc.es (S.Y.-V.); manuelantonio.gonzalez@usc.es (M.G.-G.); jose.rivas@usc.es (J.R.)

* Correspondence: cblanco@uniovi.es

Received: 22 June 2020; Accepted: 18 July 2020; Published: 22 July 2020

Abstract: Superparamagnetic iron oxide nanoflowers coated by a black carbon layer (Fe_3O_4@C) were studied as labels in lateral flow immunoassays. They were synthesized by a one-pot solvothermal route, and they were characterized (size, morphology, chemical composition, and magnetic properties). They consist of several superparamagnetic cores embedded in a carbon coating holding carboxylic groups adequate for bioconjugation. Their multi-core structure is especially efficient for magnetic separation while keeping suitable magnetic properties and appropriate size for immunoassay reporters. Their functionality was tested with a model system based on the biotin–neutravidin interaction. For this, the nanoparticles were conjugated to neutravidin using the carbodiimide chemistry, and the lateral flow immunoassay was carried out with a biotin test line. Quantification was achieved with both an inductive magnetic sensor and a reflectance reader. In order to further investigate the quantifying capacity of the Fe_3O_4@C nanoflowers, the magnetic lateral flow immunoassay was tested as a detection system for extracellular vesicles (EVs), a novel source of biomarkers with interest for liquid biopsy. A clear correlation between the extracellular vesicle concentration and the signal proved the potential of the nanoflowers as quantifying labels. The limit of detection in a rapid test for EVs was lower than the values reported before for other magnetic nanoparticle labels in the working range 0–3 × 10^7 EVs/μL. The method showed a reproducibility (RSD) of 3% (n = 3). The lateral flow immunoassay (LFIA) rapid test developed in this work yielded to satisfactory results for EVs quantification by using a precipitation kit and also directly in plasma samples. Besides, these Fe_3O_4@C nanoparticles are easy to concentrate by means of a magnet, and this feature makes them promising candidates to further reduce the limit of detection.

Keywords: superparamagnetic iron oxide nanoflowers; lateral flow immunoassays; biosensor; extracellular vesicles; exosomes

1. Introduction

The small size and the unique properties of magnetic nanoparticles have aroused considerable interest in the field of nanomedicine. In fact, they are currently being studied for biomedical applications such as hyperthermia, drug delivery, biosensors, and imaging [1,2]. Superparamagnetic nanoparticles (SPM NPs) are especially attractive due to their large saturation magnetization and initial magnetic permeability [3]. One of the most interesting features of SPM NPs is the possibility to be attracted by magnetic field gradient and then redispersed when the field is removed. This enables remote manipulation at immunoseparations for analytical determinations and other customized applications.

Magnetite and maghemite are the preferred crystal structures of iron oxide superparamagnetic nanoparticles due to their good magnetic response, biocompatibility, facile synthesis, and low-cost production. In addition, they can be modified with different surface coatings for electrosteric stabilization to prevent their uncontrolled agglomeration in solution [4]. Core/shell superparamagnetic nanoparticles are considered as a superparamagnetic core coated with other materials such as polymers, gold, silver, fatty acids, or surfactants, which increase colloidal stability and biocompatibility, preventing the oxidation of the iron oxide core and enhancing chemical versatility by linking functional groups [5]. A chemical versatile shell is desirable for the conjugation of bioreceptors such as antibodies, nucleotides, or peptides, which are used for molecular recognition of specific targets at analytical tests and devices [6].

Carbon coatings have several advantages as capping agents, since they are chemically stable, cheap, and light and allow an easy encapsulation of iron oxide cores [7–9]. Additionally, carbon films include carboxylic groups enabling the bioconjugation of nanoparticles with receptors for molecular recognition [10,11]. Carbon nanoparticles and other carbon nanomaterials have been reported as novel labels in lateral flow due to their advantages such as its strong color and its high contrast against the white background of nitrocellulose membrane [12,13]. Moreover, the limit of detection reached for these labels have been demonstrated to be better than those obtained with gold nanoparticles [14,15]. Therefore, the carbon coating can develop a strong visual signal as other carbon materials, and this could be advantageous when the nanoparticles are used in biosensors based on immunoassays.

In recent years, biosensors based on superparamagnetic nanoparticles have received considerable attention due to magnetic nanoparticles properties mentioned above, because they enhance the analytical figures of merit of biosensors such as electrochemical, optical, and piezoelectric sensors [16]. The magnetic nanoparticles are generally used to attract the analyte towards the detection surface by an external magnetic field. Sometimes, they are even integrated into the sensor transducer. Numerous biosensors based on superparamagnetic nanoparticles have been reported by using electrochemical [17–19], optical [20,21], piezoelectric [22,23], and magnetic [24–27] transducers.

Lateral flow immunoassay (LFIA) is a rapid test that meets the requirements for point of care (POC) testing. It is simple and provides results within a short time without the necessity of centralized laboratory at low cost [28]. The tests consist of different cellulosic elements (sample pad, conjugate pad, nitrocellulose membrane, and adsorption pad), which are assembled on a plastic backing to get robustness. Biorecognition elements are immobilized on the nitrocellulose membrane displaying lines (generally control and test lines), which become active upon flow of the liquid sample that contains the analyte of interest and nanoparticles. Generally, this kind of test relies on a visual signal, providing a yes/no response by means of colored nanoparticles such as gold or latex [29]. Magnetic nanoparticles are key players for the development of quantitative LFIA in combination with external magnetic transductors [30]. Additionally, magnetic nanoparticles have been used as colorimetric labels detected by naked-eye and optical readers due to their dark brown color easily distinguishable on white nitrocellulose membranes [31–33]. To date, several LFIA based on magnetic nanoparticles have been reported to detect and quantify human immunodeficiency virus (HIV) [34], allergen parvalbumin [35], and *Bacillus anthracis* spores [36,37] by means of a magnetic assays readers (MAR) system. The magnetic measurements require the excitation of the magnetic nanoparticles used as labels in the immunoassay by an oscillating magnetic field in order to quantify the magnetization of the nanoparticles. Moreover, magnetic nanoparticles can be used in order to purify and enrich analytes, enhancing the limits of detections and selectivity of conventional LFIA [38–41].

Our research group developed a novel magnetic sensor coupled to LFIA to quantify superparamagnetic nanoparticles immobilized at test and control lines. We used it to determine prostate specific antigen concentrations in the clinical range of interest by using a sandwich format [42] and histamine in wine following a competitive immunoassay [43]. For those studies, we used

10–12 nm size magnetic nanoparticles, which are stable in solution due to their electrostatic repulsion. The sensor is simpler compared with others, because the external field is not required. However, the superparamagnetism of the nanoparticles should be carefully optimized in order to produce the desired increase of impedance on a radio frequency (RF) current-carrying copper conductor [44,45]. This impedance change is directly proportional to the number of magnetic nanoparticles. Thus, this device allows an indirect detection of analyte of interest thanks to magnetic reporters.

The aim of this work was to study the use of iron oxide nanoflowers coated by a black carbon layer (Fe_3O_4@C) as labels for LFIA. The nanoflower structure consists of several superparamagnetic cores embedded with an external carbon surface. This core-shell structure was designed aiming its use at both immunoseparation and detection; multiple cores are suitable for immunoisolation, and the carbon coating provides a strong optical signal, keeping the magnetic properties unaltered.

In order to test their applicability to LFIA, we firstly studied these Fe_3O_4@C nanoparticles with a model affinity molecular recognition system. With this aim, we bioconjugated the particles to neutravidin and tested them against biotin printed across the membrane. Then, the Fe_3O_4@C nanoparticles were applied to an immunoassay for extracellular vesicles (EVs). EVs are nanovesicles produced by all cells via endocytosis processes. They carry proteins and nucleic acids from the original cell and therefore can be used to get molecular information about their parent cells. They are attracting a lot of attention in recent years as a source of non-invasive biomarkers for liquid biopsy. The bottleneck at EV research that is limiting industrial and clinical translations is the current isolation from biological fluids and the development of simple quantification methods. Our research group developed LFIA for EVs based on tetraspanin recognition [46,47]. In this work, as proof of concept, we tested the potential of the Fe_3O_4@C nanoparticles to EVs separation and quantification.

2. Materials and Methods

2.1. Chemicals and Instruments

Ferrocene ($Fe(C_5H_5)_2$, 98%), hydrogen peroxide (H_2O_2, 30%), acetone (C_3H_6O, 99.9%), bovine serum albumin (BSA), 1-ethyl-3-[3dimethylaminopropyl]-carbodiimide hydrochloride (EDC), and 2-(*N*-morpholino)ethanesulfonic acid (MES) were of analytical reagent grade purchased from Aldrich (Madrid, Spain) and were used without further purification. NeutrAvidin protein was obtained from Thermo Fischer Scientific (Waltham, MA, USA).

For biotin–neutravidin tests, glass fiber membrane (GFCP001000) used as sample pad and backing cards (HF000MC100) were purchased from Millipore (Darmstadt, Germany). Other materials used were nitrocellulose membranes (UniSart CN95, Sartorius, Spain) and absorbent pads (Whatman, Madrid, Spain). The sample buffer consisted of 10 mM phosphate buffer (PB) pH 7.4 with 0.5% Tween-20 and 1% BSA.

For EVs tests, nitrocellulose membranes (HF07504XSS) were purchased from Millipore (Germany). Other materials used for the preparation of strips were similar to the biotin-neutravidin test. Based on previous results, the sample buffer consisted of 10 mM HEPES pH 7.4 with 0.5% Tween-20 and 1% BSA. HEPES was purchased from Fisher Scientific (Madrid, Spain). Anti-tetraspanin antibodies anti-CD9 and anti-CD63 were provided by Immunostep (Salamanca, Spain). Anti-mouse IgG was purchased from Sigma-Aldrich (Spain). Lyophilized commercial exosomes purified from plasma (HBM-PEP) of healthy donors were purchased from HansaBioMed (Tallinn, Estonia).

In order to dispense the control and the detection lines, an IsoFlow reagent dispensing system (Imagene Technology, Lebanon, NH, USA) was used with a dispense rate of 0.100 µL/mm. A guillotine Fellowes Gamma (Madrid, Spain) was used to cut the strips. For quantification at the test line by reflectance measurements, a portable strip reader ESE Quant LR3 lateral flow system (Qiagen Inc., GmbH, Hilden, Germany) was used.

2.2. *Synthesis and Characterization of Carbon-Coated Nanoflowers*

2.2.1. Synthesis

The synthesis was carried out following a one-pot solvothermal method as previously reported by Wang [48]. Typically, ferrocene (m = 0.3 g) was dissolved in acetone (V = 25 mL) under vigorous magnetic stirring for 30 min; then, hydrogen peroxide (V = 1.5 mL) was slowly added into the above mixture solution and vigorously stirred for another 30 min. This precursor solution was transferred to the Teflon-lined stainless-steel autoclave (Parr Instrument, Illinois, IL, USA) with a total volume of 45.0 mL and heated to 210 °C for 96 h. Finally, the autoclave was cooled to room temperature, the reaction products were magnetically collected, and the supernatant was discarded. The precipitates were washed with acetone four times, and again, the products were magnetically separated to eliminate the acetone and were repeatedly washed with water.

2.2.2. Characterization

Hysteresis loops were recorded with a vibrating sample magnetometer (VSM, DMS, Lowell, MA, USA) at room temperature and under external magnetic fields from −10,000 to 10,000 Oe.

Room temperature X-ray diffraction (XRD) patterns of powder samples were obtained with a Philips PW1710 diffractometer (Panalytical, Callo End, UK) with a Cu Kα radiation (λ = 1.54186 Å) between 10° and 80° with steps of 0.02° and 10 s/step.

The composition was analyzed with a TGA Perkin Elmer model 7 (Perkin, Waltham, MA, USA).

Fourier transform infrared (FTIR) spectra were performed in a Thermo Nicolet Nexus spectrometer (Thermo Fisher Scientific, Madrid, Spain) using the attenuated total reflectance (ATR) method from 4000 to 400 cm^{-1}.

Scanning electron microscopy (SEM) with a Zeiss FE-SEM ULTRA Plus (5 kV) microscope (Zeiss, Oberkochen, Germany) and transmission electron microscopy (TEM) with a JEOL JEM-1011 microscope (100 kV) were employed to study the morphology.

2.3. *Bioconjugation of Superparamagnetic Iron Oxide Nanoflowers Coated by Fe_3O_4@C*

2.3.1. Functionalization

Fe_3O_4@C with carboxyl functional groups were functionalized using neutravidin to test their performance as label in immunoassays through the neutravidin–biotin interaction. Firstly, 100 µL of nanoflowers were mixed with 100 µL of neutravidin (different concentrations were studied) and 20 µL of EDC (1 mg/mL in MES 1 mM, pH 6.00) under continuous sonication for one hour. Then, 20 µL of EDC were added one hour and two hours after under sonication. After the last addition of EDC, the mixture was sonicated for 10 min.

The protocol to cover nanoflowers with neutravidin was adapted to bioconjugate the nanoflowers to antitetraspanin antibodies against EVs. Firstly, 50 µL of nanoflowers were mixed with 50 µL of anti-CD63 (1 mg/mL) and 20 µL of EDC (1 mg/mL in MES 1 mM, pH 6.00) under continuous sonication for one hour. Then, 20 µL of EDC were added one hour and two hours after under sonication. After the last addition of EDC, the mixture was sonicated for 10 min.

2.3.2. Characterization of Nanoparticles Conjugates by Dynamic Light Scattering

A Zetasizer Nano ZS ZEN3600 (Malvern Instruments, Malvern, UK) equipped with a solid-state He–Ne laser (λ = 633 nm) was used to measure size distribution and ζ-potential. In order to monitor the conjugation process, 30 measurements of the backscattered (173°) intensity were carried out at 25 °C and averaged. For data analysis, Zetasizer software version 7.03 was used.

2.4. Enrichment and Quantification of EV from Real Samples

Extracellular vesicles derived from human plasma samples were isolated using ExoQuick™ precipitation reagent (System Biosciences, Palo Alto, CA, USA), according to the manufacturer's instructions. Freshly isolated EVs were analyzed using a NanoSight LM10 instrument (Malvern, Worcestershire, UK) and NTA 3.1 software at Nanovex Biotechnologies S.L (Asturias, Spain).

2.5. Lateral Flow Assays

2.5.1. Preparation of the Strips

The immunoassay was based on a dipstick format. The strips consist of four parts: sample pad, nitrocellulose membrane, absorbent pad, and backing plastic card. The first step was to incorporate the nitrocellulose membrane into the backing plastic card to get robustness. Then, for biotin–neutravidin affinity test, a biotin-BSA test line was immobilized on the membrane with a concentration of 1 mg/mL. An IsoFlow dispenser at a rate of 0.100 µL/mm was employed.

For EVs tests, two lines of antibodies were immobilized across the nitrocellulose strip: (i) the test line gave us the result of the analysis following a sandwich format and (ii) the control line was used to validate the strip indicating that the liquid sample flowed adequately along the strip. Both lines were applied by the IsoFlow dispenser at a rate of 0.100 µL/mm with 1 mg/mL concentration of anti-CD9 and anti-IgG for test line and control line, respectively.

The nitrocellulose membrane after dispensing was kept for 20 min at 37 °C. Finally, the absorbent pad and the sample pad were stuck onto the backing card overlapping them 2 mm. Finally, individual strips of 5 mm were cut. For storage, strips were kept at room temperature and preserved with desiccant bags to avoid moisture.

2.5.2. Magnetic Quantification

Quantification of the test line in the LFIAs was provided by a Scanning MagnetoInductive Sensor (SMISensor) specifically designed in-house for this task. The sensing head consists of a double copper line printed on a rigid insulating substrate across which an alternating current is continuously flowing. A precision impedance analyzer (Agilent 4294A, Agilent Technologies, Madrid, Spain) was used to monitor the magnitude and the phase of the sensing head impedance. For this purpose, 16048G test leads and a 500 mV/40 MHz excitation voltage were used. The test lines on the strips were scanned laterally over the sensing head by a micro-positioner producing an increase of the impedance of the circuit due to the large magnetic permeability of the superparamagnetic particles present on it. This variation proved to be directly proportional to number of nanoparticles at the test line in previous studies. To account for all the particles in the test line, the signal was integrated across the position. For further information, please, see Supplementary Materials.

2.5.3. Optical Measurements

In order to quantify the color intensity of the test line by reflectance measurements, a portable strip reader ESE-Quant LR3 lateral flow system (Qiagen Inc., Hilden, Germany) was used.

2.5.4. Characterization of the Strip by SEM

The morphology of the strips was characterized by SEM in a Zeiss FE-SEM ULTRA Plus (5 kV) microscope (Zeiss, Oberkochen, Germany).

3. Results and Discussion

3.1. Characterization of the Carbon-Coated Nanoflowers before Bioconjugation

After synthesis, carbon-coated nanoflowers were characterized using various analytical techniques: VSM, X-ray diffraction, ζ potential, thermogravimetric analysis (TGA), FTIR, SEM, and TEM (Figure 1).

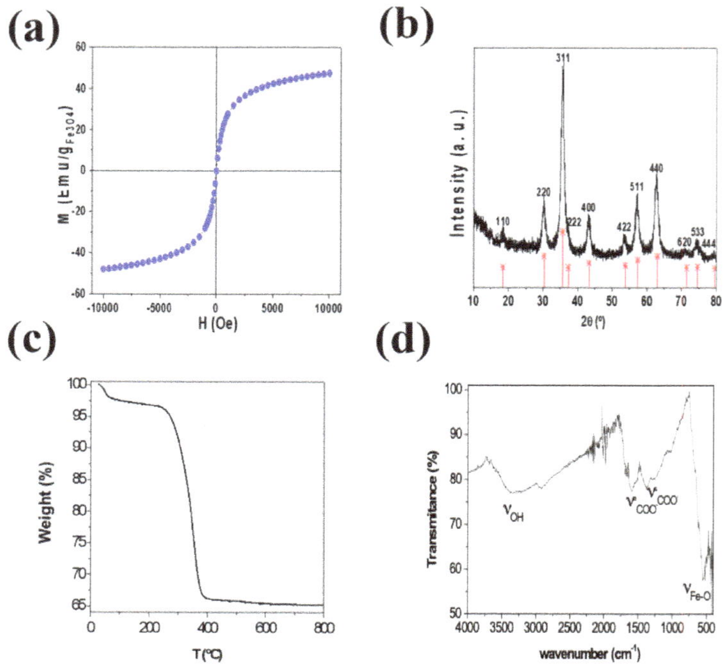

Figure 1. (**a**) Hysteresis loops of core/shell structure of magnetite/carbon performed at room temperature with a vibrating sample magnetometer (VSM) (−10, +10) KOe. (**b**) X-ray diffraction (XRD) pattern of the core/shell structure of magnetite/carbon colloidal nanoparticles compared to the XRD pattern of magnetite from JCPDS 19-0629 data base (**c**) thermogravimetric analysis (TGA) curves of the core/shell structure of magnetite/carbon colloidal nanoparticles. (**d**) FT-IR.

Magnetic characterization of the dried samples performed with a VSM at room temperature showed a magnetization normalized to the content of magnetite with a saturation at 10 KOe around 50 emu. This indicated superparamagnetic behavior (Figure 1a), in concordance with the dominant surface effects in small nanoparticles, for which the dead magnetic layer significantly decreased the magnetization. Moreover, negligible coercivity and absent remanence were observed. This can be ascribed to the superparamagnetic behavior of small magnetite nanoparticles.

Figure 1b shows X-ray diffraction spectra of the carbon coated multicore NP shown together with the theoretical diffraction peaks of magnetite (JCPDS card No. 19-0629) [49]. It can be seen that both the location and the relative intensity of m-Fe_3O_4@C nanoparticles coincided with the main theoretical 111, 220, 311, 400, 422, 511, 440 magnetite reflections (red lines in Figure 1b). This confirmed that magnetite was the crystalline phase of iron oxide present in the sample. To obtain the crystallite size, Scherrer formula [50] was applied to main reflection peak 311, providing an average Dhkl = 16.3 nm and σ = 2.8 nm. Moreover, on low diffraction angles, a broad band corresponding to the amorphous carbon coating shell could be seen.

The surface charge of Fe_3O_4@C nanoparticles was negative with a ζ potential value of -32 mV and a rate of magnetite/total mass of 0.651% (Wmagnetite/Wsample) (Figure 1c). The ζ potential value confirms the stability of Fe_3O_4@C nanoparticles in water suspension.

Figure 1d shows the FT-IR spectra of Fe_3O_4@C nanoparticles. The peak observed around 550 cm^{-1} was characteristic of Fe-O vibrations [51]. Additionally, a large band around 3400 cm^{-1} was also observed due to the -OH groups adsorbed on the nanoparticle surface. Besides, two peaks appeared around 1595 and 1384^{-1}, which corresponded to the asymmetric and the symmetric stretching vibrations of COO- groups.

The morphology of the Fe_3O_4@C nanoparticles was characterized by scanning electron microscopy (SEM) and by transmission electron microscopy (TEM), as shown in Figure 2a,b respectively. It can be seen (Figure 2a) that the obtained particles had a nearly spherical shape and uniform size with a regular core@shell structure where the carbon coating homogeneously encapsulated dozens of magnetite nanocrystals with a size 10–20 nm (Figure 2b). From TEM images, distribution of sizes was analyzed showing a mono-modal histogram with slight polydispersity, with an average diameter of 129 nm ± 19 nm (Figure 2c).

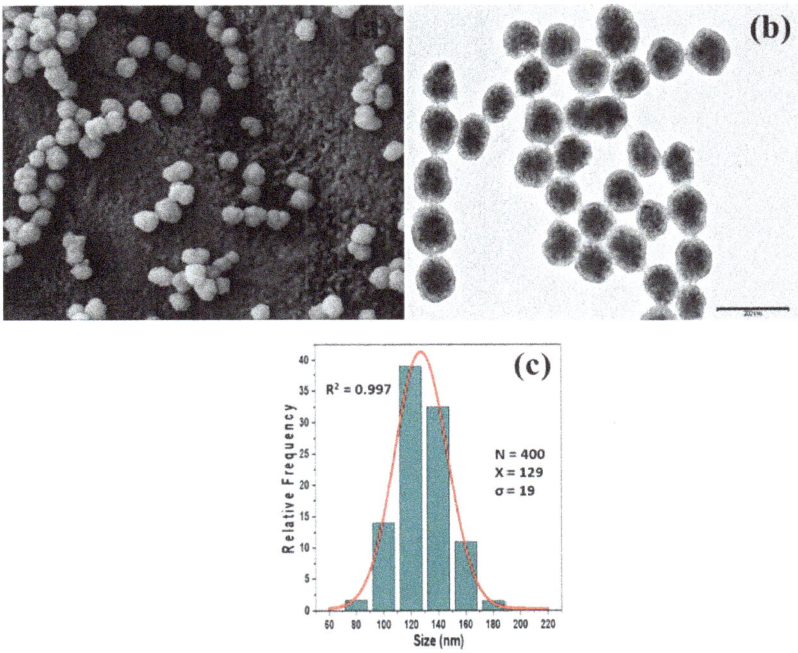

Figure 2. (**a**) SEM and (**b**) TEM images of the core/shell structure of magnetite/carbon colloidal NPs. (**c**) Analysis of size distribution obtained from (**b**).

3.2. *Study of Neutravidin Concentration during the Bioconjugation Process*

Increasing neutravidin concentrations were used to coat the nanoflowers: 0.03, 0.05, 0.07, 0.1, and 0.3 mg/mL. Phosphate buffer was used as diluent to prepare the neutravidin standards. Dynamic light scattering (DLS) measurements were carried out to compare nanoparticles hydrodynamic size before and after conjugation reaction. The hydrodynamic diameter of nanoparticles before conjugation was 178 nm (polydispersity index 0.058). The Figure 3a shows the values for Z-average of the hydrodynamic sizes for the different concentration of neutravidin used during the conjugation process. The hydrodynamic diameter of nanoparticles was higher when the concentration of neutravidin increased (Figure 3b). The bioconjugation process was confirmed by the increase of the hydrodynamic diameter of nanoparticles before and after conjugation. The mean size of neutravidin is around 2–3 nm (60 kDa). Therefore, considering the neutravidin size, it can be confirmed that this protein was bound to nanoparticles surface though the carboxyl functional groups available. The Z-average of the hydrodynamic sizes increased directly with the concentration of neutravidin. This could be because more than one molecule of neutravidin can be bound to a nanoparticle through their multiple functional groups on their surface.

(a)

Sample	Z-average	PDI
Before conjugation	177.8	0.058
Conjugated with 0.03 mg/ml of neutravidin	183.2	0.045
Conjugated with 0.05 mg/ml of neutravidin	207.7	0.039
Conjugated with 0.07 mg/ml of neutravidin	237.1	0.176
Conjugated with 0.1 mg/ml of neutravidin	210.5	0.133
Conjugated with 0.3 mg/ml of neutravidin	223.3	0.130

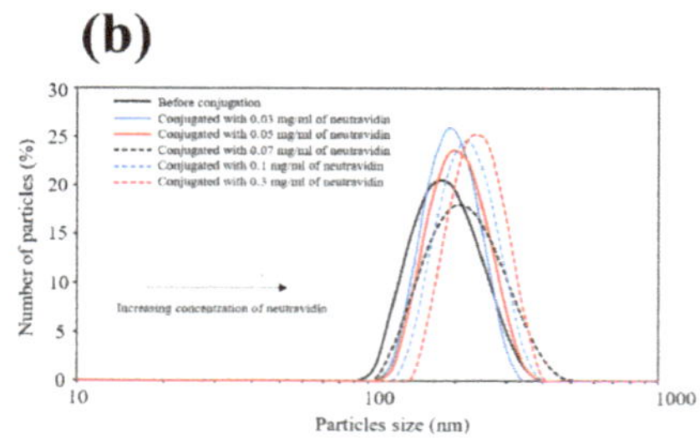

Figure 3. (**a**) Results obtained by dynamic light scattering (DLS) (Z-average and polydispersity index) for study performed with neutravidin. (**b**) Hydrodynamic size distribution profiles of Fe_3O_4@C nanoparticles before (solid black line) and after conjugation with neutravidin concentrations of 0.3 mg/mL (solid red line), 0.1 mg/mL (solid blue line), 0.07 mg/mL (dashed black line), 0.05 mg/mL (dashed red line), and 0.03 mg/mL (dashed blue line) of neutravidin.

3.3. *Lateral Flow Assay Procedure*

A biotin–neutravidin affinity test (Figura 4a) and a lateral flow immunoassay for extracellular vesicles (Figure 4b) were developed.

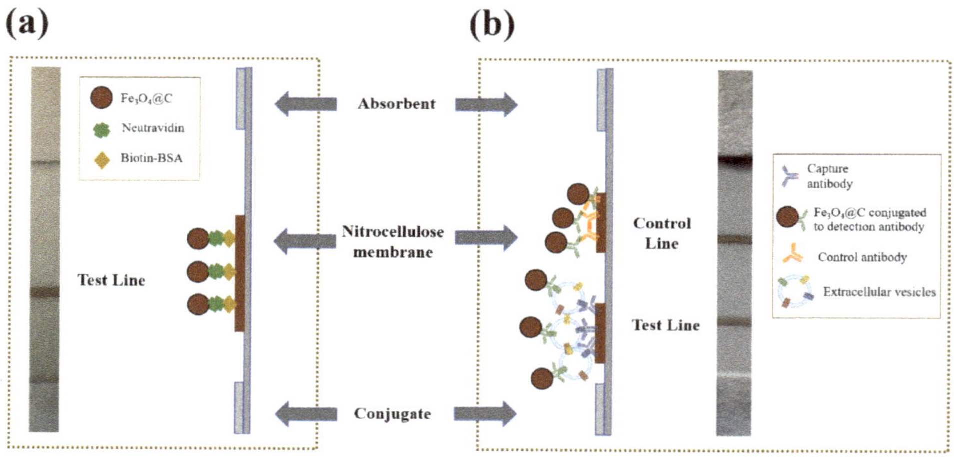

Figure 4. (**a**) Photography (front view) and schematic illustration of the biotin–neutravidin affinity test (side view). (**b**) Photography (front view) and schematic representation of the lateral flow immunoassay for extracellular vesicles (side view).

3.3.1. Biotin-Neutravidin Affinity Test

The biotin-streptavidin/neutravidin/avidin system has been widely used in immunoassays because it is a powerful non-covalent interaction with high specificity and strong affinity [52,53]. Streptavidin, neutravidin, and avidin are molecules that contain four binding sites with an extraordinarily high affinity for biotin. This system is very attractive for biosensing because of its amplification capability [54–56]. In addition, biotin can be easily covalently bonded to proteins such as antibodies, nucleotides, and enzymes, enabling a strong binding between biotinylated proteins and streptavidin, neutravidin, or avidin. In this case, the biotin–neutravidin system was used as model system for affinity interactions as a first step to study the feasibility to bioconjugate these nanoparticles.

In order to test the suitability of the NP for LFIA, 20 µL of suspensions with different concentration of neutravidin and 80 µL of running buffer were transferred into a microtube. The sample pad was introduced into the mixture, and the buffer started to flow through the strip by capillary action. Only the nanoflowers coated with neutravidin were retained at the biotin-BSA line, as shown in Figure 4. A simplified schematic representation of biotin–neutravidin interaction in LFIA is shown in Figure 4.

3.3.2. Magnetic and Optical Quantification for Biotin-Neutravidin Test

The test line, once dried, was analyzed by reflectance and magnetic measurements. Figure 5a shows the results for both measurements. The optical and the magnetic signals increased with neutravidin concentration until saturation was reached. The saturation was produced due to the depletion of free biotin molecules at the test line, thus the nanoflowers coated with neutravidin were not retained. The optical and the magnetic methods showed a reproducibility (RSD) of 1% (n = 3) for both cases. Figure 5b shows a representative example of strips for an increase in neutravidin concentration. Both curves followed the same trend, achieving a slightly better fit for the magnetic one. The signal corresponding to nanoparticles coated with 0.01 mg/mL of neutravidin was difficult to distinguish from the blank at the inductive sensor. However, the optical signal was significantly different from the blank, as Figure 5b shows. For the blank, Fe_3O_4@C nanoparticles without neutravidin coating were assayed in order to test that bare nanoparticles were not attached to biotin-BSA test line. The Figure 5b confirms that there were not non-specific interactions between bare Fe_3O_4@C nanoparticles and biotin-BSA test line. A higher concentration of neutravidin (0.03 mg/mL) on the nanoparticles could be easily detected by both magnetic and optical instruments.

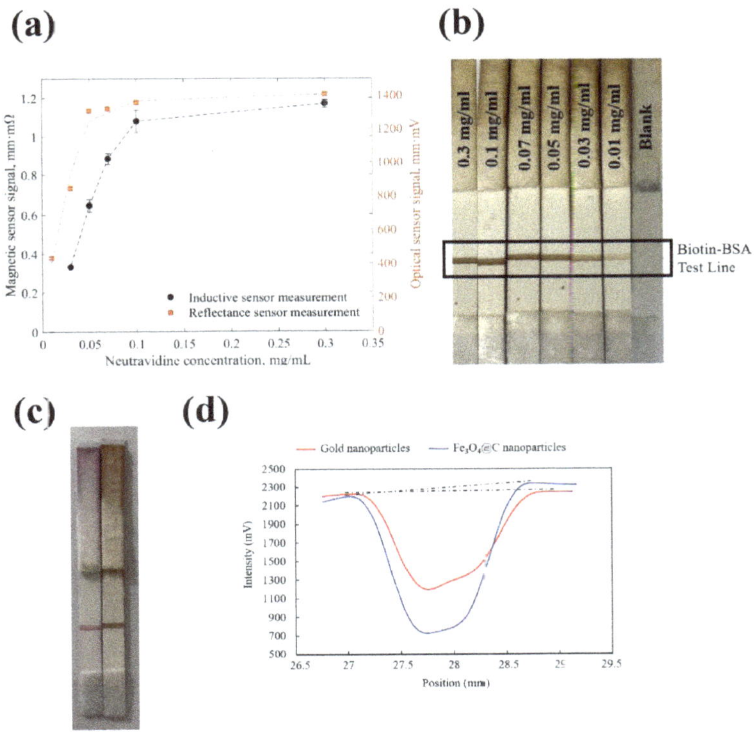

Figure 5. (**a**) Magnetic and optical signals as function of the concentration of neutravidin. (**b**) Representative example of results obtained in the strips for the different concentrations of neutravidin. (**c**) Comparison of different labels: gold nanoparticles (left) and Fe_3O_4@C nanoparticles (right). (**d**) Intensity profiles measured using ESEQuant reader for gold (red line) and Fe_3O_4@C (blue line) nanoparticles.

Colloidal gold was used as label at the same immunoassay (biotin-neutravidin interaction test) to compare the results with those of Fe_3O_4@C nanoparticles. Colloidal nanoparticles have been widely used as labels in LFIA, the 40 nm size being the most popular choice for this kind of assay due to their higher sensitivity [57]. Gold nanoparticles were conjugated to neutravidin by passive binding [58]. The optimal concentration of neutravidin to stabilize the colloidal gold was found through a titration assay as described in reference [58]. The protocol for the titration was executed as described elsewhere [58]. The results indicated that 0.4 mg/mL of neutravidin was the lower concentration needed for the conjugation with gold nanoparticles (data no shown). To perform the comparison, Fe_3O_4@C nanoparticles were conjugated to 0.3 mg/mL of neutravidin, because at this concentration, the optical signal has reached the saturation (Figure 5a), and therefore the intensity of this signal should be similar to that obtained by 0.4 mg/mL of neutravidin. Therefore, the concentrations chosen were comparable for both labels.

The test was carried out using the same protocol described above, and the optical reader was used to analyze the signals. Figure 5c shows that naked-eye signals for gold and Fe_3O_4@C nanoparticles at the test line apparently looked equivalent. However, when the test lines were quantified by optical reader, the results show that the nanoflowers displayed a significantly stronger optical signal (1405.9 mm·mV) compared with gold nanoparticles (937.3 mm·mV). Figure 5d shows the intensity profiles measured using ESEQuant reader for gold nanoparticles (red line) and Fe_3O_4@C nanoparticles (blue line). The intensity profiles were based on reflectance measurements. The ESEQuant reader scanned the test line by moving the light source over the strip. The reader was adjusted so that the light reflected from the strip was collected by a confocal detector and registered as intensity (mV). When the incident beam passed across the test line, the reflected light decreased compared with the light reflected from the membrane (base line). This was because the nanoparticles at the test line absorbed light, and this effect reduced light intensity. This was shown as a negative peak (Figure 5d). Hence, the recorded graph represents intensity (mV) versus position (mm). The result of reflectance measurements for gold and Fe_3O_4@C nanoparticles were obtained by integration of negative peak area (Figure 5d, dashed black lines). Absorption and reflection of the light were proportional to the overall intensity of the nanoparticles of the test line. Therefore, the results indicate that the density of Fe_3O_4@C nanoparticles was higher that the density of gold nanoparticles at the biotin-BSA test line for a similar concentration of neutravidin.

3.3.3. Characterization of the Strip by SEM

Morphological characterization of the strips was carried out by SEM in order to observe the distribution of the Fe_3O_4@C nanoparticles after using the strip in the detection device. Figure 6a shows an image of the porous structure of a non-colored part of the strip. Spheres with size between 3–5 μm embedded in a network of fibers were clearly observed in the region corresponding to the test line (Figure 6b). We tried several membranes with different pore sizes to let Fe_3O_4@C nanoparticles flow along membrane until the end of the strip. It could be observed that the pores of this cellulose network were large enough to allow the diffusion of the Fe_3O_4@C nanoparticles through the membrane (Figure 6b).

An energy-dispersive X-ray spectroscopy (EDS) analysis was also carried out in order to confirm that these spherical nanoparticles corresponded to Fe_3O_4@C nanoparticles agglomerates. The EDS spectrum showed the presence of Fe on the analyzed area (Figure 6c). With this analysis, we confirmed the presence of Fe_3O_4@C nanoparticles on the test line.

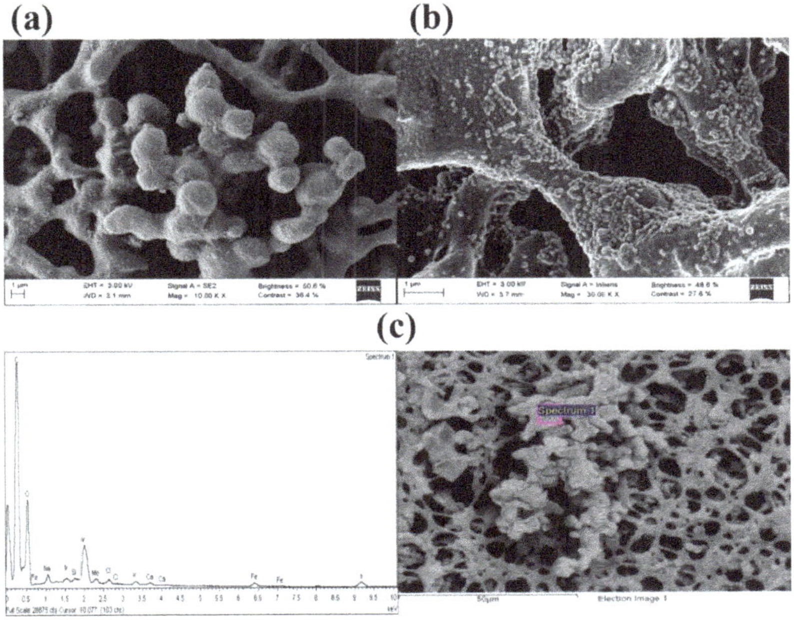

Figure 6. (**a**) SEM images of the cellulose membrane and (**b**) Fe_3O_4@C nanoparticle agglomerates attached to the cellulose network at the test line. The strips were examined after use in the detection device. (**c**) energy-dispersive X-ray spectroscopy (EDS) spectrum obtained from one particular location in a SEM image of the strips with presence of Fe_3O_4@C nanoparticles.

3.3.4. Applications of Fe_3O_4@C Nanoparticles for Detection of EVs by LFIA

Finally, as proof of concept, we developed a lateral flow immunoassay for EVs by using these nanoflowers as label. The immunoassay relies on sandwich format developed in our research group [46,47]. Anti-CD9 and anti-CD63 were used as capture and detection antibodies, respectively. Capture antibody was immobilized on the nitrocellulose membrane at the test line, and detection antibody was conjugated to the nanoflower surfaces. Figure 4b shows a schematic illustration of the immunoassay for EVs. Commercial standard exosomes were used to study these magnetic nanoparticles as labels and to evaluate the magnetic signal in this application.

In order to carry out the test, different concentrations of standard commercial exosomes (6.00×10^6, 1.50×10^7, 2.10×10^7, 3.00×10^7, 6.00×10^7, 1.05×10^8 EVs/μL) were added into a microtube that contained 10 μL nanoflowers coated with anti-CD63 and buffer until a final volume of 100 μL. HEPES was used to dilute the standard commercial exosomes

The strips were analyzed at both the magnetic sensor and the optical reader, but quantification was possible only with the optical reader. Previous results were based on the plain biotin–neutravidin interaction, but the assay used for the EVs detection was more complex; it involved a capture antibody of bulky EVs at the test line and Fe_3O_4@C bioconjugated antibody for detection. The optical signal corresponding to the highest concentration of EVs was 779 mm·mV (Figure 7). If we look at Figure 5a, the corresponding value at the magnetic sensor curve for that concentration would be below the quantification possibilities of this device. Further work is in progress to optimize the microelectronics for amplification.

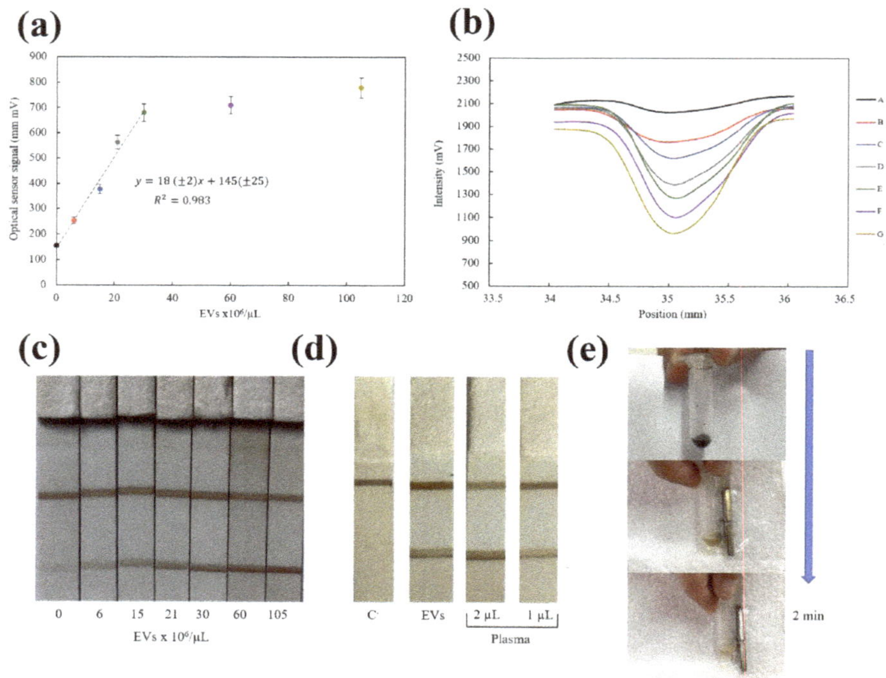

Figure 7. (**a**) Optical calibration curve for extracellular vesicles (EVs). (**b**) Intensity profiles measured using ESEQuant reader for different concentration of EVs: blank (A), 6.00 × 10^6 (B), 1.50 × 10^7 (C), 2.10 × 10^7 (D), 3.00 × 10^7 (E), 6.00 × 10^7 (F), 1.05 × 10^8 (G) EVs/µL (**c**) Representative example of results obtained in the strips with different concentrations of standard EVs. (**d**) Representative strips of lateral flow immunoassay (LFIA) tested with real samples: negative control (C-), EVs isolated from plasma sample using ExoQuick (EVs), and plasma. (**e**) Separation of Fe_3O_4@C nanoparticles using a conventional magnet.

Nevertheless, a direct relationship between the concentration of exosomes and the optical signal was found (Figure 7). Figure 7b shows the intensity profiles obtained by commercial optical reader for different concentration of EVs. Reflectance peak area increased proportionally with EVs concentration since the density of Fe_3O_4@C nanoparticles at the test line was higher. The limit of detection (LOD) achieved was 4 × 10^6 EVs/µL (calculated with the 3 S_B/m criterium), and the method showed a reproducibility (RSD) of 3% (n = 3). This LOD agrees with the values reported before with this type of tests [46,47] by using gold nanoparticles (8.5 × 10^5 and 4.5 × 10^6 EVs/µL), carbon black nanoparticles (9.2 × 10^6 EVs/µL) or other commercial magnetic nanoparticles coated with polyacrylic acid (1.0 × 10^7 EVs/µL). Figure 7c shows a representative example of the strips for standard EVs. Prompted by these results, this LFIA system was next tested with freshly isolated EV from human plasma and also directly with different volumes of plasma (Figure 7d). Quantification of isolated EVs using the optical calibration curve (Figure 7a) was compared to that obtained by NTA. Despite the different basis of these measuring principles, results were in the same range of concentration (3 × 10^9 EVs/µL determined in LFIA vs. 7 × 10^9 EVs/µL determined by NTA). Estimation of the number of EVs by NTA is based on the tracking of their Brownian displacements, whereas LFIA is a single step procedure, in which EVs are recognized by targeting their surface markers. Moreover, human plasma without previous treatment was used for the first time directly on the test strips. An excellent agreement was achieved for the number of EVs obtained in two different volumes using the calibration curve: 21.4 × 10^6 EVs/µL in 1 µL and 20.9 × 10^6 EVs/µL when a 2 µL plasma aliquot was tested. This is an indication of the absence of matrix effects. Thus, the magnetic LFIA developed in this work with Fe_3O_4@C nanoflowers was found suitable for analysis of complex samples such as plasma or serum with minimum sample preparation in a simple protocol.

In addition, another important strength of these particles is their rapid separation with a conventional magnet in 2 min (Figure 7e) while keeping their superparamagnetic behavior. Therefore, they show a great potential to easily enrich CD63+ EV for further functional or analytical assays.

4. Conclusions

Fe_3O_4@C nanoparticles were synthesized with the purpose of having the double function of separation and quantification at immunoassays. Their characteristics are summarized in Table 1. They exhibited superparamagnetic behavior, and their value of ζ potential confirms their stability in suspension.

Table 1. Characteristics of carbon-coated superparamagnetic oxide nanoflowers Fe_3O_4@C.

Carbon-Coated Superparamagnetic Oxide Nanoflowers (Fe_3O_4@C)	
Synthesis	One-pot solvothermal method
Composition (core/shell)	Magnetite cores Black carbon coating
Magnetite crystals size (d)	10–20 nm
Magnetization saturation	50 emu per g of magnetite
Mean size	129 nm
Mean hydrodynamic diameter	178 nm
Electrokinetic potential (ζ)	−32 mV
Rate magnetite/total mass	0.651%

These carbon coated nanoflowers were conjugated to neutravidin by using the carbodiimide chemistry and tested at lateral flow immunoassays with a biotin test line. The conjugations process was monitored by DLS and confirmed by LFIA based on biotin–neutravidin interaction line. Quantification was carried out by means of an optical reader and a magnetic sensor. With this basis, the nanoflowers were used to develop a lateral flow immunoassay for detection of EVs.

In summary, these Fe_3O_4@C magnetic particles proved to be effective for bioconjugation purposes due to the availability of carboxylic groups on their surface. Their carbon coating provided a strong optical signal when using them as labels in lateral flow assays. In fact, the LFIA developed in this work achieved lower LOD values than those obtained by means of other commercial magnetic nanoparticle labels for EV detection. Therefore, the Fe_3O_4@C magnetic particles employed in this study have a great potential of application as labels in EV enrichment and detection.

Supplementary Materials: The following are available online at http://www.mdpi.com/2079-6374/10/8/80/s1, Figure S1: (a) Photograph of the sensing planar coil. (b) Image of the device integrating the planar coil and the micropositioner. (c) 40 MHz Impedance variation measured in a scan of a sample of nanoparticles containing 32 µg of Fe_3O_4. (d) Impedance recording observed for the scan of an LFIA with 1 mg/mL of neutravidin in two consecutive scans in the sensor.

Author Contributions: Methodology, A.M., E.S.-P., S.Y.-V., and M.G.-G.; investigation, A.M., E.S.-P., M.S., J.C.M.-G., S.Y.-V., and M.G.-G.; writing—original draft preparation, A.M. and S.Y.-V.; writing—review and editing, E.S.-P., Y.P., J.R., M.R., and M.C.B.-L.; supervision, Y.P., J.R., M.R., and M.C.B.-L.; project administration, Y.P., J.R., M.R., and M.C.B.-L.; funding acquisition, Y.P., J.R., M.R., and M.C.B.-L. All authors have read and agreed to the published version of the manuscript.

Funding: This work was supported in part by Spanish Ministry of Economy and Competitiveness under projects MAT2017-84959-C2-1-R and the Principality of Asturias (Spain) under project IDI/2018/000185 and the Consellería de Educación Program for Development of a Strategic Grouping in Materials (AEMAT) at the University of Santiago de Compostela under Grant No. ED431E208/08, Xunta de Galicia. Amanda Moyano was supported by a "Severo Ochoa" fellowship (Consejería de Educación y Cultura del Gobierno del Principado de Asturias, grant BP17-152).

Conflicts of Interest: The authors declare that they have no conflict of interest.

References

1. Colombo, M.; Carregal-Romero, S.; Casula, M.F.; Gutierrez, L.; Morales, M.P.; Boehm, I.B.; Heverhagen, J.T.; Prosperi, D.; Parak, W.J. ChemInform abstract: Biological applications of magnetic nanoparticles. *Chem. Soc. Rev.* **2012**, *41*, 4306–4334. [CrossRef] [PubMed]
2. Zhang, L.; Dong, W.-F.; Sun, H.-B. Multifunctional superparamagnetic iron oxide nanoparticles: Design, synthesis and biomedical photonic applications. *Nanoscale* **2013**, *5*, 7664–7684. [CrossRef] [PubMed]
3. Knobel, M.; Nunes, W.C.; Socolovsky, L.M.; De Biasi, E.; Vargas, J.M.; Denardin, J.C. Superparamagnetism and other magnetic features in granular materials: A review on ideal and real systems. *J. Nanosci. Nanotechnol.* **2008**, *8*, 2836–2857. [CrossRef] [PubMed]
4. Wu, W.; Wu, Z.; Yu, T.; Jiang, C.; Kim, W.S. Recent progress on magnetic iron oxide nanoparticles: Synthesis, surface functional strategies and biomedical applications. *Sci. Technol. Adv. Mater.* **2015**, *16*. [CrossRef] [PubMed]
5. McNamara, K.; Tofail, S.A.M. Nanoparticles in biomedical applications. *Adv. Phys. X* **2017**, *2*, 54–88. [CrossRef]
6. Salvador, M.; Moyano, A.; Martínez-García, J.C.; Blanco-López, M.C.; Rivas, M. Synthesis of superparamagnetic iron oxide nanoparticles: SWOT analysis towards their conjugation to biomolecules for molecular recognition applications. *J. Nanosci. Nanotechnol.* **2019**, *19*, 4839–4856. [CrossRef]
7. Rafiee, E.; Khodayari, M. Starch as a green source for Fe_3O_4@carbon core-shell nanoparticles synthesis: A support for 12-tungstophosphoric acid, synthesis, characterization, and application as an efficient catalyst. *Res. Chem. Intermed.* **2016**, *42*, 3523–3536. [CrossRef]
8. Mendes, R.G.; Koch, B.; Bachmatiuk, A.; El-Gendy, A.A.; Krupskaya, Y.; Springer, A.; Klingeler, R.; Schmidt, O.; Büchner, B.; Sanchez, S.; et al. Synthesis and toxicity characterization of carbon coated iron oxide nanoparticles with highly defined size distributions. *Biochim. Biophys. Acta Gen. Subj.* **2014**, *1840*, 160–169. [CrossRef]
9. Prajapat, C.L.; Sharma, P.; Gonal, M.R.; Vatsa, R.K.; Singh, M.R.; Ravikumar, G. Synthesis and magnetic study of carbon coated iron oxide nanoparticles by laser ablation in solution. *AIP Conf. Proc.* **2016**, *1731*, 130011.
10. Taylor, A.; Krupskaya, Y.; Costa, S.; Oswald, S.; Kramer, K.; Füssel, S.; Klingeler, R.; Büchner, B.; Borowiak-Palen, E.; Wirth, M.P. Functionalization of carbon encapsulated iron nanoparticles. *J. Nanoparticle Res.* **2010**, *12*, 513–519. [CrossRef]
11. Bonanni, A.; Ambrosi, A.; Pumera, M. On oxygen-containing groups in chemically modified graphenes. *Chem. A Eur. J.* **2012**, *18*, 4541–4548. [CrossRef]
12. Zhang, X.; Zhao, F.; Sun, Y.; Mi, T.; Wang, L.; Li, Q.; Li, J.; Ma, W.; Liu, W.; Zuo, J.; et al. Development of a highly sensitive lateral flow immunoassay based on receptor-antibody-amorphous carbon nanoparticles to detect 22 β-lactams in milk. *Sens. Actuators B Chem.* **2020**, *321*, 128458. [CrossRef]
13. Guoa, J.; Chenb, S.; Jinhong Guoa, J.; Ma, X. Nanomaterial Labels in lateral flow immunoassays for point-of-care-testing. *J. Mater. Sci. Technol.* **2020**, *60*, 90–104. [CrossRef]
14. Zhang, X.; Yu, X.; Wen, K.; Li, C.; Mujtaba Mari, G.; Jiang, H.; Shi, W.; Shen, J.; Wang, Z. Multiplex lateral flow immunoassays based on amorphous carbon nanoparticles for detecting three fusarium mycotoxins in maize. *J. Agric. Food Chem.* **2017**, *65*, 8063–8071. [CrossRef]
15. Liu, B.; Wang, L.; Tong, B.; Zhang, Y.; Sheng, W.; Pan, M.; Wang, S. Development and comparison of immunochromatographic strips with three nanomaterial labels: Colloidal gold, nanogold-polyaniline-nanogold microspheres (GPGs) and colloidal carbon for visual detection of salbutamol. *Biosens. Bioelectron.* **2016**, *85*, 337–342. [CrossRef] [PubMed]
16. Rocha-Santos, T.A.P. Sensors and biosensors based on magnetic nanoparticles. *TrAC Trends Anal. Chem.* **2014**, *62*, 28–36. [CrossRef]
17. Hervás, M.; López, M.Á.; Escarpa, A. Simplified calibration and analysis on screen-printed disposable platforms for electrochemical magnetic bead-based inmunosensing of zearalenone in baby food samples. *Biosens. Bioelectron.* **2010**, *25*, 1755–1760. [CrossRef]
18. Yang, Z.; Zhang, C.; Zhang, J.; Bai, W. Potentiometric glucose biosensor based on core-shell Fe_3O_4-enzyme-polypyrrole nanoparticles. *Biosens. Bioelectron.* **2014**, *51*, 268–273. [CrossRef]
19. Yola, M.L.; Eren, T.; Atar, N. A novel and sensitive electrochemical DNA biosensor based on Fe@Au nanoparticles decorated graphene oxide. *Electrochim. Acta* **2014**, *125*, 38–47. [CrossRef]

20. Wang, J.; Sun, Y.; Wang, L.; Zhu, X.; Zhang, H.; Song, D. Surface plasmon resonance biosensor based on Fe_3O_4/Au nanocomposites. *Colloids Surf. B Biointerfaces* **2010**, *81*, 600–606. [CrossRef]
21. Wang, Y.; Dostalek, J.; Knoll, W. Magnetic nanoparticle-enhanced biosensor based on grating-coupled surface plasmon resonance. *Anal. Chem.* **2011**, *83*, 6202–6207. [CrossRef] [PubMed]
22. Li, D.; Wang, J.; Wang, R.; Li, Y.; Abi-Ghanem, D.; Berghman, L.; Hargis, B.; Lu, H. A nanobeads amplified QCM immunosensor for the detection of avian influenza virus H5N1. *Biosens. Bioelectron.* **2011**, *26*, 4146–4154. [CrossRef] [PubMed]
23. Gan, N.; Wang, L.; Li, T.; Sang, W.; Hu, F.; Cao, Y. A Novel signal-amplified immunoassay for myoglobin using magnetic core-shell Fe_3O_4@Au-multi walled carbon nanotubes composites as labels based on one piezoelectric sensor. *Integr. Ferroelectr.* **2013**, *144*, 29–40. [CrossRef]
24. Haun, J.B.; Yoon, T.J.; Lee, H.; Weissleder, R. Magnetic nanoparticle biosensors. *Wiley Interdiscip. Rev. Nanomed. Nanobiotechnol.* **2010**, *2*, 291–304. [CrossRef]
25. Srinivasan, B.; Li, Y.; Jing, Y.; Xing, C.; Slaton, J.; Wang, J.P. A three-layer competition-based giant magnetoresistive assay for direct quantification of endoglin from human urine. *Anal. Chem.* **2011**, *83*, 2996–3002. [CrossRef]
26. Zu, P.; Chan, C.C.; Koh, G.W.; Lew, W.S.; Jin, Y.; Liew, H.F.; Wong, W.C.; Dong, X. Enhancement of the sensitivity of magneto-optical fiber sensor by magnifying the birefringence of magnetic fluid film with Loyt-Sagnac interferometer. *Sens. Actuators B Chem.* **2014**, *191*, 19–23. [CrossRef]
27. Hathaway, H.J.; Butler, K.S.; Adolphi, N.L.; Lovato, D.M.; Belfon, R.; Fegan, D.; Monson, T.C.; Trujillo, J.E.; Tessier, T.E.; Bryant, H.C.; et al. Detection of breast cancer cells using targeted magnetic nanoparticles and ultra-sensitive magnetic field sensors. *Breast Cancer Res.* **2011**, *13*, 1–13. [CrossRef]
28. Sajid, M.; Kawde, A.N.; Daud, M. Designs, formats and applications of lateral flow assay: A literature review. *J. Saudi Chem. Soc.* **2015**, *19*, 689–705. [CrossRef]
29. Huang, X.; Aguilar, Z.P.; Xu, H.; Lai, W.; Xiong, Y. Membrane-based lateral flow immunochromatographic strip with nanoparticles as reporters for detection: A review. *Biosens. Bioelectron.* **2015**, *75*, 166–180. [CrossRef]
30. Moyano, A.; Serrano-pertierra, E.; Salvador, M.; Martínez-garcía, J.C.; Rivas, M.; Blanco-López, M.C. Magnetic lateral flow immunoassays. *Diagnostics* **2020**, *10*, 288. [CrossRef]
31. Panferov, V.G.; Safenkova, I.V.; Zherdev, A.V.; Dzantiev, B.B. Setting up the cut-off level of a sensitive barcode lateral flow assay with magnetic nanoparticles. *Talanta* **2017**, *164*, 69–76. [CrossRef] [PubMed]
32. Liu, C.; Jia, Q.; Yang, C.; Qiao, R.; Jing, L.; Wang, L.; Xu, C.; Gao, M. Lateral flow immunochromatographic assay for sensitive pesticide detection by using Fe_3O_4 nanoparticle aggregates as color reagents. *Anal. Chem.* **2011**, *83*, 6778–6784. [CrossRef] [PubMed]
33. Wu, J.; Dong, M.; Zhang, C.; Wang, Y.; Xie, M.; Chen, Y. Magnetic lateral flow strip for the detection of cocaine in urine by naked eyes and smart phone camera. *Sensors* **2017**, *17*, 1286. [CrossRef] [PubMed]
34. Workman, S.; Wells, S.K.; Pau, C.P.; Owen, S.M.; Dong, X.F.; LaBorde, R.; Granade, T.C. Rapid detection of HIV-1 p24 antigen using magnetic immuno-chromatography (MICT). *J. Virol. Methods* **2009**, *160*, 14–21. [CrossRef] [PubMed]
35. Zheng, C.; Wang, X.; Lu, Y.; Liu, Y. Rapid detection of fish major allergen parvalbumin using superparamagnetic nanoparticle-based lateral flow immunoassay. *Food Control* **2012**, *26*, 446–452. [CrossRef]
36. Wang, D.B.; Tian, B.; Zhang, Z.P.; Deng, J.Y.; Cui, Z.Q.; Yang, R.F.; Wang, X.Y.; Wei, H.P.; Zhang, X.E. Rapid detection of Bacillus anthracis spores using a super-paramagnetic lateral-flow immunological detection system. *Biosens. Bioelectron.* **2013**, *42*, 661–667. [CrossRef]
37. Wang, D.B.; Tian, B.; Zhang, Z.P.; Wang, X.Y.; Fleming, J.; Bi, L.J.; Yang, R.F.; Zhang, X.E. Detection of bacillus anthracis spores by super-paramagnetic lateral-flow immunoassays based on "Road Closure". *Biosens. Bioelectron.* **2015**, *67*, 608–614. [CrossRef]
38. Sharma, A.; Tok, A.I.Y.; Lee, C.; Ganapathy, R.; Alagappan, P.; Liedberg, B. Magnetic field assisted preconcentration of biomolecules for lateral flow assaying. *Sens. Actuators B Chem.* **2019**, *285*, 431–437. [CrossRef]
39. Razo, S.C.; Panferov, V.G.; Safenkova, I.V.; Varitsev, Y.A.; Zherdev, A.V.; Dzantiev, B.B. Double-enhanced lateral flow immunoassay for potato virus X based on a combination of magnetic and gold nanoparticles. *Anal. Chim. Acta* **2018**, *1007*, 50–60. [CrossRef]
40. Huang, Z.; Xiong, Z.; Chen, Y.; Hu, S.; Lai, W. Sensitive and matrix-tolerant lateral flow immunoassay based on fluorescent magnetic nanobeads for the detection of clenbuterol in swine urine. *J. Agric. Food Chem.* **2019**, *67*, 3028–3036. [CrossRef]

41. Huang, Z.; Hu, S.; Xiong, Y.; Wei, H.; Xu, H.; Duan, H.; Lai, W. Application and development of superparamagnetic nanoparticles in sample pretreatment and immunochromatographic assay. *TrAC Trends Anal. Chem.* **2019**, *114*, 151–170. [CrossRef]
42. Lago-Cachón, D.; Oliveira-Rodríguez, M.; Rivas, M.; Blanco-López, M.C.; Martínez-García, J.C.; Moyano, A.; Salvador, M.; García, J.A. Scanning magneto-inductive sensor for quantitative assay of prostate-specific antigen. *IEEE Magn. Lett.* **2017**, *8*. [CrossRef]
43. Moyano, A.; Salvador, M.; Martínez-García, J.C.; Socoliuc, V.; Vékás, L.; Peddis, D.; Alvarez, M.A.; Fernández, M.; Rivas, M.; Blanco-López, M.C. Magnetic immunochromatographic test for histamine detection in wine. *Anal. Bioanal. Chem.* **2019**, *411*, 6615–6624. [CrossRef] [PubMed]
44. Lago-Cachón, D.; Rivas, M.; Martínez-García, J.C.; García, J.A. Cu impedance-based detection of superparamagnetic nanoparticles. *Nanotechnology* **2013**, *24*, 245501. [CrossRef] [PubMed]
45. Rivas, M.; Lago-Cachón, D.; Martínez-García, J.C.; García, J.A.; Calleja, A.J. Eddy-current sensing of superparamagnetic nanoparticles with spiral-like copper circuits. *Sens. Actuators A Phys.* **2014**, *216*, 123–127. [CrossRef]
46. Oliveira-Rodríguez, M.; López-Cobo, S.; Reyburn, H.T.; Costa-García, A.; López-Martín, S.; Yáñez-Mó, M.; Cernuda-Morollón, E.; Paschen, A.; Valés-Gómez, M.; Blanco-López, M.C. Development of a rapid lateral flow immunoassay test for detection of exosomes previously enriched from cell culture medium and body fluids. *J. Extracell. Vesicles* **2016**, *5*, 31803. [CrossRef]
47. Oliveira-Rodríguez, M.; Serrano-Pertierra, E.; García, A.C.; Martín, S.L.; Mo, M.Y.; Cernuda-Morollón, E.; Blanco-López, M.C. Point-of-care detection of extracellular vesicles: Sensitivity optimization and multiple-target detection. *Biosens. Bioelectron.* **2017**, *87*, 38–45. [CrossRef]
48. Wang, H.; Sun, Y.B.; Chen, Q.W.; Yu, Y.F.; Cheng, K. Synthesis of carbon-encapsulated superparamagnetic colloidal nanoparticles with magnetic-responsive photonic crystal property. *Dalton Trans.* **2010**, *39*, 9565–9569. [CrossRef]
49. Nakagiri, N.; Manghnani, M.H.; Ming, L.C.; Kimura, S. Crystal structure of magnetite under pressure. *Phys. Chem. Miner.* **1986**, *13*, 238–244. [CrossRef]
50. Zsigmondy, R.; Scherrer, P. Bestimmung der inneren struktur und der größe von kolloidteilchen mittels röntgenstrahlen. In *Kolloidchemie Ein Lehrbuch*; Springer: Berlin/Heidelberg, Germany, 1912; Volume 277, pp. 387–409.
51. Wang, L.; Bao, J.; Wang, L.; Zhang, F.; Li, Y. One-pot synthesis and bioapplication of amine-functionalized magnetite nanoparticles and hollow nanospheres. *Chem. A Eur. J.* **2006**, *12*, 6341–6347. [CrossRef]
52. Lakshmipriya, T.; Gopinath, S.C.B.; Tang, T.H. Biotin-streptavidin competition mediates sensitive detection of biomolecules in enzyme linked immunosorbent assay. *PLoS ONE* **2016**, *11*, 16–20. [CrossRef]
53. Lin, Z.; Wang, X.; Li, Z.J.; Ren, S.Q.; Chen, G.N.; Ying, X.T.; Lin, J.M. Development of a sensitive, rapid, biotin-streptavidin based chemiluminescent enzyme immunoassay for human thyroid stimulating hormone. *Talanta* **2008**, *75*, 965–972. [CrossRef] [PubMed]
54. Sai, N.; Chen, Y.; Liu, N.; Yu, G.; Su, P.; Feng, Y.; Zhou, Z.; Liu, X.; Zhou, H.; Gao, Z.; et al. A sensitive immunoassay based on direct hapten coated format and biotin-streptavidin system for the detection of chloramphenicol. *Talanta* **2010**, *82*, 1113–1121. [CrossRef] [PubMed]
55. Liu, N.; Nie, D.; Zhao, Z.; Meng, X.; Wu, A. *Ultrasensitive Immunoassays Based on Biotin-Streptavidin Amplified System for Quantitative Determination of Family Zearalenones*; Elsevier Ltd.: Amsterdam, The Netherlands, 2015; Volume 57, ISBN 8621549220.
56. Liu, R.; Liu, J.; Xie, L.; Wang, M.; Luo, J.; Cai, X. A fast and sensitive enzyme immunoassay for brain natriuretic peptide based on micro-magnetic probes strategy. *Talanta* **2010**, *81*, 1016–1021. [CrossRef] [PubMed]
57. Kim, D.S.; Kim, Y.T.; Hong, S.B.; Kim, J.; Huh, N.S.; Lee, M.K.; Lee, S.J.; Kim, B., II; Kim, I.S.; Huh, Y.S.; et al. Development of lateral flow assay based on size-controlled gold nanoparticles for detection of hepatitis B surface antigen. *Sensors* **2016**, *16*, 2154. [CrossRef] [PubMed]
58. Thobhani, S.; Attree, S.; Boyd, R.; Kumarswami, N.; Noble, J.; Szymanski, M.; Porter, R.A. Bioconjugation and characterisation of gold colloid-labelled proteins. *J. Immunol. Methods* **2010**, *356*, 60–69. [CrossRef]

biosensors

MDPI

Article

Lipid–Polymer Hybrids Encapsulating Iron-Oxide Nanoparticles as a Label for Lateral Flow Immunoassays

Shayesteh Bazsefidpar [1], Amanda Moyano [1], Gemma Gutiérrez [2], María Matos [2,*] and María Carmen Blanco-López [1,*]

[1] Department of Physical and Analytical Chemistry & Institute of Biotechnology of Asturias, University of Oviedo, c/Julián Clavería 8, 33006 Oviedo, Spain; bazsefidparshayesteh@uniovi.es (S.B.); moyanoamanda@uniovi.es (A.M.)

[2] Department of Chemical and Environmental Engineering & Institute of Biotechnology of Asturias, University of Oviedo, 33006 Oviedo, Spain; gutierrezgemma@uniovi.es

* Correspondence: matosmaria@uniovi.es (M.M.); cblanco@uniovi.es (M.C.B.-L.)

Abstract: The feasibility of using Superparamagnetic Iron Oxide Nanoparticles (SPIONs) encapsulated by lipid–polymer nanoparticles as labels in lateral flow immunoassays (LFIA) was studied. First, nanoparticles were synthesized with average diameters between 4 and 7 (nm) through precipitation in W/O microemulsion and further encapsulated using lipid–polymer nanoparticles. Systems formulated were characterized in terms of size and shape by DLS (Nanozetasizer from Malvern) and TEM. After encapsulation, the average size was around (≈20 and 50 nm). These controlled size agglomerates were tested as labels with a model system based on the biotin–neutravidin interaction. For this purpose, the encapsulated nanoparticles were conjugated to neutravidin using the carbodiimide chemistry, and the LFIA was carried out with a biotin test line. The encapsulated SPIONs showed that they could be promising candidates as labels in LFIA test. They would be useful for immunomagnetic separations, that could improve the limits of detection by means of preconcentration.

Keywords: SPIONs; encapsulation; PLGA; lipid; lipid–polymer hybrid nanoparticles; lateral flow immunoassays (LFIA)

Citation: Bazsefidpar, S.; Moyano, A.; Gutiérrez, G.; Matos, M.; Blanco-López, M.C. Lipid–Polymer Hybrids Encapsulating Iron-Oxide Nanoparticles as a Label for Lateral Flow Immunoassays. *Biosensors* **2021**, *11*, 218. https://doi.org/10.3390/bios11070218

Received: 27 April 2021
Accepted: 28 June 2021
Published: 1 July 2021

1. Introduction

Point-of-care (POC) tests are diagnostic tools for rapid detection and fast analysis [1,2]. Lateral flow immunoassays are among the most popular devices for POC diagnostics [3,4]. Lateral flow immunoassays (LFIA), also called immunochromatographic tests, have been established as POC tests in recent years due to the rapid, low-cost, simple detection [5]. They are the basis for the well-known pregnancy test, but they can be designed for different applications in biomedicine, toxicology, food, agriculture, and environmental fields [6,7]. They consist of a membrane of nitrocellulose, where bioreceptors are immobilized. As the sample flows through the membrane, the analyte is captured at the test line. The molecular recognition event is revealed by means of labeled bioreagents. For the actual pandemic situation, LFIA tests are attracting a lot of interest because they are the basis of antigen and antibody tests for COVID-19 diagnostics [8].

However, their usage is still limited due to the poor sensitivity, specificity, and limited stability [3,9]. The actual challenges include the need to achieve low limits of detection, with direct use or simplified protocols of sample preparation, or the possibility for quantification. The nanopaticles used in LFIAs play an important role in the sensitivity of the system [5]. The most common are gold or latex nanoparticles, which are good for visual detection. In recent years, nanoparticle research has been a cutting edge field in biosensing, and other particles with fluorescent, magnetic or electrical properties are currently investigated in this field [1,10,11]. Among nanoparticles, Superparamagnetic Iron

Oxide Nanoparticles (SPIONs) have received considerable attention based on their properties such as potent magnetic, superparamagnetic properties, high surface-area-to-volume ratio, and quick magnetophoretic response for the development of a new generation of LFIA [11–13]. However, SPIONs have several limitations such as aggregation, lack of normal distribution in the suspension in water, and lack of maintenance of the long-term stability of functionalized SPIONs which limit their usage in point-of-care diagnosis [1]. Therefore, encapsulation of SPIONs could be one of the most useful methods for controlled agglomeration of nanoparticles, improving the stability of the suspension in water, preventing oxidation, reducing the toxicity of nanoparticles, and facilitating the bioconjugation of nanoparticles with biological molecules [14–16].

Lipid–polymer hybrid nanoparticles (LPHNPs) are hybrid systems conceptually made of both liposomes and polymeric nanoparticles. In this system, polymer as a core is covered by a lipid layer. This system strongly enhances the encapsulation efficiency [17,18]. Encapsulation by LPHNPs offers several advantages such as enhancing device sensitivity as well as the possibility to modify the properties of the surface to provide a better interaction with biological molecules [19]. Poly (lactic-co-glycolic acid) (PLGA) presents a promising alternative to other conventional nanocolloids, such as nanoemulsions or liposomes. PLGA is one of the most successfully developed biodegradable hydrophobic polymers and hence presents biodegradability and biocompatibility. Moreover, it offers not only the possibility to encapsulate various types of drugs e.g., hydrophilic or hydrophobic small molecules or macromolecules, but also allows the possibility to modify surface properties to provide stealthiness and/or better interaction with biological materials [18,19]. Phosphatidylcholine (PC) is a neutral phospholipid that adsorbs and self-assembles onto the surface of the hydrophobic polymer through hydrophobic interactions with the goal of reducing the free energy of the system [18].

In this study, SPIONs through precipitation in W/O microemulsion were synthesized and further encapsulated in lipid–polymer hybrid system as labels for LFIA. Biodegradable PLGA used as a polymer core for entrapping SPIONs, PVA as a non-ionic surfactant for the stabilization, and PC as the lipid layer surrounding the polymer core (SPIONs in PLGA–PVA/PC). The emulsion/solvent method was used for the encapsulation of SPIONs in PLGA and PC. To test their application in LFIA, we studied these encapsulated SPIONs in PLGA–PVA/PC as labels in LFIA. The encapsulated SPIONs in PLGA–PVA/PC were bioconjugated to neutravidin and tested against biotin in the strips.

2. Materials and Methods

2.1. Chemicals

Ferric Chloride Hexahydrate ($FeCl_3 \cdot 6H_2O$) and ($FeCl_2 \cdot 4H_2O$) was supplied from Sigma-Aldrich (Madrid, Spain). Ammonia 30% (v/v) (NH_3) were supplied by Panreac AppliChem (Barcelona, Spain). Ferrous Chloride Tetrahydrate. Cetyl Trimethyl Ammonium Bromide 99% (CTAB), 1-butanol (min. 99%) and hydrochloric acid 38% (HCl), ethanol (95%), and phosphotungstic acid hydrate (99.995%), 1-Hexanol, and Sodium Hydroxide (NaOH) were supplied by Sigma-Aldrich (Madrid, Spain). Nitric Acid, min. 69.5% (HNO_3) was supplied by Scharlau (Barcelona, Spain).

PLGA (LG 50:50, Mw 24–38 kDa) was purchased from Sigma Chemical Co. (Steinheim, Germany). Polyvinyl alcohol (PVA) (Mw 30–70 kDa), phosphatidylcholine (PC) (predominant species: C42H80NO8P, MW = 775.04 g/mol) from soybean (Phospholipon 90G) was obtained from Lipoid (Koln, Germany). Cholesteryl hemisuccinate (MW = 486.73 g/mol) was obtained from Sigma Chemical Co. Sepharose^{TM} CL-4B was purchased from GE Healthcare company in (Stockholm, Sweden).

N-Hydroxysuccinimide (NHS), 1-ethyl-3-[3-dimethylaminopropyl]–carbodiimide hydrochloride (EDC), 2-(N-morpholino) ethanesulfonic acid (MES), bovine serum albumin (BSA) were purchased by Sigma-Aldrich. Neutravidin protein was provided from Thermo Fischer Scientific (Waltham, MA, USA).

For biotin–neutravidin tests, glass fiber membrane (GFCP001000) was employed as sample pad and backing cards (HF000MC100) were obtained from Millipore (Darmstadt, Germany). Other materials used were nitrocellulose membranes (UniSart CN95, Sartorius, Spain) and absorbent pads (Whatman, Madrid, Spain). 4-(2-hydroxyethyl)-1-piperazineethanesulfonic acid (HEPES) was used as a buffer for conjugation. The sample buffer consisted of 10 mM phosphate buffer (PB) pH 7.4 with 0.5% Tween-20 and 1% BSA.

2.2. Synthesis and Characterization of SPIONs

2.2.1. Synthesis of SPIONs

SPIONs were produced based on two different formulations following the *W/O* microemulsion method reported in previous studies [20]. *W/O* microemulsions were formulated using CTAB as the main surfactant, 1-butanol as a cosurfactant, and 1-hexanol as the continuous oily phase. CTAB and 1-butanol were used using a constant weight ratio of 3:2 (CTAB: 1-butanol). The water phase consisted of a solution that contained Fe^{2+}/Fe^{3+} in a 2:1 molar ratio, being 0.7 and 1.4 M, respectively. It was prepared by dissolving an appropriate amount of the chloride salts aforementioned being homogenized by magnetic stirring. The 0.01 M HCl solution was added to avoid Fe (II) oxidation. Both the formulation of the *W/O* microemulsions used for the synthesis of the SPIONs and the resulting mean diameters are shown in Table 1.

Table 1. W/O formulations used for the synthesis of SPIONs and mean sizes obtained by DLS.

Sample	Microemulsion Formulation (% *w/w*)				Size (nm)
	CTAB	1-Butanol	1-Hexanol	Water Phase	
1	24	16	45	15	5.4
2	15	10	57	18	6.6

First of all, the microemulsions were prepared and left to rest for a while until the appearance was totally translucid. Then, the synthesis of the SPIONs was performed through the co-precipitation of the iron salts present in the microemulsion water droplets by the dropwise addition of ammonia (30% (*v/v*) solution) upon vigorous stirring with the Silent Crusher M Homogenizer (Heidolph 8F) (Atlanta, GA, USA) (set at 6500 rpm. Once a black precipitate appeared, the solution was left for two hours under magnetic stirring. Finally, all the samples were washed five times using a solution consisting of ethanol and water in a ratio of 90:10 (% *v/v*) and dispersed in ethanol or in water.

2.2.2. Preparation of Hybrid Nanoparticles Encapsulating SPIONs

SPIONs were encapsulated by a single emulsion/solvent method. The two formulations of SPIONs synthesized (1 and 2) were previously dispersed in water or ethanol and therefore 4 sets of experiments were carried out.

For encapsulation, 400 µL of SPIONs dispersed (in ethanol or water) were added into 2 mL of organic phase consisting of 12.5% (*v/v*) methanol in chloroform solution. Then, 30 mg of PLGA and cholesteryl hemisuccinate 1% (*w/v*) of the total membrane compounds (PC) was dissolved in the solution. To prepare the hybrid nanoparticles (SPIONs in PLGA–PVA/PC), the organic solution was emulsified in 6 mL of aqueous phase containing 10 mg PC and 2% PVA (*w/v*) to form an oil-in-water (*O/W*) emulsion under continuous sonication with the amplitude of 70% for 5 min on an ice bath. After sonication, 8 mL of PVA 5% (*w/v*) were added and the sonication continued for 5 min. Then, the emulsions were stirred using a mechanical stirrer (Teflon type) (Fenteer Brand in Shenzhen, China) to allow evaporation of the organic solvent and form the particles for 24 h. Nanoparticles were purified using Size Exclusion Chromatography (SEC) including 13.5 cc of Sepharose^{TM} CL-4B and 20 cc of milli-Q water.

2.3. Bioconjugation of Lipid–Polymer SPIONs

Encapsulated SPIONs with carboxyl functional groups were functionalized using neutravidin to test their function as the label in LFIA through the interaction of neutravidin–biotin. Firstly, the carboxyl groups of the nanoparticles were activated using carbodiimide chemistry. For this, 0.0015 g of EDC and 0.0030 g of NHS were dissolved in 100 μL of MES buffer (pH 5.73) and were mixed with 1000 μL of nanoparticles for 30 min. After shaking, nanoparticles were separated from supernatant through a centrifuge at 5000× g for 4 min and washed with 200 μL of MES buffer (pH 5.73). Then, 100 μL of MES buffer (pH 7.4) and 100 μL of neutravidin (1 mg/mL) were added to nanoparticles. After shaking for 3 h, the excess EDC and NHS were removed via centrifuge at 5000× g for 4 min and washed with 200 μL of MES buffer (pH 7.4). In the next step, the residual carboxyl groups on the surfaces were blocked by adding 100 μL of MES buffer (pH 7.4) and 100 μL of BSA (0.010 g/L mL MES buffer (pH 7.4)) and reacted through shaking for 30 min. Then, the mixture was centrifuged at 5000× g for 4 min and washed with MES buffer (pH 7.4)). Finally, the supernatant was discarded, and nanoparticles were dispersed in 100 μL of PVA and PB buffer separately for comparison of their movement on the strips. After dispersing of nanoparticles 20 μL were added to 80 μL of running buffer containing Tween 20%, BSA, and PB buffer (pH = 7.4).

2.4. Characterization of Nanoparticles Conjugates

2.4.1. Dynamic Light Scattering and Zeta Potential

Encapsulated SPIONs were characterized in terms of measurement of particle size distribution as well as zeta potential by dynamic light scattering (DLS) analysis using Zetasizer Nano ZS ZEN3600 (Malvern Instruments, Malvern, UK) equipped with a solid-state He–Ne laser (λ = 633 nm). Measurements were performed at 25 °C. Additionally, this instrument was used to monitor the conjugation process. Zetasizer software version 7.03 was used for data analysis.

2.4.2. Transmission Electron Microscopy

Particle morphology, size, and the structure of aggregation of the encapsulated SPIONs were determined by TEM. The aqueous dispersion was drop-cased onto the former-coated copper grid and placed in the TEM for analysis with a JEOL-2000 Ex II TEM (Saint-Herblain, France).

2.5. Lateral Flow Assays

Preparation of the Strips

The LFIA was based on a dipstick format. The test strips were composed of a sample pad, nitrocellulose membrane, absorbent pad, and backing plastic card. Firstly, the nitrocellulose membrane (25 mm wide) was attached to a backing plastic card to get robustness. Then, for the test of biotin–neutravidin affinity, a test line of biotin-BSA was immobilized across the membrane by the IsoFlow dispenser (Imagene Technology, Lebanon, NH, USA) at a rate of 0.100 μL/mm. In the next step, the nitrocellulose membrane after the immobilization of the biotin-BSA test line was kept for 20 min at 37 °C. Finally, the sample pad and the absorbent pad were stuck onto the backing card with an overlap between them of 2 mm. The complete card was cut into 5 mm wide strips.

3. Results and Discussion

3.1. Characterization of the Lipid–Polymer SPIONs before Bioconjugation

Figure 1 indicates the scheme of the preparation and structure of the SPIONs loading in lipid–polymer nanoparticle by emulsion/solvent method. In this procedure, PLGA was used as a carrier to entrap the nanoparticles and coat them with lipid layers. PVA (a non-ionic surfactant) was used as a stabilizer in the formulation.

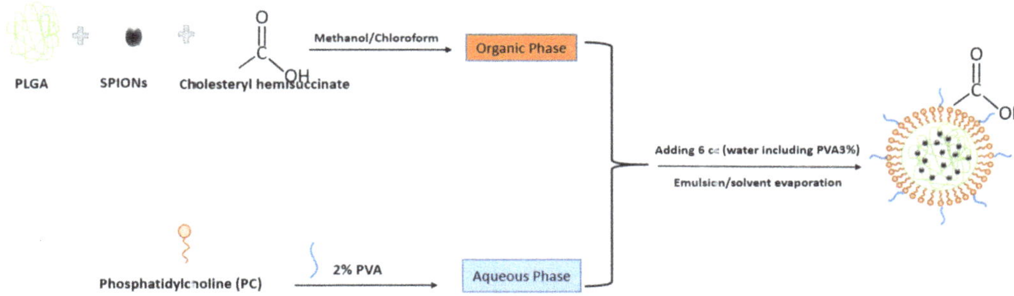

Figure 1. Scheme showing the preparation and structure of the SPIONs in PLGA–PVA/PC by emulsion/solvent evaporation method.

Transmission electron microscopy (TEM) was used to investigate the state of aggregation of SPIONs in the lipid–polymer nanoparticles and to assess their size distribution. Figure 2a,b show that the SPIONs agglomerated under controlled size aggregates when they were dispersed in ethanol (organic phase). The average size of the agglomerates was around 20–50 nm, whereas that of single SPION was 4–7 nm. These micrographs show that the encapsulation of SPIONs in PLGA–PVA/PC was successful. However, when SPIONS were dispersed in the water phase, the encapsulation was not size-controlled (Figure 2c,d). The size obtained by DLS and zeta potential for SPIONs in PLGA–PVA/PC are summarized in Table 2. It was observed that after the addition of SPIONs, the hydrodynamic size of lipid–polymer nanoparticles was slightly larger than empty lipid–polymer nanoparticles ($\approx$80 nm) in all samples. Additionally, the zeta potential of SPIONs in PLGA–PVA/PC nanoparticle was higher than empty lipid–polymer nanoparticles (≈ -0.3 mV) that relied on the presence of cholesteryl hemisuccinate.

3.2. Biotin–Neutravidin Affinity Test

The biotin–neutravidin complex is one of the strongest non-covalent interactions that due to the high specificity and strong affinity is widely used in immunoassays [21,22]. In this study, the biotin–neutravidin system was applied as model system in order to study the feasibility of using SPIONs in PLGA–PVA/PC as labels for LFIA. Figure 3 shows the steps of bioconjugation of encapsulated SPIONs with neutravidin.

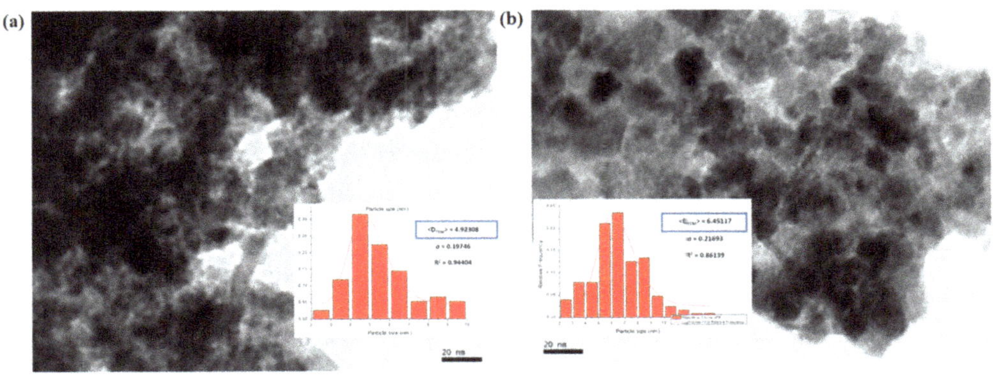

Figure 2. *Cont.*

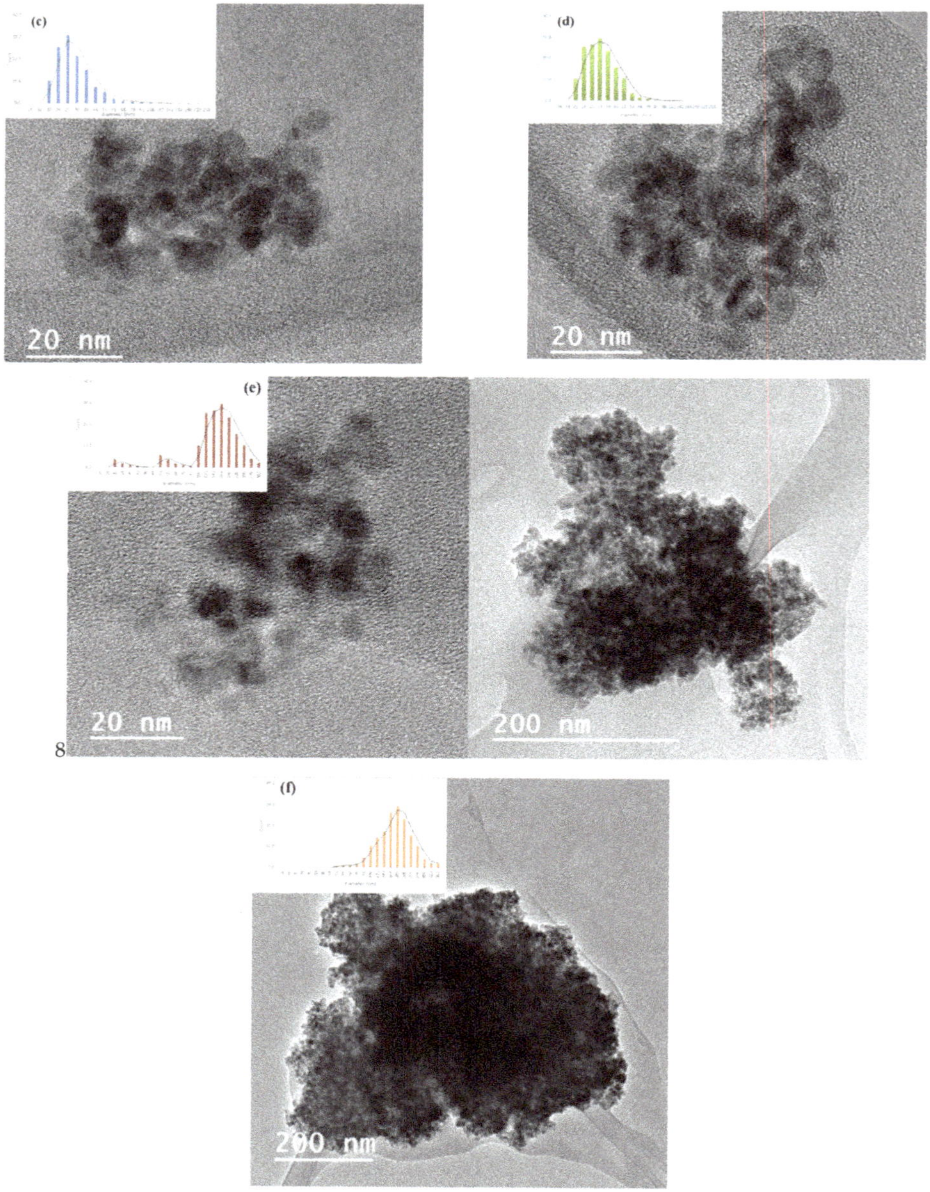

Figure 2. TEM images of the SPIONs in PLGA–PVA/PC. (**a**) SPIONs F1 before encapsulation (**b**) SPIONs F2 before encapsulation. (**c**) SPIONs F1 in the organic phase, (**d**) SPIONs F2 in the organic phase (**e**) SPIONs F1 in the water phase, (**f**) SPIONs F2 in the water phase.

Table 2. Results of the hydrodynamic size of SPIONs after encapsulation in lipid–polymer nanoparticles DLS (Z-average and polydispersity index).

Formulation	Size (nm)	PDI	Zeta Potential (mV)
Empty	80	0.243	−0.3
F1 (dispersed in organic phase)	82.34	0.263	−12.6
F2 (dispersed in organic phase)	89.45	0.345	−13.3
F1 (dispersed in water phase)	83.59	0.230	−13.43
F2 (dispersed in water phase)	99.87	0.229	−7.3

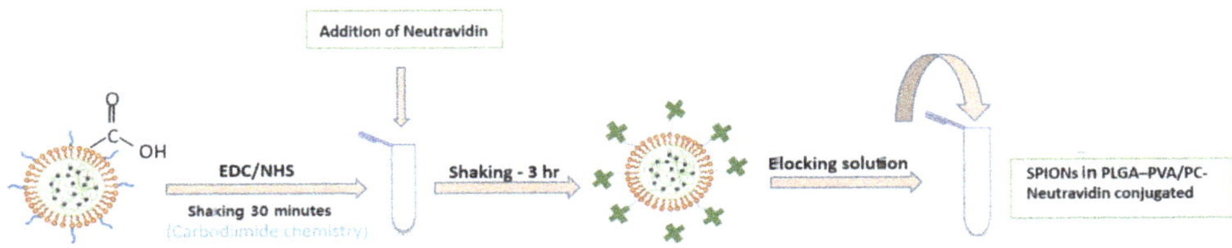

Figure 3. A schematic of the protocol of bioconjugation of encapsulated SPIONs with carboxyl functional groups with neutravidin.

After bioconjugation of the four different types of encapsulated SPIONs in PLGA–PVA/PC with neutravidin separately, bioconjugated nanoparticles were dispersed in PVA (1%, 3%, and 5%) and in PB separately to compare the movement of conjugated nanoparticles on the strips. The results showed that the bioconjugated nanoparticles dispersed in PVA 100 µL could flow through the strips, while the bioconjugated nanoparticles dispersed in PB did not. The movement of nanoparticles on strips in PVA 1% was weak but when using PVA 3% it was observed that nanoparticles moved better on the strips to make an optical signal, but it was not strong. The visual signal was optimized when using PVA 5% and this was the medium chosen for dispersing nanoparticles after bioconjugation.

DLS was used to compare the hydrodynamic size of nanoparticles before and after the bioconjugation reaction (Table 3 and Figure 4). The results indicate that the hydrodynamic diameter of nanoparticles was higher after bioconjugation, confirming the success of this process.

Table 3. DLS results of hydrodynamic diameter of SPIONs in PLGA–PVA/PC before and after conjugation with neutravidin.

Formulation	Size (nm) (Before Conjugation)	Size (nm) (After Conjugation)	PDI
F1 (dispersed in organic phase)	82.34	136.68	0.253
F2 (dispersed in organic phase)	89.45	150.35	0.225
F1 (dispersed in water phase)	83.59	164	0.230
F2 (dispersed in water phase)	99.87	142	0.234

In order to investigate the performance of four different formulations of encapsulated SPIONs for LFIA, 20 µL of suspensions and 80 µL of running buffer were transferred into a microtube. The strip was introduced into the microtube and the buffer including encapsulated SPIONs started to flow through the strips. As it can be seen in Figure 5, the SPIONs (1 and 2) dispersed in the organic phase could flow through the strips and make an optical signal in the test line. However, there was no signal when the SPIONs dispersed in the water phase were used. The unsuccessful encapsulation of SPIONs dispersed in the water phase in a controlled size could be the reason because they did not flow through the strips. In addition, encapsulated nanoparticles showed rapid separation with a conventional magnet in 1 min (Figure 5), whereas the free nanoparticles were very stable in colloidal suspension and they were not attracted by the magnet. On the other side, nanoparticles with larger size might not redisperse after removal of the magnetic field, but these lipid hybrids keep the ability to redisperse when the magnetic field is removed, and they have demonstrated good flowing behavior. This is very important for their successful use as labels when they are conjugated with the detection bioreagent.

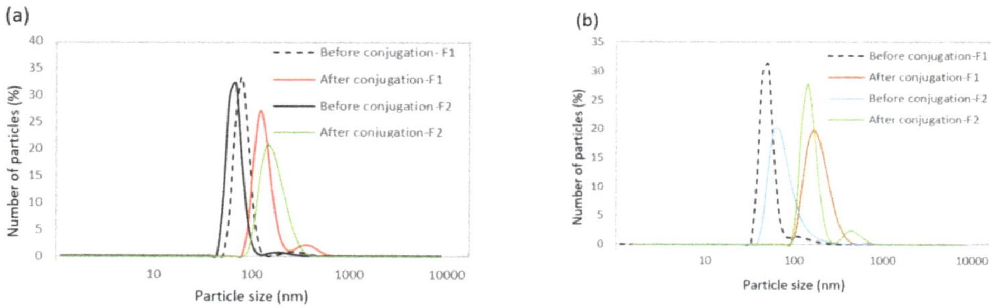

Figure 4. (**a**) Hydrodynamic diameter distribution profiles of SPIONs dispersed in the organic phase and (**b**) dispersed in the water phase before and after bioconjugation.

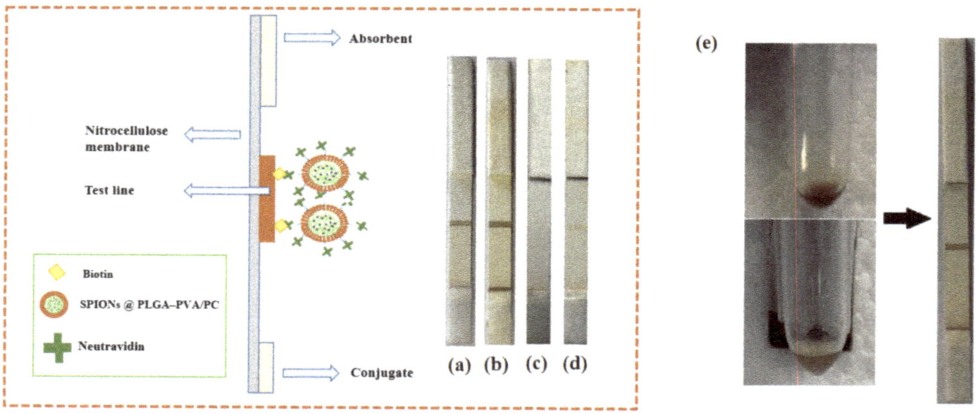

Figure 5. Scheme showing the preparation and structure of biotin–neutravidin affinity test. (**a**) SPIONs dispersed in the organic phase (F1), (**b**) SPIONs dispersed in the organic phase (F2), (**c**) SPIONs dispersed in the water phase (F1), (**d**) SPIONs disperses in the water phase (F2). (**e**) Separation of SPIONs in PLGA–PVA/PC nanoparticles using a conventional magnet.

Figure 6 shows a scheme with the steps involved.

The intensity profiles of the nanoparticles at the biotin-BSA test line were measured for SPIONs in PLGA–PVA/PC and gold nanoparticles using ESEQuant Lateral Flow Reader (ESEQuant LRS). Table 4 shows the results obtained. It can be observed that the instrumental optical density reading of SPIONs in PLGA–PVA/PC nanoparticles was higher than that of gold nanoparticles (height and peak area).

Table 4. Comparison of the intensity profiles measured for gold nanoparticles and SPIONs in PLGA–PVA/PC nanoparticles.

Type of Nanoparticles	X-Pos (mm)	Intensity (mV)	Peak Start (mm)	Peak End (mm)	Height (mV)	Area (mm × mv)
SPIONs in PLGA–PVA/PC	25.48	1164	24.48	26.48	752.84	702.15
Gold	24.96	255	24.04	26.04	619.68	469.85

This work shows proof of concept of the efficiency of the new labeling system and opens the path for applications in the health and environmental fields. The limits of detection could be greatly improved by using the capacity for magnetic separation of this lipid hybrid nanoparticle system.

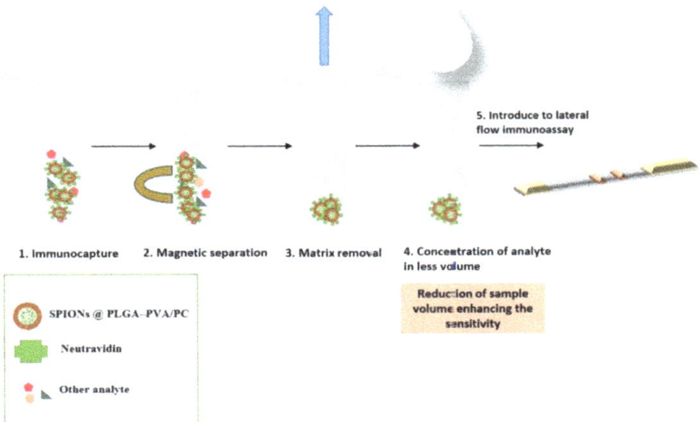

Figure 6. Scheme showing the immunomagnetic separation and preconcentration.

4. Conclusions

In this work, the feasibility of using SPIONs encapsulated in lipid–polymer nanoparticles as labels for LFIA was investigated. SPIONs were synthesized with average diameters between 4 and 7 nm by precipitation using the W/O microemulsion method and were encapsulated in lipid–polymer nanoparticles by the emulsion/solvent method to obtain multinanoparticle systems based on SPIONs with controlled size. The results showed that SPIONs dispersed in the organic phase could be encapsulated with controlled size by lipid–polymer nanoparticles (SPIONsinPLGA–PVA/PC) providing an optical signal when using them as labels in lateral flow assays after bioconjugation with neutravidin. By using biotin-BSA test lines, this system has shown advantages such as good optical density intensity readings, as compared with gold nanoparticles. These multilabel systems have a great potential of application for magnetic separation and analyte preconcentration for these types of rapid tests.

Author Contributions: Conceptualization, M.M., G.G. and M.C.B.-L.; methodology, A.M., M.M., G.G. and M.C.B.-L.; investigation, S.B., M.M., G.G.; resources, M.M., G.G. and M.C.B.-L.; writing—original draft preparation, S.B.; writing—review and editing, M.M. and M.C.B.-L.; supervision, M.M. and M.C.B.-L.; project administration, M.C.B.-L.; funding acquisition, M.C.B.-L. All authors have read and agreed to the published version of the manuscript.

Funding: This work is part of a project that has received funding from the European Union's Horizon 2020 research and innovation program under the Marie Sklodowska-Curie Grant Agreement No. 813439 (Break Biofilms). The work was also supported by the project MAT2017-84959-C2-1-R and the project Nanotechnology in translational hyperthermia (HIPERNANO) RED2019-102626T.

Institutional Review Board Statement: Not applicable.

Informed Consent Statement: Not applicable.

Data Availability Statement: Not applicable.

Conflicts of Interest: The authors declare no conflict of interest.

References

1. Xianyu, Y.; Wang, Q.; Chen, Y. Magnetic particles-enabled biosensors for point-of-care testing. *TrAC Trends Anal. Chem.* **2018**, *106*, 213–224. [CrossRef]
2. Connolly, R.; O' Kennedy, R. Magnetic lateral flow immunoassay test strip development—Considerations for proof of concept evaluation. *Methods* **2017**, *116*, 132–140. [CrossRef] [PubMed]
3. Liu, L.; Yang, D.; Liu, G. Signal amplification strategies for paper-based analytical devices. *Biosens. Bioelectron.* **2019**, *136*, 60–75. [CrossRef] [PubMed]
4. Antiochia, R. Paper-Based Biosensors: Frontiers in Point-of-Care Detection of COVID-19 Disease. *Biosensors* **2021**, *11*, 110. [CrossRef]
5. Sajid, M.; Kawde, A.N.; Daud, M. Designs, formats and applications of lateral flow assay: A literature review. *J. Saudi Chem. Soc.* **2015**, *19*, 689–705. [CrossRef]
6. Guo, J.; Chen, S.; Guo, J.; Ma, X. Nanomaterial Labels in Lateral Flow Immunoassays for Point-of-Care-Testing. *J. Mater. Sci. Technol.* **2021**, *60*, 90–104. [CrossRef]
7. Moyano, A.; Serrano-pertierra, E.; Salvador, M.; Martínez-garcía, J.C. Magnetic Lateral Flow Immunoassays. *Diagnostics* **2020**, *10*, 288. [CrossRef]
8. Chen, Z.; Zhang, Z.; Zhai, X.; Li, Y.; Lin, L.; Zhao, H.; Bian, L.; Li, P.; Yu, L.; Wu, Y.; et al. Rapid and Sensitive Detection of anti-SARS-CoV-2 IgG, Using Lanthanide-Doped Nanoparticles-Based Lateral Flow Immunoassay. *Anal. Chem.* **2020**, *92*, 7226–7231. [CrossRef]
9. Yamada, K.; Shibata, H.; Suzuki, K.; Citterio, D. Toward practical application of paper-based microfluidics for medical diagnostics: State-of-the-art and challenges. *Lab Chip* **2017**, *17*, 1206–1249. [CrossRef] [PubMed]
10. Hu, C.; Yue, W.; Yang, M. Nanoparticle-based signal generation and amplification in microfluidic devices for bioanalysis. *Analyst* **2013**, *138*, 6709–6720. [CrossRef]
11. Estelrich, J.; Escribano, E.; Queralt, J.; Busquets, M.A. Iron oxide nanoparticles for magnetically-guided and magnetically-responsive drug delivery. *Int. J. Mol. Sci.* **2015**, *16*, 8070–8101. [CrossRef] [PubMed]
12. Zia, M.; Phull, A.R.; Ali, J.S. Synthesis, characterization, applications, and challenges of iron oxide nanoparticles. *Nanotechnol. Sci. Appl.* **2016**, *9*, 49–67. [CrossRef]
13. Ha, Y.; Ko, S.; Kim, I.; Huang, Y.; Mohanty, K.; Huh, C.; Maynard, J.A. Recent Advances Incorporating Superparamagnetic Nanoparticles into Immunoassays. *ACS Appl. Nano Mater.* **2018**, *1*, 512–521. [CrossRef]
14. Salvador, M.; Moyano, A.; Martínez-García, J.C.; Blanco-López, M.C.; Rivas, M. Synthesis of Superparamagnetic Iron Oxide Nanoparticles: SWOT Analysis Towards Their Conjugation to Biomolecules for Molecular Recognition Applications. *J. Nanosci. Nanotechnol.* **2019**, *19*, 4839–4856. [CrossRef] [PubMed]
15. Farhanian, D.; De Crescenzo, G.; Tavares, J.R. Large-Scale Encapsulation of Magnetic Iron Oxide Nanoparticles via Syngas Photo-Initiated Chemical Vapor Deposition. *Sci. Rep.* **2018**, *8*, 1–11. [CrossRef]
16. Inozemtseva, O.A.; German, S.V.; Navolokin, N.A.; Bucharskaya, A.B.; Maslyakova, G.N.; Gorin, D.A. *Encapsulated Magnetite Nanoparticles: Preparation and Application as Multifunctional Tool for Drug Delivery Systems*; Elsevier Inc.: Amsterdam, The Netherlands, 2018; pp. 175–192. ISBN 9780128138854.
17. Mukherjee, A.; Waters, K.; Kalyan, P.; Achrol, A.S. Lipid—Polymer hybrid nanoparticles as a next- generation drug delivery platform: State of the art, emerging technologies, and perspectives. *Int. J. Nanomed.* **2019**, *14*, 1937–1952. [CrossRef] [PubMed]
18. Bose, R.J.C.; Lee, S.H.; Park, H. Lipid-based surface engineering of PLGA nanoparticles for drug and gene delivery applications. *Biomater. Res.* **2016**, *20*, 1–9. [CrossRef]
19. Danhier, F.; Ansorena, E.; Silva, J.M.; Coco, R.; Le Breton, A.; Préat, V. PLGA-based nanoparticles: An overview of biomedical applications. *J. Control. Release* **2012**, *161*, 505–522. [CrossRef] [PubMed]
20. Salvador, M.; Gutiérrez, G.; Noriega, S.; Moyano, A.; Blanco-López, M.C.; Matos, M. Microemulsion synthesis of superparamagnetic nanoparticles for bioapplications. *Int. J. Mol. Sci.* **2021**, *22*, 427. [CrossRef] [PubMed]
21. Moyano, A.; Serrano-Pertierra, E.; Salvador, M.; Martínez-García, J.C.; Piñeiro, Y.; Yañez-Vilar, S.; Gónzalez-Gómez, M.; Rivas, J.; Rivas, M.; Carmen Blanco-López, M. Carbon-coated superparamagnetic nanoflowers for biosensors based on lateral flow immunoassays. *Biosensors* **2020**, *10*, 80. [CrossRef]
22. Haley, W.E. The Principles and Applications of Avidin-Based Nanoparticles in Drug Delivery and Diagnosis. *Physiol. Behav.* **2017**, *176*, 139–148. [CrossRef]

biosensors

MDPI

Article

DNA–Gold Nanozyme-Modified Paper Device for Enhanced Colorimetric Detection of Mercury Ions

Min-Xin Mao [1,2,†], Rong Zheng [2,†], Chi-Fang Peng [1,2,*] and Xin-Lin Wei [3]

1 State Key Laboratory of Dairy Biotechnology, Shanghai Engineering Research Center of Dairy Biotechnology, Dairy Research Institute, Bright Dairy & Food Co., Ltd., Shanghai 200436, China; 6190112080@jiangnan.edu.cn
2 School of Food Science and Technology, Jiangnan University, Wuxi 214122, China; 6170112120@stu.jiangnan.edu.cn
3 School of Agriculture and Biology, Shanghai Jiaotong University, Shanghai 200240, China; weixinlin@sjtu.edu.cn
* Correspondence: pcf@jiangnan.edu.cn
† These authors contributed equally to this work.

Received: 9 December 2020; Accepted: 16 December 2020; Published: 18 December 2020

Abstract: In this work, a paper device consisted of a patterned paper chip, wicking pads, and a base was fabricated. On the paper chip, DNA–gold nanoparticles (DNA–AuNPs) were deposited and Hg^{2+} ions could be adsorbed by the DNA–AuNPs. The formed DNA–AuNP/Hg^{2+} nanozyme could catalyze the tetramethylbenzidine (TMB)–H_2O_2 chromogenic reaction. Due to the wicking pads, a larger volume of Hg^{2+} sample could be applied to the paper device for Hg^{2+} detection and therefore the color response could be enhanced. The paper device achieved a cut-off value of 50 nM by the naked eye for Hg^{2+} under optimized conditions. Moreover, quantitative measurements could be implemented by using a desktop scanner and extracting grayscale values. A linear range of 50–2000 nM Hg^{2+} was obtained with a detection limit of 10 nM. In addition, the paper device could be applied in the detection of environmental water samples with high recoveries ranging from 85.7% to 105.6%. The paper-device-based colorimetric detection was low-cost, simple, and demonstrated high potential in real-sample applications.

Keywords: paper device; signal enhancement; mercury ion; colorimetric detection

1. Introduction

Mercury ions (Hg^{2+}) are one of toxic heavy metals. They are widely found in the environment [1], are a serious threat to human health [2]. In order to control the risk of Hg^{2+}, the US Environmental Protection Agency (EPA) and the World Health Organization (WHO) set a maximum contents of Hg^{2+} in drinking water which are 2.0 μg/L (10 nM) and 6.0 μg/L (30 nM), respectively [3]. In practice, numerous conventional lab-dependent techniques such as inductively-coupled plasma mass spectrometry (ICP-MS) [4], atomic fluorescence spectrometry (AFS) [5], and high performance liquid chromatography (HPLC) [6,7] have been well-established for detection of Hg^{2+}. However, their operations are highly dependent on time-consuming sample pretreatments, expensive instrumentation, and skilled technicians, making them unsuitable for rapid and on-site detection of target Hg^{2+} ions [8,9].

In recent years, many researchers have established a large number of methods for the detection of Hg^{2+} based on nanomaterials, such as fluorescent [10], colorimetric [11], chemiluminescent [12], surface-enhanced raman spectroscopy (SERS), and electrochemical methods [13,14]. The above methods have demonstrated many advantages, such as rapidness and high sensitivity. Hg^{2+} can strongly interact with many nanomaterials such as gold nanoparticles (AuNPs), gold nanorods, and silver nanoparticles. When these nanoparticles adsorb Hg^{2+}, Au–Hg nano-alloys or Ag–Hg nano-alloys could be formed [15]. Some researchers have reported that Au–Hg nano-alloys possess peroxidase-like property and could catalyze H_2O_2-mediated oxidation of tetramethylbenzidine (TMB) [16]. We reported that the DNA–AuNP complex could capture Hg^{2+} and form DNA–Au–Hg nano-alloys [17]. These DNA–Au–Hg nano-alloys demonstrated much stable peroxidase-like activity and could achieve highly-sensitive colorimetric detection of Hg^{2+}.

Since Whiteside's group first proposed a paper-based device for the detection of biochemicals in blood, this device has received considerable attention by many researchers [18,19]. Paper is an excellent substrate material for sample filtration and preconcentration [20,21] due to its high surface-to-volume ratio, low-cost, and portability [22,23]. It has also been widely used in medical diagnosis [24,25], environmental monitoring [26], and food quality analysis [27], etc. Paper can be modified by various nanomaterials, such as ceria nanoparticles [28], AuNPs [29], silver nanoparticles [30,31], and carbon nanotubes [32–34] in order to develop assays for various targets or improve colorimetric homogeneity and intensity. For example, He et al. successfully developed an ultrasensitive nucleic acid biosensor based on HRP–AuNP dual labels and a lateral flow strip biosensor [35]. Qiao et al. developed a fluorometric Hg^{2+} test strip using Au–Ag nanoclusters as fluorescent probes combined with suppressing "coffee stains" by a bio-inspired fabrication strategy [36]. Zhang et al. used Cy5-labeled functional ssDNA toward multiple analytes, graphene oxide, and paper substrate to fabricate a paper device to report the presence of the Hg^{2+} and Ag^{+} ions and aminoglycoside antibiotics in food [37]. Li et al. prepared three kinds of doped carbon quantum dots and fabricated a smartphone-based three-channel ratio fluorescence device for simultaneous determination of Hg^{2+}, Fe^{3+}, and Cu^{2+} ions in environmental samples [38]. Zhou et al. developed a rapid and sensitive paper-based analytical device (PAD) to detect the total tetracyclines in environmental water based on a paper channel by field amplification stacking and fluorescent imaging [39].

In this paper, DNA–AuNPs were deposited onto filter paper and a nanozyme-based colorimetric detection of Hg^{2+} was carefully optimized on the filter paper. The detection was eventually carried out on a paper chip, which had detection zones modified with DNA–AuNPs and connected to a substrate reservoir by multiple channels [40]. Layers of filter paper as a wicking pad were placed under the detection zones to facilitate Hg^{2+} enrichment. This paper device demonstrated advantages including being simple, low-cost, and sensitive.

2. Experimental

2.1. Reagents and Instruments

Chloroauric acid ($HAuCl_4$) and sodium citrate were purchased from Sigma-Aldrich (Shanghai, China) and 3,3′,5,5′-tetramethylbenzidine (TMB) and hydrogen peroxide were purchased from Aladdin Reagent Company (Shanghai, China). All metal ion standard solutions were purchased from the National Institute of Metrology P. R. China. All other reagents were of analytical grade. Whatman No. 1 filter paper was obtained from GE Healthcare (Shanghai, China). Ultra-pure water was prepared with a Milli-Q pure system for all the experiments.

UV–visible (UV–vis) absorption spectra were measured with an Agilent Cary 60 UV–vis spectrophotometer (Crawford Scientific, Strathaven, UK) at room temperature. Absorption values of reaction solutions were obtained with a microplate reader (Bio-Tek, Elx800, Winooski, VT, USA). Transmission electron microscopy (TEM) images were obtained on a JEOL JEM-2100 at an accelerating voltage of 200 kV.

2.2. Fabrication of Gold Nanozyme Paper Device

The pattern of paper chip was designed using CorelDRAW software. As shown in Figure S1a, the paper chip had a substrate reservoir, which connected with eight detection zones through eight channels. The pattern of the paper device base was similar to the pattern of the paper chip, except for having eight smaller holes located at the center of each detection zone (Figure 1).

The filter paper was cut into the paper chip according to the designed pattern by a CO_2 laser engraving machine (Golden, CO 80403 USA). Then, the obtained paper cuttings were immersed into ultrapure water, rinsed for 30 s, and then dried at 40 °C for later use.

A wood board was engraved by the CO_2 laser to produce the pattern (Figure S1b). The depth of the groove for paper chip was set at 1.5 mm and other sizes are shown in the pattern. The engraved wood board was immersed in 1% paraffin solution (dissolved in n-hexane) for 5 min, and then baked at 80 °C for 10 min. The fabrication of the paper device is shown in Figure 1. Firstly, Scotch tape was attached to back side of the pretreated wood base. Layers of round filter paper as a wicking pad were filled into the holes of the wood board. Then the paper chip was fixed closely to the patterned wood board. The images of the paper device are shown in Figure S2.

2.3. Colorimetric Detection of Hg^{2+} on Paper Device

The DNA–AuNP complex were prepared according to our previous reports [17,41]. To each detection zone on paper chip, 2 μL of the DNA–AuNPs (0.6 nM) was added. After being dried at room temperature for 5 min, 20–100 μL of standard Hg^{2+} solution or sample was added to each sample detection zone. After being incubated for 20 min, 300 μL of substrate (0.4 mM TMB and 3.0% H_2O_2 in 0.1 M citric buffer) was added to the reagent reservoir. After the substrates were distributed to each detection zone, chromogenic reaction was initiated and continued for 20 min. The color development was recorded by mobile phone and desktop scanning, and the colorimetric signal was analyzed using Image J software.

2.4. Validation of the Colorimetric Detection

Tap water and lake water samples from Li Lake (Wuxi, China) were spiked with different concentrations (200, 500, and 1000 nM) of Hg^{2+}, filtered twice through 0.22 μm membrane, and then measured by the paper device. The lake water samples were filtered with filter paper modified with graphene oxide, and then filtered with a 0.22 μm membrane to carry out the next detection.

3. Results and Discussion

3.1. Fabrication of Paper Device

Generally, the sensitivity of paper-based assays is negatively affected by small volumes of sample loaded onto a small-size detection zone [20,35,42]. The volume of loaded sample could be increased significantly through adopting water adsorbent, thereby improving the detection sensitivity [22]. In order to achieve enhanced sensitivity in our designed paper chip for Hg^{2+} colorimetric detection, we used a base to hold the paper chip and wicking pad. The patterns of paper chip and wood base were both easily produced. The cost of one paper chip is about 15 cents (CNY). The cost of the wood base is about twenty cents and it could be reusable. Thus, the device is low-cost. The wood base and paper chip were obtained through laser engraving as shown in Figure 1. The size of the paper chip and wood base could be easily controlled to match with each other. On the wood base, it was much simpler to cut a hole than engrave a well. In order to fix the wicking pad, scotch tape was used to seal the holes.

In order to prevent rapid sample leakage along the wood base surface, the inner surface of the base was hydrophobically modified by coating with paraffin. As shown in Figure S3, the contact angle to the waterdrop on the wood board surface was over 90 degree and the water drop on the base surface could be kept stable for over 60 min. These results confirmed the good hydrophobicity of the paraffin-modified base, which facilitated stable sample flow vertical from paper chip to the wicking pad and Hg^{2+} absorption by the DNA–AuNPs on the detection zone (Figure 2).

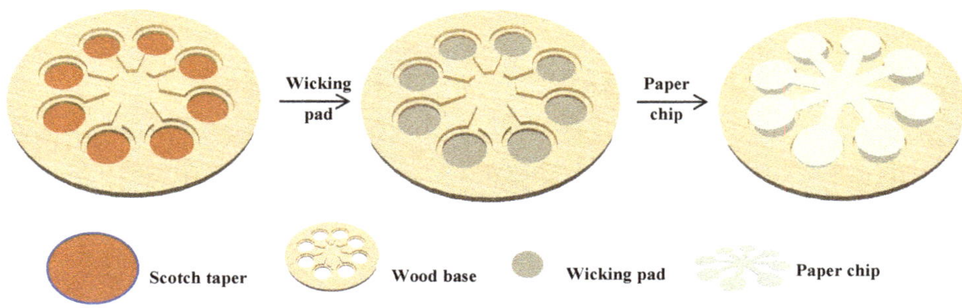

Figure 1. Schematic of paper device fabrication.

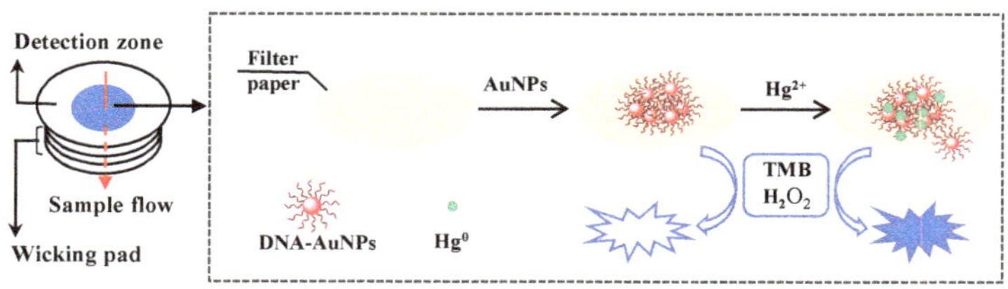

Figure 2. Schematic diagram of Hg^{2+} detection.

3.2. Colorimetric Detection of Hg^{2+}

The DNA–AuNPs had peroxidase-like activity and could catalyze the chromogenic reaction of TMB–H_2O_2, but the catalytic activity was weak. The peroxidase activity of the DNA–AuNPs could be significantly enhanced after the DNA–AuNPs adsorbed Hg^{2+} [15] and produced a very strong peak of TMB–H_2O_2 at 650 nm (Figure S4).

In order to obtain sensitive detection on the paper chip, the effect of H_2O_2 concentration and DNA–AuNP concentration were investigated. The optimal conditions were evaluated by the colorimetric intensity difference, $\Delta I = I - I_0$ (I and I_0 refer to the gray value obtained with and without Hg^{2+}). As shown in Figure 3a,b, the highest color intensity could be obtained with 3% H_2O_2 and 0.6nM DNA–AuNPs, respectively.

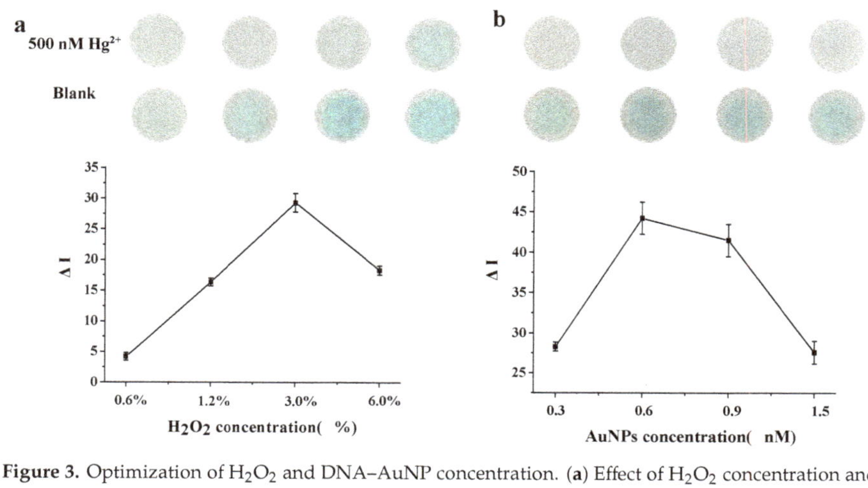

Figure 3. Optimization of H_2O_2 and DNA–AuNP concentration. (**a**) Effect of H_2O_2 concentration and (**b**) effect of DNA–AuNP concentration.

The Hg^{2+} volume was also investigated. When more than 20 μL was applied onto the detection zone, Hg^{2+} solution would overflow to the substrate reservoir, resulting in uncontrolled color development. With the superimposed wicking pad under paper chip, the volume of Hg^{2+} solution could be increased linearly with the increasing layers of wicking pad. In order to simplify the operation, five layers of wicking pad and 100 μL of Hg^{2+} solution at most were investigated. As shown in Figure 4a, it was found that darkest blue appeared when 60 μL of Hg^{2+} solution was used. Compared with 20 μL of Hg^{2+} solution, 60 μL was suitable for paper chip alone, and the ΔI increased four-fold. Unfortunately, no higher signal increasement was found when over 60 μL of Hg^{2+} solution was dropped onto the paper chip. These results were probably due to the fact that part of the DNA–AuNPs could be washed away by excessive Hg^{2+} solution. After incubating with Hg^{2+} for 15–20 min, the highest colorimetric intensity could be obtained when 60 μL of Hg^{2+} solution was used (Figure 4b).

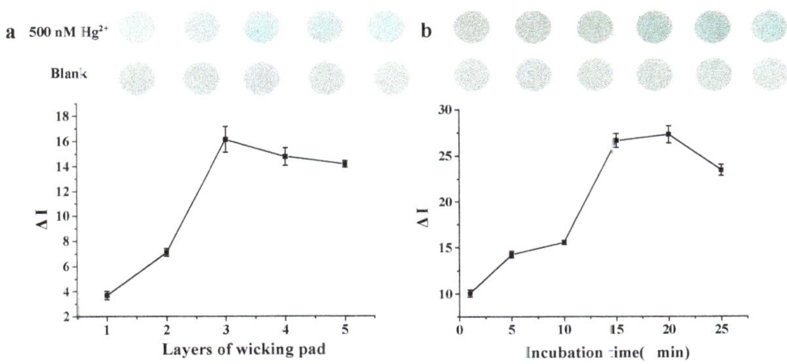

Figure 4. Optimization of the Hg^{2+} volume and adsorption time. (**a**) Effect of layers of wicking pad and (**b**) effect of incubation time.

As shown in Figure 5, the color intensity increased with the increased Hg^{2+} concentration on the paper device, and 50 nM Hg^{2+} could be distinguished by the naked eye. With the desktop scanning, quantitative determination could be implemented. A linear relationship between the gray intensity and logarithm of Hg^{2+} concentration could be obtained in the range of 0.05–2 μM. A detection limit of 10 nM was achieved, based on a 3σ/slope, where σ was the standard deviation of blank samples. Compared with some typical nanomaterial-modified papers or test strips for Hg^{2+} colorimetric detection, the above paper-device-based detection demonstrated comparable sensitivity (Table S1).

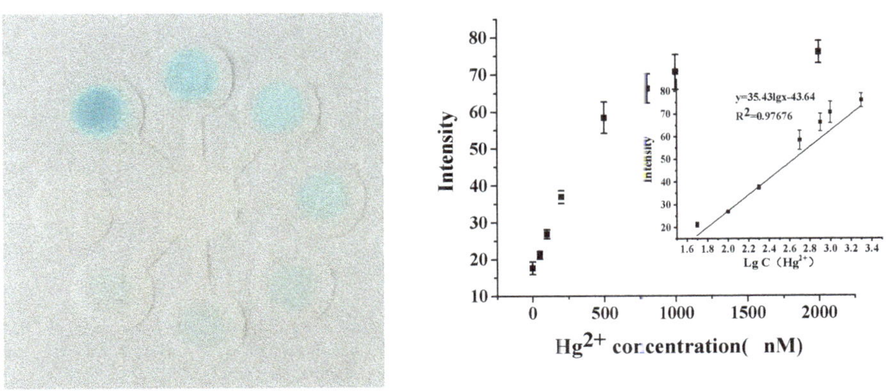

Figure 5. The image of detection of Hg^{2+} and calibration curve of colorimetric detection.

To explore the selectivity of this colorimetric detection, various common metal ions including $MeHg^{+}$, Mn^{2+}, Cu^{2+}, Ni^{2+}, Ba^{2+}, Cd^{2+}, Al^{3+}, Zn^{2+}, Fe^{3+}, Cr^{3+}, Co^{2+}, Sr^{2+}, and Bi^{3+} were tested. As shown in Figure 6, Hg^{2+} ions (1 μM) showed a deep blue color in the paper and negligible color responses were observed toward the other metal ions (10 μM), indicating that the high selectivity of this method was toward Hg^{2+}.

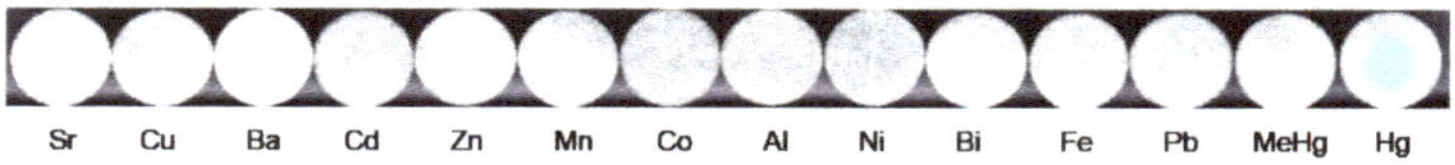

Figure 6. Selectivity of the method toward heavy metal ions.

3.3. Application in Real Samples

To verify the feasibility of this paper device in detecting real samples, tap water and lake water samples were spiked with Hg^{2+} and applied to the paper device. The results were obtained as shown in Table 1. The recoveries ranged from 85.7% to 105.6% when water samples spiked with 200, 500, and 1000 nM Hg^{2+} were measured. These results showed the great potential of this paper device for Hg^{2+} detection in practical applications.

Table 1. Determination of Hg^{2+} in tap water and lake water samples. (n = 3).

Sample	Added (nM)	Detected (nM)	Recovery (%)	RSD (%)
	200	189.8	94.9	3.6
Tap water	500	506.8	101.4	2.9
	1000	856.9	85.7	2.0
	200	197.6	98.8	4.7
Lake water	500	483.7	96.7	4.2
	1000	1056.0	105.60	2.7

4. Conclusions

In conclusion, a paper device consisting of a patterned paper chip and a base were successfully fabricated. The designed paper chip and wicking pad on the paper device facilitated the operation of DNA–gold nanozyme-based colorimetric detection of Hg^{2+} and enhanced the sensitivity. The color development of 50 nM Hg^{2+}on the paper device could be distinguished by the naked eye. Moreover, quantitative analysis of the color could be implemented by desktop scanner and gray intensity extracting. The colorimetric detection of Hg^{2+} was a low-cost, simple operation that demonstrated great potential in real sample detection. In addition, the paper device could be extended to combine with other nanosensors for more applications.

Supplementary Materials: The following are available online at http://www.mdpi.com/2079-6374/10/12/211/s1, Figure S1: Designed pattern of (**a**) paper chip and (**b**) base of the paper device; Figure S2: Photographs of paper-based device. (**a**) paper chip; (**b**) wood base; (**c**) wood base filled with wicking pad and (**d**) paper device; Figure S3: Photograph of a water drop on wood board surface; Figure S4: UV- vis absorption spectra of TMB-H_2O_2 reaction. (**a**) DNA-AuNPs without Hg^{2+}; (**b**) DNA-AuNPs with Hg^{2+}. Table S1: Comparison of paper-based devices for the detection of Hg^{2+} reported in literatures.

Author Contributions: Investigation, manuscript writing (M.-X.M.); Investigation, data treatment (R.Z.); Experiment design, result discussion, manuscript revision (C.-F.P.); Result discussion, manuscript revision (X.-L.W.). All authors have read and agreed to the published version of the manuscript.

Funding: This research was funded by the MOST, PRC (National Key Research and Development Program 2018YFC1604400), the Open Project Program of State Key Laboratory of Dairy Biotechnology, Bright Dairy & Food Co. Ltd. (SKLDB2017-00) and the National Natural Science Foundation of China (31871879).

Conflicts of Interest: The authors declare no conflict of interest.

References

1. Li, X.; Zhang, Y.; Chang, Y.; Xue, B.; Kong, X.; Chen, W. Catalysis-reduction strategy for sensing inorganic and organic mercury based on gold nanoparticles. *Biosens. Bioelectron* **2017**, *92*, 328–334. [CrossRef] [PubMed]
2. Deng, L.; Li, Y.; Yan, X.; Xiao, J.; Ma, C.; Zheng, J.; Liu, S.; Yang, R. Ultrasensitive and Highly Selective Detection of Bioaccumulation of Methyl-Mercury in Fish Samples via Ag-0/Hg-0 Amalgamation. *Anal. Chem.* **2015**, *87*, 2452–2458. [CrossRef] [PubMed]
3. Kan, C.; Shao, X.; Song, F.; Xu, J.; Zhu, J.; Du, L. Bioimaging of a fluorescence rhodamine-based probe for reversible detection of Hg (II) and its application in real water environment. *Microchem. J.* **2019**, *150*, 104142. [CrossRef]
4. Rofouei, M.K.; Rezaei, A.; Masteri-Farahani, M.; Khani, H. Selective extraction and preconcentration of ultra-trace level of mercury ions in water and fish samples using Fe3O4-magnetite-nanoparticles functionalized by triazene compound prior to its determination by inductively coupled plasma-optical emission spectrometry. *Anal. Methods* **2012**, *4*, 959–966.
5. Carneado, S.; Pero-Gascon, R.; Ibanez-Palomino, C.; Lopez-Sanchez, J.F.; Sahuquillo, A. Mercury(II) and methylmercury determination in water by liquid chromatography hyphenated to cold vapour atomic fluorescence spectrometry after online short-column preconcentration. *Anal. Methods* **2015**, *7*, 2699–2706. [CrossRef]
6. Zhou, Q.; Xing, A.; Zhao, K. Simultaneous determination of nickel, cobalt and mercury ions in water samples by solid phase extraction using multiwalled carbon nanotubes as adsorbent after chelating with sodium diethyldithiocarbamate prior to high performance liquid chromatography. *J. Chromatogr. A* **2014**, *1360*, 76–81. [CrossRef]
7. Wang, L.; Zhou, J.-B.; Wang, X.; Wang, Z.-H.; Zhao, R.-S. Simultaneous determination of copper, cobalt, and mercury ions in water samples by solid-phase extraction using carbon nanotube sponges as adsorbent after chelating with sodium diethyldithiocarbamate prior to high performance liquid chromatography. *Anal. Bioanal. Chem.* **2016**, *408*, 4445–4453. [CrossRef]
8. Anand, T.; Sankar, M. A dual colorimetric chemosensor for Hg(II) and cyanide ions in aqueous media based on a nitrobenzoxadiazole (NBD)-antipyrine conjugate with INHIBIT logic gate behaviour. *Anal. Methods* **2020**, *12*, 4526–4533. [CrossRef]
9. Feng, X.; Zhang, J.; Wang, J.; Han, A.; Fang, G.; Liu, J.; Wang, S. The stabilization of fluorescent copper nanoclusters by dialdehyde cellulose and their use in mercury ion sensing. *Anal. Methods* **2020**, *12*, 3130–3136. [CrossRef]
10. Chen, L.; Fu, X.; Lu, W.; Chen, L. Highly Sensitive and Selective Colorimetric Sensing of Hg^{2+} Based on the Morphology Transition of Silver Nanoprisms. *ACS Appl. Mater. Interfaces* **2013**, *5*, 284–290. [CrossRef]
11. Long, Y.J.; Li, Y.F.; Liu, Y.; Zheng, J.J.; Tang, J.; Huang, C.Z. Visual observation of the mercury-stimulated peroxidase mimetic activity of gold nanoparticles. *Chem. Commun.* **2011**, *47*, 11939–11941. [CrossRef] [PubMed]
12. Cai, S.; Lao, K.; Lau, C.; Lu, J. "Turn-On" Chemiluminescence Sensor for the Highly Selective and Ultrasensitive Detection of Hg^{2+} Ions Based on Interstrand Cooperative Coordination and Catalytic Formation of Gold Nanoparticles. *Anal. Chem.* **2011**, *83*, 9702–9708. [CrossRef] [PubMed]
13. Song, C.; Yang, B.; Yu, Z.; Yang, Y.; Wang, L. Ultrasensitive sliver nanorods array SERS sensor for mercury ions. *Biosens. Bioelectron.* **2017**, *15*, 59–65. [CrossRef] [PubMed]
14. He, Z.-J.; Kang, T.-F.; Lu, L.-P.; Cheng, S.-Y. An electrochemiluminescence sensor based on CdSe@CdS-functionalized MoS2 and a GOD-labeled DNA probe for the sensitive detection of Hg(ii). *Anal. Methods* **2020**, *12*, 491–498. [CrossRef]
15. Long, F.; Zhu, A.; Shi, H. Recent Advances in Optical Biosensors for Environmental Monitoring and Early Warning. *Sensors* **2013**, *13*, 13928–13948.
16. Tan, L.; Zhang, Y.; Qiang, H.; Li, Y.; Sun, J.; Hu, L.; Chen, Z. A sensitive Hg(II) colorimetric sensor based on synergistic catalytic effect of gold nanoparticles and Hg. *Sens. Actuator B Chem.* **2016**, *229*, 686–691. [CrossRef]
17. Peng, C.-F.; Pan, N.; Xie, Z.-J.; Wu, L.-L. Highly sensitive and selective colorimetric detection of Hg^{2+} based on the separation of Hg^{2+} and formation of catalytic DNA–gold nanoparticles. *Anal. Methods* **2016**, *8*, 1021–1025. [CrossRef]

18. Martinez, A.W.; Phillips, S.T.; Whitesides, G.M. Three-dimensional microfluidic devices fabricated in layered paper and tape. *Proc. Natl. Acad. Sci. USA* **2008**, *105*, 19606–19611. [CrossRef]
19. Fu, E.; Downs, C. Progress in the development and integration of fluid flow control tools in paper microfluidics. *Lab Chip* **2017**, *17*, 614–628. [CrossRef]
20. Nilghaz, A.; Lu, X. Detection of antibiotic residues in pork using paper-based microfluidic device coupled with filtration and concentration. *Anal. Chim. Acta* **2019**, *1046*, 163–169. [CrossRef]
21. Pena-Pereira, F.; Lavilla, I.; Bendicho, C. Paper-based analytical device for instrumental-free detection of thiocyanate in saliva as a biomarker of tobacco smoke exposure. *Talanta* **2016**, *147*, 390–396. [CrossRef] [PubMed]
22. Feng, L.; Li, X.; Li, H.; Yang, W.; Chen, L.; Guan, Y. Enhancement of sensitivity of paper-based sensor array for the identification of heavy-metal ions. *Anal. Chim. Acta* **2013**, *780*, 74–80. [CrossRef] [PubMed]
23. Evans, E.; Moreira Gabriel, E.F.; Benavidez, T.E.; Tomazelli Coltro, W.K.; Garcia, C.D. Modification of microfluidic paper-based devices with silica nanoparticles. *Analyst* **2014**, *139*, 5560–5567. [CrossRef] [PubMed]
24. Jeong, S.-G.; Lee, S.-H.; Choi, C.-H.; Kim, J.; Lee, C.-S. Toward instrument-free digital measurements: A three-dimensional microfluidic device fabricated in a single sheet of paper by double-sided printing and lamination. *Lab Chip* **2015**, *15*, 1188–1194. [CrossRef] [PubMed]
25. Yetisen, A.K.; Akram, M.S.; Lowe, C.R. Paper-based microfluidic point-of-care diagnostic devices. *Lab Chip* **2013**, *13*, 2210–2251. [CrossRef] [PubMed]
26. Liu, H.; Crooks, R.M. Three-Dimensional Paper Microfluidic Devices Assembled Using the Principles of Origami. *J. Am. Chem. Soc.* **2011**, *133*, 17564–17566. [CrossRef]
27. Ishii, S.; Segawa, T.; Okabe, S. Simultaneous Quantification of Multiple Food- and Waterborne Pathogens by Use of Microfluidic Quantitative PCR. *Appl. Environ. Microbiol.* **2013**, *79*, 2891–2898. [CrossRef]
28. Ornatska, M.; Sharpe, E.; Andreescu, D.; Andreescu, S. Paper Bioassay Based on Ceria Nanoparticles as Colorimetric Probes. *Anal. Chem.* **2011**, *83*, 4273–4280. [CrossRef]
29. Lee, Y.-F.; Huang, C.-C. Colorimetric Assay of Lead Ions in Biological Samples Using a Nanogold-Based Membrane. *ACS Appl. Mater. Interfaces* **2011**, *3*, 2747–2754. [CrossRef]
30. Ratnarathorn, N.; Chailapakul, O.; Henry, C.S.; Dungchai, W. Simple silver nanoparticle colorimetric sensing for copper by paper-based devices. *Talanta* **2012**, *99*, 552–557. [CrossRef]
31. Chaiyo, S.; Siangproh, W.; Apilux, A.; Chailapakul, O. Highly selective and sensitive paper-based colorimetric sensor using thiosulfate catalytic etching of silver nanoplates for trace determination of copper ions. *Anal. Chim. Acta* **2015**, *866*, 75–83. [CrossRef] [PubMed]
32. Esmaeili, N.; Rakhtshah, J.; Kolvari, E.; Shirkhanloo, H. Ultrasound assisted-dispersive-modification solid-phase extraction using task-specific ionic liquid immobilized on multiwall carbon nanotubes for speciation and determination mercury in water samples. *Microchem. J.* **2020**, *154*, 104632. [CrossRef]
33. Figueredo, F.; Garcia, P.T.; Corton, E.; Coltro, W.K.T. Enhanced Analytical Performance of Paper Microfluidic Devices by Using Fe3O4 Nanoparticles, MWCNT, and Graphene Oxide. *ACS Appl. Mater. Interfaces* **2016**, *8*, 11–15. [CrossRef]
34. Wang, P.; Ge, L.; Yan, M.; Song, X.; Ge, S.; Yu, J. Paper-based three-dimensional electrochemical immunodevice based on multi-walled carbon nanotubes functionalized paper for sensitive point-of-care testing. *Biosens. Bioelectron.* **2012**, *32*, 238–243. [CrossRef] [PubMed]
35. He, Y.; Zhang, S.; Zhang, X.; Baloda, M.; Gurung, A.S.; Xu, H.; Zhang, X.; Liu, G. Ultrasensitive nucleic acid biosensor based on enzyme-gold nanoparticle dual label and lateral flow strip biosensor. *Biosens. Bioelectron.* **2011**, *26*, 2018–2024. [CrossRef] [PubMed]
36. Qiao, Y.; Shang, J.; Li, S.; Feng, L.; Jiang, Y.; Duan, Z.; Lv, X.; Zhang, C.; Yao, T.; Dong, Z. Fluorimetric Mercury Test Strips with Suppressed "Coffee Stains" by a Bio-inspired Fabrication Strategy. *Sci. Rep.* **2016**, *6*, 36494. [CrossRef]
37. Zhang, Y.; Zuo, P.; Ye, B.C. A low-cost and simple paper-based microfluidic device for simultaneous multiplex determination of different types of chemical contaminants in food. *Biosens. Bioelectron.* **2015**, *68*, 14–19. [CrossRef]
38. Li, D.; Sun, Y.; Shen, Q.; Zhang, Q.; Huang, W.; Kang, Q.; Shen, D. Smartphone-based three-channel ratiometric fluorescent device and application in filed analysis of Hg^{2+}, Fe^{3+} and Cu^{2+} in water samples. *Microchem. J.* **2020**, *152*, 104423. [CrossRef]

39. Zhou, T.; Liu, J.-J.; Xu, Y.; Wu, Z.-Y. Fast and sensitive screening detection of tetracyclines with a paper-based analytical device. *Microchem. J.* **2019**, *145*, 703–707. [CrossRef]
40. Han, K.N.; Choi, J.S.; Kwon, J. Gold nanozyme-based paper chip for colorimetric detection of mercury ions. *Sci. Rep.* **2017**, *7*, 7. [CrossRef]
41. Xie, Z.-J.; Bao, X.-Y.; Peng, C.-F. Highly Sensitive and Selective Colorimetric Detection of Methylmercury Based on DNA Functionalized Gold Nanoparticles. *Sensors* **2018**, *18*, 2679. [CrossRef] [PubMed]
42. Mei, Z.; Chu, H.; Chen, W.; Xue, F.; Liu, J.; Xu, H.; Zhang, R.; Zheng, L. Ultrasensitive one-step rapid visual detection of bisphenol A in water samples by label-free aptasensor. *Biosens. Bioelectron.* **2013**, *39*, 26–30. [CrossRef] [PubMed]

Publisher's Note: MDPI stays neutral with regard to jurisdictional claims in published maps and institutional affiliations.

biosensors

MDPI

Communication

A Cellulose Paper-Based Fluorescent Lateral Flow Immunoassay for the Quantitative Detection of Cardiac Troponin I

Satheesh Natarajan [1], Joseph Jayaraj [1,2] and Duarte Miguel F. Prazeres [3,*]

[1] Healthcare Technology Innovation Centre, Indian Institute of Technology, Madras, Chennai, Tamil Nadu 600113, India; satheesh@htic.iitm.ac.in (S.N.); jayaraj@htic.iitm.ac.in (J.J.)
[2] Department of Electrical Engineering, Indian Institute of Technology, Chennai, Tamil Nadu 600113, India
[3] IBB—Institute for Bioengineering and Biosciences, Department of Bioengineering, Instituto Superior Técnico, Universidade de Lisboa, 1049-001 Lisboa, Portugal
* Correspondence: miguelprazeres@tecnico.ulisboa.pt

Abstract: This paper presents a lateral flow assay (LFA) for the quantitative, fluorescence-based detection of the cardiac biomarker troponin I (cTnI) that features an analytical strip made of cellulose filter paper. The results show that the wicking and test time are comparable to those obtained with conventional nitrocellulose (NC)-based LFAs. Further, the cellulose paper provides an excellent background with no auto-fluorescence that is very adequate in detecting fluorescent lines. While fluorescence that was generated with cellulose strips was lower when compared to that generated in NC strips, signals could be improved by layering carbon nanofibers (CNF) on the cellulose. A nonlinear behavior of the concentration–response relationship was observed for the LFA architectures with NC, cellulose, and cellulose-CNF in the 0 to 200 ng/mL cTnI concentration range. The measurements were consistent and characterized by coefficients of variation lower than 2.5%. Detection and quantitation limits that were in the range 1.28–1.40 ng/mL and 2.10–2.75 ng/mL were obtained for LFA with cellulose and cellulose CNF strips that are equivalent to the limits obtained with the standard NC LFA. Overall, we showed that commercially available filter paper can be used in the analytical strip of LFA.

Keywords: biomarker; carbon nanofiber; cellulose; diagnostics; immunoassay; lateral flow assays; paper; point-of-care testing; troponin I

Citation: Natarajan, S.; Jayaraj, J.; Prazeres, D.M.F. A Cellulose Paper-Based Fluorescent Lateral Flow Immunoassay for the Quantitative Detection of Cardiac Troponin I. *Biosensors* **2021**, *11*, 49. https://doi.org/10.3390/bios11020049

Received: 5 January 2021
Accepted: 11 February 2021
Published: 14 February 2021

1. Introduction

Lateral flow assays (LFA) are the dominant segment in the Point-Of-Care (POC) testing market. These portable devices are designed to perform diagnostics at the time and place of patient care [1–3]. The moving of screening, diagnosis, and monitoring testing from a laboratory setting to the field could be particularly useful in the context of (i) emergencies that require fast results for clinical and healthcare decision making, (ii) diagnosis of people in remote areas, (iii) regular monitoring of chronic patients, (iv) the testing of patients during primary-care appointments, and (v) auto-monitoring [1–3]. The target applications for LFA include infectious disease testing (e.g., influenza, HIV, hepatitis C); glucose, cholesterol, urine, haematology, pregnancy, and fertility monitoring; cardiac and tumor/cancer marker testing; coagulation and activated clotting time analysis; and, the control of drugs-of-abuse, among others [4]. The current SARS-CoV-2 coronavirus pandemic provides an excellent example of the advantages and complementarity of LFA diagnostics. In particular, and as the disease spreads, LFA tests are being extensively used to detect anti-viral antibodies (e.g., IgG, IgM) and, thus, determine who has been infected and what is the seroprevalence in the population [5,6]. Further, LFA that detect SARS-CoV-2 antigens have been developed for the rapid diagnosis of infection [7,8]. Apart from human diagnostics, testing at the point-of-contact with LFA is also being pursued in the veterinary, environmental, agro-food, forensics, and bio-defense areas [9,10].

The LFA concept and the underlying technologies that are required for the manufacturing of the LFA hand-held cartridges at scale are well established [9–11]. At the heart of a conventional LFA, we find a series of overlapping rectangular strips of different components (sample pad, conjugate release pad, analytical strip, and absorbent pad) that are mounted on a backing card and combined with specific reagents for analyte recognition (Figure 1a). While changes to this architecture have been proposed (see, for example, Parolo et al., [12]), this simple design is predominant. Each material in the LFA accomplishes a specific function: (i) the backing card provides support, (ii) the sample pad receives the liquid sample, (iii) the release pad contains reagents that are required for the test, (iv) the analytical strip contains test and control areas where signals are generated and detected, and (v) the absorbent pad acts as a sink to receive the liquid that runs through the LFA [13,14].

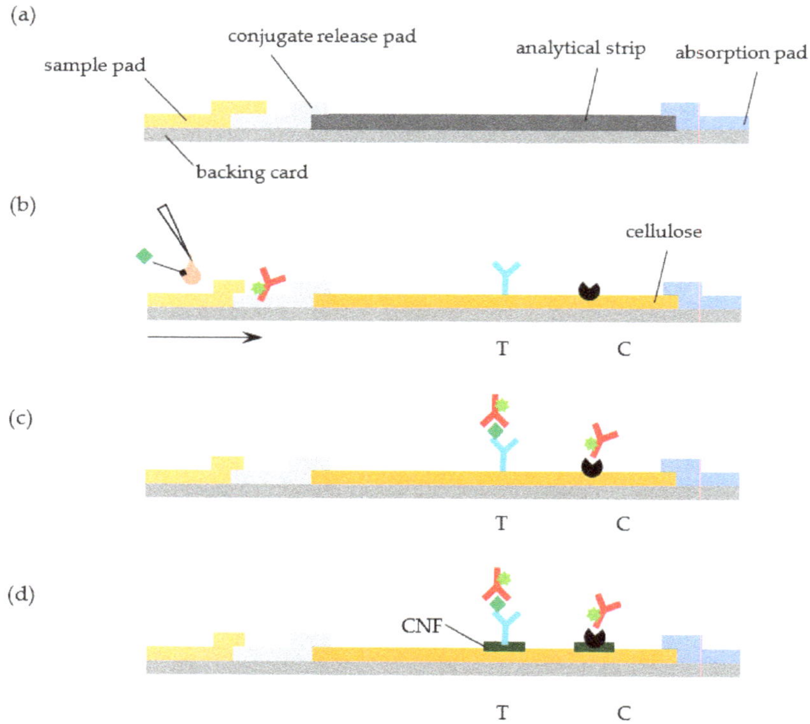

Figure 1. Lateral flow assays (LFA) with cellulose paper strip for detection of cardiac troponin I. (**a**) A standard LFA cartridge is made of overlapping materials mounted on a backing card: an analytical strip, usually made of nitrocellulose (NC), and sample, conjugate release and absorbent pads. (**b**) The LFA tested uses cellulose paper in the analytical strip instead of NC. Capture anti-cTnI IgG antibody (blue) and protein A (black) are adsorbed on the test (T) and control (C) lines, and Alexa Fluor-labeled anti-cTnI antibody (red) is impregnated in the conjugate pad. Upon addition of a sample of serum (~50 μL) spiked with cTnI (green lozenge), conjugates are released from the conjugate pad and form complexes with cTnI. (**c**) Anti-cTnI:cTnI complexes formed are captured by anti-cTnI antibodies in the test line. Unbound, residual complexes continue to move through the strip and are captured in the control line by protein A. (**d**) To improve detection, cellulose nanofibers (CNF) are layered over cellulose paper in the test and control line regions.

Porous nitrocellulose (NC) membranes are the material of choice for the analytical strip in most cases [13,14]. NC is prepared by incorporating nitro groups in the glucose units of cellulose chains via nitration. This NC polymer can then be cast in the form of membranes with thicknesses of the order 100–150 μm by controlling its precipitation from a solvent system. The resulting material strongly interacts with proteins via hydrophobic and hydrogen bonds interactions, with binding capacities of more than 100 μg of IgG per cm^2 [14]. This is one of the characteristics that justify the popularity of NC as a substrate for molecular detection. While NC is hydrophobic, it is rendered hydrophilic during casting by adding surfactants [14]. This confers it the ability to move fluids by capillarity that is critical for LFA. Other key characteristics of NC that make it ubiquitous in LFA include high brightness, wide availability at low cost, and with a range of porosities and pore sizes (3–20 μm) that are compatible with capillary migration, and ease of handling [13]. Nevertheless, some authors have pointed out that NC is not necessarily the best matrix for an LFA. For example, the inability to control the orientation of most proteins following adsorption usually translates into a loss of recognition activity [13] and, hence, in the need to use excess protein. Other notable shortcomings include, for example, changes and inconsistencies in the flow rates over time, lot-to-lot variability, flammable nature, sensitivity to humidity, and an inherent brittleness [13,14].

Cellulose constitutes an attractive and popular material for biosensors and LFA [15]. It is often found in the sample and absorbent pads due to its absorptive capacity, low cost, tensile strength, wide availability, and suitability for rapid roll-to-roll manufacture. However, few reports describe the use of cellulose in the analytical strip of LFA. This is somewhat surprising, given the widespread use of cellulose in many analytical (bio) chemistry applications and the fact that cellulose strips could cost one order of magnitude less than NC strips. For example, the increasing popularity of microfluidic paper-based devices provides a clear demonstration that cellulose could potentially be used in the analytical strip of LFA [16–19]. In one very comprehensive study, Lappalainen et al. investigated the adequateness of paper that is derived from different pulps as a material for the analytical strip of LFA by examining properties, like brightness, wet strength, and lateral flow speed [20]. The authors demonstrated that a functioning hemoglobin LFA could be set up by using analytical strips made of 80 g/m^2 paper derived from unbeaten, bleached eucalyptus pulp, and modified with a resin [20]. In a follow up study, the same group successfully implemented a direct sandwich assay for the detection of morphine in an LFA. where the NC, analytical membrane was substituted by paper manufactured in the house [21].

A couple more recent studies also demonstrate how LFA sensitivity to protein and nucleic detection can be enhanced by layering cellulose nanofibers (CNF) on the test line regions of NC strips [22,23]. The core idea is to promote the penetration of CNF into the top pores of NC and, thus, contribute to increase the amount of capture antibodies close to the surface of the strips. As a result, the density of selectively bound gold nanoparticles in the top part of test lines increases, enhancing the LFA sensitivity by 36.6% in the case of the detection of human IgG [22] and by 20-fold in the case of the detection of *Staphylococcus aureus* nucleic acid [23].

The goal of this work is to test whether the standard NC analytical strip in a LFA can be replaced by a commercially available cellulose paper strip without a significant loss of performance. We further study the effect of layering CNF on the test line regions of the cellulose strip in the signals generated (Figure 1d). As a model system, we consider the quantitative detection of cardiac troponin I (cTnI), a biomarker of myocardial cell damage, by a fluorescent LFA [24]. The system relies on the recognition of the target cTnI molecule by fluorescently labelled anti-cTnI detection antibodies and subsequent capture of the complex at the test line by an anti-cTnI capturing antibody (Figure 1b,c). The fluorescent signal that is generated at the LFA lines is then quantified while using a portable immunoanalyzer [25,26].

2. Materials and Methods

2.1. Materials

The cellulose paper strip (chromatography paper Whatman N.1), sample pad (CF4), conjugate pad (Fusion 5), absorbent pad (CF6), and Sephadex G20 were procured from

Cytiva (Bangalore, India). The NC membrane (Hi-Flow Plus HF180) was purchased from Merck Millipore (Burlington, MA, USA). The mouse anti-cTnI capturing antibody (clone 4C2 for cTnI), the mouse anti-cTnI detecting antibody (clone 19C7 for cTnI), recombinant troponin I, protein A, and Alexa Fluor 647 were from Abcam (Cambridge, UK). PBS (137 mm NaCl, 2.7 mm KCl, 10 mm Na_2HPO_4, 1.8 mm KH_2PO_4) and PB (75.4 mm Na_2HPO_4-$7H_2O$, 24.6 mm $NaH_2PO_4H_2O$) buffers, NaOH, $NaHCO_3$, NaN3, BSA, and Tween-20, N-methylmorpholine N-oxide (NMMO) were purchased from Sigma–Aldrich (St. Louis, MO, USA). CNF (ref. NG01NC0201, 10–20 nm width, 2–3 μm length) were bought from Nanografi Nano Teknoloji (Ankara, Turkey).

An Easy Printer Model LPM-02 from MDI-Advanced Microdevices Pvt. Ltd. (Ambala, India) was used to dispense lines over the analytical strip. The fluorescence signals that were generated at the LFA test and control lines were evaluated using ImageQuant (Figure S1 (Supplementary Materials)), a portable immunoanalyzer developed and designed at the Healthcare Technology Innovation Center (IIT, Madras, India) [25,26]. This instrument relies on a laser based confocal optics system to capture the fluorescence of the test and control lines of the LFA strips. The captured images are analyzed with LabVIEW™ software (National Instruments, Austin, TX, USA) to obtain signal data from test and control lines. The system uses intelligent image-analytics techniques that identify the reaction kinematics from a sequence of images, track the reaction progress and development of the test and control lines, identify the stabilization of the reaction, and calculate the test and control line areas and area ratios [25,26].

2.2. Antibody Labeling

The anti-cTnI detection antibody was labelled with the fluorescent dye Alexa Fluor 647 following the manufacturer's instructions. The Alexa Fluor 647 dye binds to the primary amine group of proteins at high molar ratios without self-quenching, forming stable dye–protein conjugates. The detection antibody (1 mg/mL in PBS buffer) was incubated with a 20 molar excess of the dye at room temperature for one hour under constant stirring. The fluorescent conjugates were purified by size exclusion chromatography on a Sephadex G20 gel column run with PBS buffer. Following purification, the conjugates were mixed with 0.02% NaN_3 and stored at −20 °C.

2.3. LFA Strip Assembly

The LFA strip was assembled by sequentially joining and partially overlapping four types of pads/materials: a sample pad (9.5 mm length) for the analyte application, a polyester fiber membrane (6 mm length) that holds the detection antibody-dye conjugate, a cellulose paper strip (27.8 mm length) to generate signals, and an absorbent pad (13.0 mm length). LFA with plain cellulose paper strips or with cellulose paper strips that were layered with CNF were also used. Layering of CNF was performed by dispensing a 0.5% (*w*/*w*) suspension of CNF in Milli Q water over the cellulose paper in the form of lines on the test and control region using the Easy Printer. The suspension was repeatedly dispensed on the same position to increase the concentration of CNF within cellulose pores (from one to six times, i.e., 0.5–2.5%, respectively). The cellulose strips were dried overnight at room temperature. The sample pad was soaked with PBS that was supplemented with 0.15% Tween-20, 1% sucrose, 0.5% BSA, and dried for one hour at RT. The conjugate pad was immersed in a 0.3 mg/mL solution of antibody-dye conjugate that was diluted in 100 mm PB buffer with 0.1% Triton, 0.1% BSA, 20% sucrose, and subsequently dried for one hour at 40 °C. The test zone of the LFA strips was observed by scanning electron microscopy (SEM) using a Quanta 200 SEM from FEI (Hillsboro, OR, USA) located at International Centre for Clean Water (ICCW), IIT Madras. Prior to analysis, the samples were coated with CNF that were dissolved in an NMMO solution using the Easy Printer instrument.

Capture anti-cTnI antibody and protein A were dispensed over NC, cellulose, or CNF-layered cellulose strips (1 μL/cm) at a rate of 0.2 mg/mL in 1 × PBS using the Easy Printer. The analytical strips were then kept at 37 °C for one hour. Finally, the pads and analytical strips were laminated with a partial overlapping of 2 mm and then cut with a width of 3.2 mm. The assembled LFA strips were kept at 4 °C until used.

2.4. Analysis of Troponin I Samples

Standard samples (0, 5, 25, 100, and 200 ng/mL) were prepared by mixing serum with solutions of cTnI that was prepared in 0.1 M PB buffer at a volumetric ratio of 1:49 v/v. For analysis, 50 µL of each standard were added to the LFA cartridge, which was then inserted into the ImageQuant analyzer. The run button was pressed and the process was monitored for about 15 min. The images of the test and control lines of the cellulose paper strip were captured and analyzed to quantify the fluorescence intensity of the generated signals. Assays were performed in triplicate for each sample.

All of the quantitative data were assessed with GraphPad Prism 6.0 (GraphPad Software, La Jolla, CA, USA). Fluorescence intensity data were used to calculate the pixel volume of the test, V_T, and control, V_C, lines, which correspond to the two-dimensional summation of all pixel intensities within each line [26]. The corresponding mean volume ratio, V_R, defined as the ratio V_T/V_C, was plotted versus the cTnI concentration to generate calibration curves. The standard deviation, SD, across triplicates was used to calculate the coefficient of variation (CoV) according to CoV = SD/mean × 100%. Calibration curve data were fitted to a power function, as described in Supplementary Materials. The limits of detection (LOD) and quantitation (LOQ) were determined based on the residual standard deviation and slope of the calibration curves obtained [27] (see Supplementary Materials).

3. Results and Discussion

Sandwich type assays for cTnI have been implemented in LFA while using detection strategies that rely on fluorophores [24], quantum dots [28], gold nanoparticles [29], or Raman tags [30], to name a few. On all of these systems, as in most LFA devices, NC is used in the analytical strip. Here, we examine whether the commercially available Whatman N. 1 chromatographic paper can be used instead of NC in a LFA for cTnI detection.

The cTnI LFA system used here relies on the recognition of the target cTnI molecule by fluorescently labelled anti-cTnI detection antibodies. Anti-cTnI capture antibody and protein A are immobilized on the test and control lines of the strip and conjugates of Alexa Fluor and anti-cTnI detection antibody are impregnated in the conjugate pad, as illustrated in Figure 1b,c. Upon sample addition and migration through the conjugate pad, conjugates are released, and they bind to the target analyte. The formed complexes then move through the cellulose paper strip and are captured by the anti-cTnI antibody adsorbed in the test line. Unbound, residual complexes continue to move through the strip, and they are captured in the control line by the adsorbed protein A.

The experiments were performed using LFA that was assembled with three different materials as analytical strips: (i) cellulose-LFA, (ii) cellulose-CNF-LFA, and (iii) NC-LFA. Cellulose-LFA were assembled using 27.8 mm length strips of Whatman N. 1 paper, a cellulose based (>98%) porous material defined by 15 µm fibers that has a basis weight of 87 g/m^2 (Figure S2 (Supplementary Materials)). This paper features pores with a size distribution centered around 5 µm and spanning the 1–19 µm range [31–34]. Aqueous solutions wick through Whatman N. 1 cellulose strips with flow times of the order of ~484 ± 69 s/4 cm, which are larger than the flow times that were obtained with typical NC membranes (e.g., 75–240 s/4 cm, [14]). Cellulose-CNF-LFA were assembled with paper strips that were pre-layered with CNF on the test zone. Control NC-LFA were also mounted while using conventional NC strips (Figure S2 (Supplementary Materials)).

In a typical experiment, a 50 µL sample is dispensed on the sample pad of the LFA cartridges. The cartridge is incubated and subsequently analyzed in the ImageQuant instrument. A 15-min incubation was found to be sufficient for the completion of the cellulose-LFA and cellulose-CNF-LFA, which is five minutes more than the time that is required to run the equivalent NC-LFA [24]. This is consistent with the difference in flow times for cellulose and NC reported above. No dimensional changes (curling, waving, and cockling) in the cellulose paper strip were observed following sample addition and completion of the test. Fluorescence images of the cellulose, cellulose-CNF, and NC strips in the LFA cartridges were then captured by the ImageQuant camera. Positive samples run in the cellulose and cellulose-CNF-LFA display well-defined and sharp fluorescent lines in the test and control zones that contrast significantly with the dark background that is provided by the cellulose (Figure S3 (Supplementary Materials)). This lack of background fluorescence indicates that Whatman paper is compatible with fluorescence detection. Following image processing by the instrument inbuilt software, the pixel volumes of the test (V_T), and control (V_C) lines were calculated and then used to determine the pixel volume ratio, V_R (see Shah et al. 2018 for details [26]).

Sets of experiments were performed with the three LFA types (NC, cellulose, cellulose-CNF) in triplicate while using serum samples that were spiked with known cTnI concentrations ranging from 0 to 200 ng/mL (Figure 2).

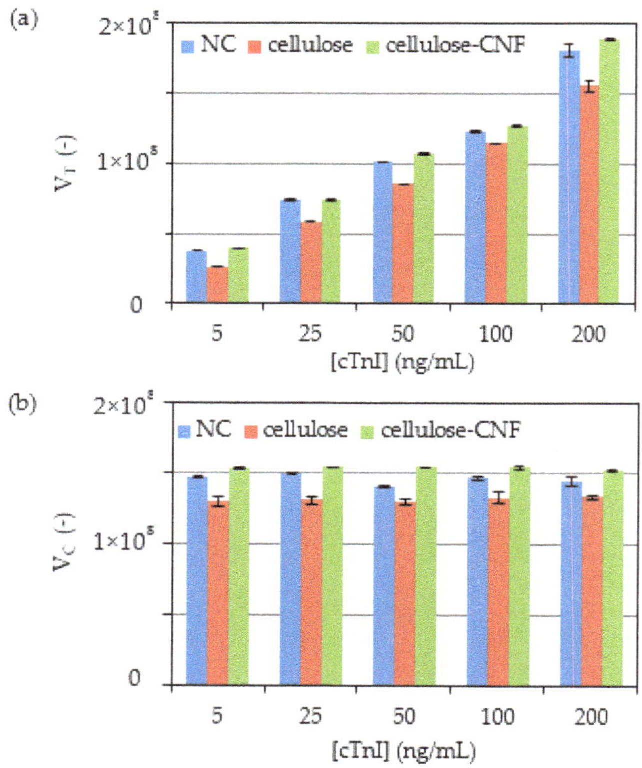

Figure 2. Effect of the concentration of samples containing cTnI on the immunofluorescence signals generated in the NC, cellulose and cellulose-CNF LFA. Pixel volumes of the (**a**) test and (**b**) control lines are shown. Experiments were run in triplicate.

Fluorescence signals were obtained in the test lines of the LFA with a pixel volume, V_T that increases with cTnI concentration, as expected for the sandwich assay implemented (Figure 2a), whereas the pixel volume of the control line, V_C, remained essentially constant (Figure 2b). A comparison of the signals that are generated across the cTnI concentration range shows that the replacement of NC for cellulose leads to a decrease in the fluorescence intensity of 7% to 31% and 7% to 13% for the test and control lines, respectively. These results are likely the consequence of paper properties that affect the distribution of capture and detection biomolecules over the test and control line volume. In particular, paper is thicker than NC (180 μm vs. 100 μm of the NC membrane used) and, thus, captured molecules will distribute over a larger volume of material, thus leading to lower densities at the line surface. Because the detection of fluorescence is mainly sensitive to complexes that are present on the surface of the strip, which is opaque, lower intensities are expected [29]. Nevertheless, we envisaged that fluorescence signals from cellulose strips could be improved by layering CNF on the test line regions of the paper strips, as described in the literature for the case of gold nanoparticle-based signals in NC strips [22,23]. Thus, LFA were prepared by depositing CNF generated from wood-derived fibrils with lengths in the micrometer and width in the nanometric range (Figure S4 (Supplementary Materials)), in the test and control regions of the strip. Photos of the strips with layered CNF obtained before and after running of tests provide evidence for the successful modification of CNF and confirm that the fibers are not washed away during the analysis (see Figure S5a–c (Supplementary Materials)). A comparison of the signals generated across the cTnI concentration range shows that the replacement of cellulose for cellulose-CNF results in a recovery of the intensity of signals generated at both the test and control lines, which even slightly surpassed the intensity of signals generated with NC strips (Figure 2).

A SEM observation of the test strips shows that the layering of the very thin and long CNF (Figure S4 (Supplementary Materials)) over cellulose significantly alters its microstructure (Figure 3 and Figure S5) by penetrating into and closing the pores of cellulose. The contrast between the CNF-layered cellulose and plain cellulose is quite evident in Figure 3a,b (also in Figure S5d,e (Supplementary Materials)), which captures the boundary region between the two zones at increasing magnification. It is also apparent from the figures that the deposition of CNF results in the covering of the cellulose microfibrils with a smoother mesh of material with a significantly lower porosity (Figure 3a). Given their dimensions, the numerous nanofibers deposited increase the cellulose surface area close to the top face of the strips. As a result, a larger amount of captured biomolecules will adsorb to the cellulose material close to the surface of the strips. This adsorption of antibodies to the cellulose fibers is likely to involve different types of interactions, e.g., hydrogen binding, van der Waals, electrostatic, and aromatic stacking interactions [35]. Subsequently, this translates into a higher density of fluorescence complexes at the surface and, hence, to an increase in the fluorescence signals.

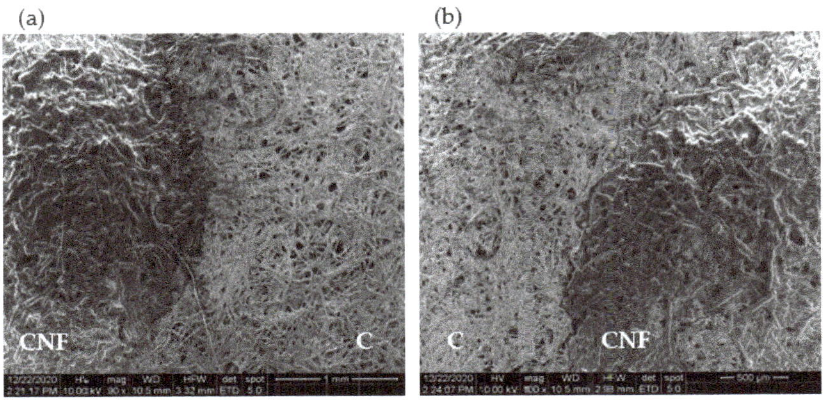

Figure 3. Scanning electron microscopy (SEM) analysis of cellulose strips with layered CNF at (**a**) 90×, and (**b**) 100× magnification. The boundary region between cellulose (marked C) and cellulose with layered CNF (marked CNF) is clearly visible.

The V_T and V_C data were used to calculate V_R, a relative measured used as the response of the LFA devices (Tables S1–S3 (Supplementary Materials)). Triplicate V_R data were further used to compute the individual CoV, which were then averaged to yield the intra-assay CoV (see Tables S1–S3 (Supplementary Materials)). The values of average CoV of 0.48%, 2.43% and 0.69% were obtained for the NC-LFA, cellulose-LFA, and cellulose-CNF-LFA, respectively (Table 1). This provides a good indication that measurements of cTnI concentration in the devices are reliable and consistent.

Table 1. Fitting of the experimental response of the three LFA types (NC, cellulose, cellulose-CNF) to a power-function response of the type $V_R = a[cTnI]^b$. The intra-assay CoV and the limits of detection and quantitation are also provided.

Parameter	NC	Cellulose	Cellulose-CNF
a	0.1320	0.0963	0.1320
B	0.4198	0.4785	0.4144
R	0.9958	0.9977	0.9957
CoV (%)	0.4771	2.430	0.6943
LOD (ng/mL)	1.39	1.28	1.40
LOQ (ng/mL)	2.73	2.10	2.75

The calibration curves were constructed next by plotting the replicate V_R data as a function of cTnI concentration for the three LFA types (NC, cellulose, cellulose-CNF). A nonlinear behavior of the concentration–response relationship was observed for the three different LFA architectures, which closely resembles a power-function response of the form:

$$V_R = a[cTnI]^b \quad (1)$$

where a and b are constants.

Equation (1) was linearized and fitted to the experimental data using the regression function of Microsoft Excel (2010) to extract the values of parameters a and b for the three LFA architectures (see Figures S6–S8 (Supplementary Materials)). The regression statistics data (Table 1) showed that Equation (1) fitted the experimental data very well, as can be judged by Figure 4.

In healthy individuals, the cTnI concentration is recorded at 0.1–0.3 ng/mL. However, with the onset of AMI, the cTnI values increase considerably and remain high for several hours [36]. A cTnI cutoff level of 6 ng/mL at one hour has been defined as appropriate for an accurate and rapid exclusion and identification of patients with suspected AMI [35]. The LOD and LOQ of the three LFA types tested were determined based on the residual standard deviation and slope of the linearized calibration curves [27] (see S1 (Supplementary Materials)). The obtained values were equivalent across the three LFA types, with values in the range 1.28–1.40 ng/mL and 2.10–2.75 ng/mL obtained for the LOD and LOQ, respectively (Table 1). Thus, these limits are compatible with a use in the context of emergencies, where rapid triage is required to exclude/identify patients with AMI.

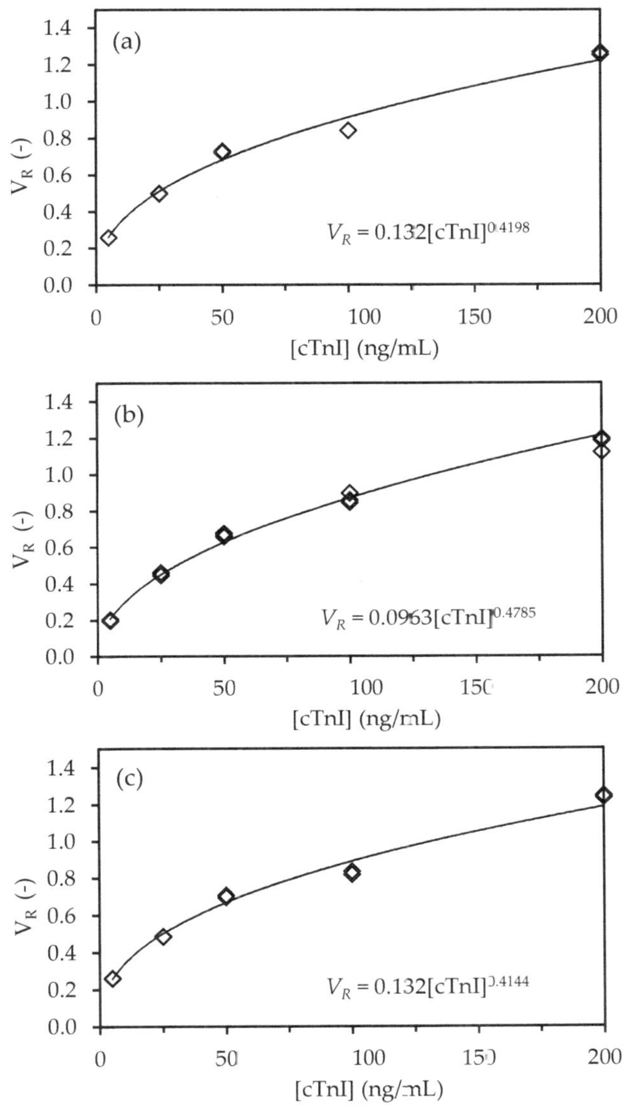

Figure 4. Calibration curves for the detection of cTnI with NC, cellulose and cellulose-CNF-LFA. The ratio of pixel volume of test line and control line ($V_R = V_T/V_C$) is plotted as a function of the concentration of cTnI ([cTnI]) for the (**a**) NC, (**b**) cellulose, and (**c**) cellulose-CNF-LFA. The experiments were run in triplicate and data were fitted with a power function of the type $V_R = a[\mathrm{cTnI}]^b$

4. Conclusions

We show that commercially available cellulose filter paper can be used as the analytical strip in a LFA for the quantitative, fluorescence-based detection of cTnI while using a sandwich type assay. While the flow and test times were slightly larger than those obtained with conventional NC-based LFAs, analysis was complete within 15 min. Further, fluorescence signals from test and control lines could be read with an image-based analyzer and then used to produce quantitative results. CNF were successfully layered on the test line regions in order to increase the amount of capture antibodies close to the surface of the cellulose strips. This resulted in the generation of immunofluorescence signals that were identical to those that were obtained with NC strips. The concentration–response relationship in the 0 to 250 ng/mL cTnI concentration range displayed a non-linear behavior for the three LFA architectures that could be described by a power law function. Measurements of cTnI concentration across the devices were reliable and consistent, as judged by a CoV lower than 2.5%. The LFA with cellulose and cellulose CNF strips displayed detection and quantitation limits that were in the range 1.28–1.40 ng/mL and 2.10–2.75 ng/mL, which were equivalent to the limits obtained with the standard NC LFA. Overall, we provide evidence that commercially available filter paper possesses adequate characteristics to replace nitrocellulose as the material of choice for the analytical strip in LFA.

Supplementary Materials: The following are available online at https://www.mdpi.com/2079-6374/11/2/49/s1, Figure S1: Photo of ImageQuant, the Image-based Quantitative Immunoassay Analyzer developed at HITC and used to evaluate the fluorescence signals generated at the LFA test and control lines, Figure S2: Photo showing the assembled nitrocellulose (a), cellulose (b), and cellulose with deposited CNF strips before running the tests. In (c) the deposited CNF in the test and control zones is clearly visible, Figure S3: Representative black and white (a) and color (b) fluorescence images of cellulose paper strips with test (T) and control (C) lines in LFA cartridges as captured by the ImageQuant camera. The cTnI concentrations of samples run in the LFA are displayed next to each photo. (c) Profiles of the fluorescence intensity alongside the cellulose strips in (a). The gray value along the axis of the images of strips shown in (a) was measured using the Analyze/Plot profile tool of the Image J software (NIH, National Institutes of Health), Figure S4: TEM image of the NG01NC0201 (Nanografi Nano Teknoloji, Turkey) carbon nanofibers used, Figure S5: High-resolution photos of the cellulose strips (a), cellulose strips with deposited CNF before running the tests (b) and cellulose strips with deposited CNF after running the tests (c). SEM analysis of cellulose strips with layered CNF at (a) 200× and (b) 500× magnification. In (a) the boundary region between cellulose (marked C) and cellulose with layered CNF (marked CNF) is clearly visible, Figures S6–S8: Linear regression of the log V_R vs. log [cTnI] data for the NC, cellulose and cellulose-CNF LFA. Table S1–S3: Response of the NC, cellulose and cellulose-CNF-LFA to serum samples with different concentrations of cTnI and calculation of the corresponding intra-assay coefficient of variation.

Author Contributions: Conceptualization, S.N. and D.M.F.P.; methodology, S.N. and J.J.; formal analysis, S.N., J.J. and D.M.F.P.; investigation, S.N. and D.M.F.P.; resources, S.N. and J.J.; writing—original draft preparation, D.M.F.P.; writing—review and editing, S.N. and D.M.F.P. All authors have read and agreed to the published version of the manuscript.

Funding: This research was funded by FCT-Fundação para a Ciência e a Tecnologia, grant number UID/BIO/04565/2020 and PTDC/CTM-CTM/30790/2017, and by the Healthcare Technology Innovation Center, Indian Institute of Technology, Madras.

Institutional Review Board Statement: Not applicable.

Informed Consent Statement: Not applicable.

Data Availability Statement: Not applicable.

Conflicts of Interest: The authors declare no conflict of interest.

References

1. Sharma, S.; Zapatero-Rodríguez, J.; Estrela, P.; O'Kennedy, R. Point-of-Care diagnostics in low resource settings: Present status and future role of microfluidics. *Biosensors* **2015**, *5*, 577–601. [CrossRef] [PubMed]
2. Gubala, V.; Harris, L.F.; Ricco, A.J.; Tan, M.X.; Williams, D.E. Point of care diagnostics: Status and future. *Anal. Chem.* **2012**, *84*, 487–515. [CrossRef] [PubMed]
3. St John, A.; Price, C.P. Existing and emerging technologies for point-of-care testing. *Clin. Biochem. Rev.* **2014**, *35*, 155–167.
4. Reports and Data, Point of Care (PoC) Diagnostics Market Analysis, By Product Type (Glucose Monitoring, Cardiometabolic Testing, Infectious Disease), by Platform Type (Lateral Flow Assays, Biosensors), by Prescription Mode, by End User, Forecasts to 2026. Available online: https://www.reportsanddata.com/report-detail/point-of-care-poc-diagnostics-market (accessed on 13 February 2021).

5. Adams, E.; Ainsworth, M.; Anand, R.; Andersson, M.I.; Auckland, K.; Baillie, J.K.; Barnes, E.; Beer, S.; Bell, J.I.; Berry, T.; et al. Antibody testing for COVID-19: A report from the National COVID Scientific Advisory Panel. *Wellcome Open Res.* **2020**, *5*, 139. [CrossRef]
6. Carter, L.J.; Garner, L.V.; Smoot, J.W.; Li, Y.; Zhou, Q.; Saveson, C.J.; Sasso, J.M.; Gregg, A.C.; Soares, D.J.; Beskid, T.R.; et al. Assay techniques and test development for COVID-19 diagnosis. *ACS Central Sci.* **2020**, *6*, 591–605. [CrossRef] [PubMed]
7. Grant, B.D.; Anderson, C.E.; Williford, J.R.; Alonzo, L.F.; Glukhova, V.A.; Boyle, D.S.; Weigl, B.H.; Nichols, K.P. A SARS-CoV-2 Coronavirus antigen-detecting half-strip lateral flow assay towards the development of point of care tests using commercially available reagents. *Anal. Chem.* **2020**, *92*, 11305–11309. [CrossRef]
8. Linares, M.; Pérez-Tanoira, R.; Romanyk, J.; Pérez-García, F.; Gómez-Herruz, P.; Arroyo, T.; Cuadors, J. Panbio antigen rapid test is reliable to diagnose SARS-CoV-2 infection in the first 7 days after the onset of symptoms. *J. Clin. Virol.* **2020**, *133*, 104659A. [CrossRef]
9. Wong, R.C.; Tse, H.Y. (Eds.) *Lateral Flow Immunoassay*; Humana Press: Amsterdam, The Netherlands, 2009.
10. O'Farrell, B. Lateral flow technology for field-based applications—Basics and advanced developments. *Top. Compan. An. M.* **2015**, *30*, 139–147. [CrossRef] [PubMed]
11. O'Farrell, B. Lateral flow immunoassay systems: Evolution from the current state of the art to the next generation of highly sensitive, quantitative rapid assays. In *The Immunoassay Handbook*; Wild, D.G., Ed.; Elsevier Science: Amsterdam, The Netherlands, 2013; pp. 89–107.
12. Parolo, C.; Medina-Sánchez, M.; Escosura-Muñiz, A.; Merkoç, A. Simple paper architecture modifications lead to enhanced sensitivity in nanoparticle based lateral flow immunoassays. *Lab. Chip.* **2013**, *13*, 386–390. [CrossRef] [PubMed]
13. O'Farrell, B. Evolution in lateral flow–based immunoassay systems. In *Lateral Flow Immunoassay*; Wong, R.C., Tse, H.Y., Eds.; Humana Press: Amsterdam, The Netherlands, 2009; pp. 1–34.
14. Mansfield, M.A. Cellulose membranes for lateral flow immunoassays: A technical treatise. In *Lateral Flow Immunoassay*; Wong, R.C., Tse, H.Y., Eds.; Humana Press: Amsterdam, The Netherlands, 2009; pp. 95–114.
15. Kamel, S.; Khattab, T.A. Recent advances in cellulose-based biosensors for medical diagnosis. *Biosensors* **2020**, *10*, 67. [CrossRef]
16. Nery, E.W.; Kubota, L.T. Sensing approaches on paper-based devices: A review. *Anal. Bioanal. Chem.* **2013**, *405*, 7573–7595. [CrossRef] [PubMed]
17. Hu, J.; Wang, S.; Wang, L.; Li, F.; Pingguan-Murphy, B.; Lu, T.J.; Xu, F. Advances in paper-based point-of-care diagnostics. *Biosens. Bioelect.* **2014**, *54*, 585–597. [CrossRef] [PubMed]
18. Rosa, A.M.M.; Louro, A.F.; Martins, S.A.M.; Silva, J.J.I.; Azevedo, A.M.; Prazeres, D.M.F. Capture and detection of DNA hybrids on paper via the anchoring of antibodies with fusions of carbohydrate binding modules and ZZ-domains. *Anal. Chem.* **2014**, *86*, 4340–4347. [CrossRef]
19. Ruivo, S.; Azevedo, A.M.; Prazeres, D.M.F. Colorimetric detection of D-dimer in a paper-based immunodetection device. *Anal. Biochem.* **2017**, *538*, 5–12. [CrossRef]
20. Lappalainen, T.; Teerinen, T.; Vento, P.; Hakalahti, L.; Erho, T. Cellulose as a novel substrate for lateral flow assay. *Nor. Pulp Pap. Res. J.* **2019**, *25*, 536–550. [CrossRef]
21. Teerinen, T.; Lappalainen, T.; Erho, T. A paper-based lateral flow assay for morphine. *Anal. Bioanal. Chem.* **2014**, *406*, 5955–5965. [CrossRef]
22. Quesada-González, D.; Stefani, C.; González, I.; Escosura-Muñiza, A.; Domingo, N.; Mutjé, P.; Merkoçi, A. Signal enhancement on gold nanoparticle-based lateral flow tests using cellulose nanofibers. *Biosens. Bioelec.* **2019**, *141*, 111407. [CrossRef]
23. Tang, R.H.; Liu, L.N.; Zhang, S.F.; Li, A.; Li, Z. Modification of a nitrocellulose membrane with cellulose nanofibers for enhanced sensitivity of lateral flow assays: Application to the determination of Staphylococcus aureus. *Microchim. Acta* **2019**, *186*, 831. [CrossRef]
24. Natarajan, S.; Su, F.; Jayaraj, J.; Shah, M.I.I.; Huang, Y. A paper microfluidics-based fluorescent lateral flow immunoassay for point-of-care diagnostics of non-communicable diseases. *Analyst* **2019**, *144*, 6291–6303. [CrossRef] [PubMed]
25. Joseph, J.; Vasan, J.K.; Shah, M.; Sivaprakasam, M.; Mahajan, L. iQuant™ Analyser: A rapid quantitative immunoassay reader. *Conf. Proc. IEEE Eng. Med. Biol. Soc.* **2017**, 3732–3736.
26. Shah, M.I.; Rajagopalan, A.; Joseph, J.; Sivaprakasam, M. An improved system for quantitative immunoassay measurement in ImageQuant. In Proceedings of the 2018 IEEE Sensors, New Delhi, India, 28–31 October 2018; pp. 1–4.
27. *Note for Guidance on Validation of Analytical Procedures: Text and Methodology (CPMP/ICH/381/95), ICH Topic Q 2 (R1)*; EMEA: London, UK, 2006.
28. Wu, M.; Zhang, X.; Wu, R.; Wang, G.; Li, J.; Chai, Y.; Shen, H.; Li, L.S. Sensitive and quantitative determination of cardiac troponin I based on silica-encapsulated CdSe/ZnS quantum dots and a fluorescence lateral flow immunoassay. *Anal. Lett.* **2020**, *53*, 1757–1773. [CrossRef]
29. Choi, D.H.; Lee, S.K.; Oh, Y.K.; Bae, B.W.; Lee, S.D.; Kim, S.; Shin, Y.-B.; Kim, M.-G. A dual gold nanoparticle conjugate-based lateral flow assay (LFA) method for the analysis of troponin I. *Biosen. Bioelect.* **2010**, *25*, 1999–2002. [CrossRef] [PubMed]
30. Khlebtsov, B.N.; Bratashov, D.N.; Byzova, N.A.; Dzantiev, B.B.; Khlebtsov, N.G. SERS-based lateral flow immunoassay of troponin I by using gap-enhanced Raman tags. *Nano Res.* **2019**, *12*, 413–420. [CrossRef]
31. Evans, E.; Gabriel, E.F.M.; Coltro, W.K.T.; Garcia, C.D. Rational selection of substrates to improve color intensity and uniformity on microfluidic paper-based analytical devices. *Analyst* **2014**, *139*, 2127–2132. [CrossRef] [PubMed]

32. Prazeres, D.M.F.; Rosa, A.M.M.; Microscopic Structure of Chromatographic Paper (100X). Figshare. 2015. Available online: https://figshare.com/articles/figure/Microscopic_structure_of_paper_100X_/1608695/3 (accessed on 13 February 2021).
33. Prazeres, D.M.F.; Rosa, A.M.M.; Microscopic Structure of Chromatographic Paper (700X). Figshare. 2015. Available online: https://figshare.com/articles/figure/Microscopic_structure_of_paper_700X_/1608696/2 (accessed on 13 February 2021).
34. Cummins, B.M.; Chinthapatla, R.; Ligler, F.S.; Walker, G.M. Time-dependent model for fluid flow in porous materials with multiple pore sizes. *Anal. Chem.* **2017**, *89*, 4377–4381. [CrossRef]
35. Lombardo, S.; Thielemans, W. Thermodynamics of adsorption on nanocellulose surfaces. *Cellulose* **2019**, *26*, 249–279. [CrossRef]
36. Neumann, J.T.; Sörensen, N.A.; Schwemer, T.; Ojeda, F.; Bourry, R.; Sciacca, V.; Schaefer, S.; Waldeyer, C.; Sinning, C.; Renné, T.; et al. Diagnosis of myocardial infarction using a high-sensitivity. *JAMA Cardiol.* **2016**, *1*, 397–404. [CrossRef] [PubMed]

biosensors

MDPI

Article

Detection of Cardiovascular CRP Protein Biomarker Using a Novel Nanofibrous Substrate

Isaac Macwan [1,*], Ashish Aphale [2], Prathamesh Bhagvath [3], Shalini Prasad [4] and Prabir Patra [2,5]

1 Department of Electrical and Bioengineering, Fairfield University, Fairfield, CT 06824, USA
2 Department of Biomedical Engineering, University of Bridgeport, Bridgeport, CT 06604, USA; aaphale@my.bridgeport.edu (A.A.); ppatra@bridgeport.edu (P.P.)
3 Department of Biomedical Engineering, Saint Louis University, St. Louis, MO 63103, USA; pbhagvath@slu.edu
4 Department of Bioengineering, University of Texas at Dallas, Richardson, TX 75080, USA; shalini.prasad@utdallas.edu
5 Department of Mechanical Engineering, University of Bridgeport, Bridgeport, CT 06604, USA
* Correspondence: imacwan@fairfield.edu; Tel.: +1-203-254-4000

Received: 1 June 2020; Accepted: 23 June 2020; Published: 24 June 2020

Abstract: It is known that different diseases have characteristic biomarkers that are secreted very early on, even before the symptoms have developed. Before any kind of therapeutic approach can be used, it is necessary that such biomarkers be detected at a minimum concentration in the bodily fluids. Here, we report the fabrication of an interdigitated sensing device integrated with polyvinyl alcohol (PVA) nanofibers and carbon nanotubes (CNT) for the detection of an inflammatory biomarker, C-reactive protein (CRP). The limit of detection (LOD) was achieved in a range of 100 ng mL^{-1} and 1 fg mL^{-1} in both phosphate buffered saline (PBS) and human serum (hs). Furthermore, a significant change in the electrochemical impedance from 45% to 70% (hs) and 38% to 60% (PBS) over the loading range of CRP was achieved. The finite element analysis indicates that a non-redox charge transduction at the solid/liquid interface on the electrode surface is responsible for the enhanced sensitivity. Furthermore, the fabricated biosensor consists of a large electro-active surface area, along with better charge transfer characteristics that enabled improved specific binding with CRP. This was determined both experimentally and from the simulated electrochemical impedance of the PVA nanofiber patterned gold electrode.

Keywords: biosensors; C-reactive protein; carbon nanotubes; electrospinning; electrochemical impedance

1. Introduction

Detection of biomarkers specific to a particular disease is a standard method for diagnosis and can be used for successful detection of various diseases [1]. A number of proteins considered as biomarkers are secreted at an early onset of a disease when the symptoms are yet to develop. A typical source to detect such biomarkers can be a patient's blood sample. C-reactive protein (CRP) has long been used as a marker of systemic inflammation, where the level of CRP increases several hundred fold within hours of an inflammatory occurrence [2] and it is found to be over-expressed in heart disease [3]. Previous research has shown that CRP is dominant in the instigation of several pathogenic pathways that may cause atherosclerosis, a precursor to cardiovascular disease [4]. According to the Center for Disease Control and Prevention, heart attack is the leading cause of death in the US with an estimated 785,000 Americans having a new coronary attack and approximately 470,000 with a recurrent attack [5]. Thus a rapid and ultra-sensitive detection of CRP from a patient's serum sample may be clinically significant in diagnosing cardiac events such as acute myocardial infarction. In recent years, the trend in healthcare has been towards developing devices that support personalized medicine with benefits such as faster diagnosis and therapeutic turnaround time.

Nanotechnology offers a wide range of materials that can be used both as biological recognition elements and transducers for a typical biosensor with exceptional levels of sensitivity that can be used to monitor biomarkers from different diseases. Novel properties of nanoscale materials, especially in the form of nanofibers, such as larger surface to volume ratios and size-based confinement of biomolecules, give them the potential to increase the sensitivity of the diagnostic devices by several orders of magnitude [6–11]. This property can be used to detect protein biomarkers such as CRP at very low concentrations and also give an early indication of a heart condition [12,13].

Some of the previous efforts in developing biosensors based on nanofibers involved citric acid-decorated nylon nanofibers for the detection of 3-phenoxybenzoic acid (3-PBA), a common human urinary metabolite [14], and polyvinyl alcohol-co-ethylene (PVA-co-PE)-based nanofibrous membranes for the detection of chloramphenicol (CAP) residues in milk [15]. One of the aspects of polyvinyl alcohol (PVA) and carbon nanotube (CNT)-based nanofibrous biological recognition elements is that they offer spatial confinement similar to the intracellular environment through porosity. The binding affinities and the rate at which a typical self-assembly process takes place can benefit from several physical constraints during the biosensing of macromolecules in vitro [16–18]. The nanotexturing of the sensor platform enables minimization of excluded volume of the biomolecules not participating in the interaction with the biomarker, which is also evident from the increase in the concentration that is analyzed through finite element analysis of diluted species [19]. Furthermore, even though individual fibers may have been laid on the electrode, there are still multiple fibers exhibiting a certain thickness similar to a membrane-like matrix but taking advantage of a higher specific surface area.

One of the major challenges faced by nanomaterial-based biosensors is the sensitivity and reproducibility of the results, which arises mainly because of the extreme difficulty in detecting a small quantity of protein biomarkers in serum samples [20–22]. Hence, determining the concentration of trace biomarkers in a complex mixture is a challenge in patient diagnosis.

Amongst several diagnostic biosensors, most of the conventional ones rely on fluorescent labeling or dyes for gathering the data [23]. Various sensing techniques that are applied to detect biomolecules at low concentrations are electrochemical analysis, surface plasmon resonance, electromagnetic measurements or mechanical actuation. Electrochemical sensors allow a label free detection of biomolecules by detecting and measuring the electrical signal [24–26]. A variety of electrochemical techniques based on integrated devices have been employed for label free and ultra-sensitive detection of different biomarkers. These methods are based on the principles of charge transfer [27–29], radiofrequency [30], complementary metal oxide semiconductors (CMOS) [31], capacitance [32], or impedance [33]. CMOS-based sensors are limited by the complexity of the fabrication technique, which leads to an extremely high cost of the overall biosensor device. In contrast, impedance and capacitance-based measurements are potential techniques for sensing a variety of biomolecules, mainly due to the low power consumption, ease of miniaturization and relatively low cost [34]. It is known that the performance of any biosensor depends on the immobilization of biomolecules on a biocompatible electrode surface. Therefore, the specificity of a biosensor can be increased by increasing the overall specific binding sites on the surface of the electrode. Hence, a biocompatible electrode, with an extremely high surface area that can engender spatial confinement for favorable binding events, would be a preferred platform for antigen–antibody interactions to achieve high sensitivity and specificity.

Here, we report, an extremely sensitive diagnosing platform with an ability to detect CRP concentrations up to fg mL^{-1} in both phosphate buffer saline (PBS) and human serum (hs). A microchip with seven interdigitated micro-comb capacitors is integrated with a randomly aligned electrospun PVA nanofibrous mesh. Owing to the random alignment of the nanofibers, the mesh showed high porosity for macromolecule confinement and an extremely small individual fiber diameter (~250 nm), along with a larger surface area to volume ratio, providing efficient antigen–antibody interactions. Furthermore, the signal transduction is found to be chemo-electro ionic, with the protein binding resulting in a modulation of the electrical double layer at the nanofibrous surface. A 45% change in the measured impedance from the antibody saturation baseline was observed while detecting the lowest detectable dose of 1 fg mL^{-1}. Finally, a change in the oxidation current for a gold substrate with and without PVA/CNT/anti-CRP functionalization provides crucial insights on the charge transfer resistance at the double layer capacitance interface between the metal electrode and the bulk electrolyte. The results obtained through a finite element model are in close agreement with the experimental data.

2. Materials and Methods

The nanoweb was fabricated using an electrospinning process (Figure 1a). The electrospinning set up comprises a high voltage power supply and a dual syringe pump unit. A positive electrode from the high voltage power supply is connected to the needle. The nanoweb is collected on a grounded metal collector covered with insulating polyethylene. To produce the PVA-CNT nanocomposite fibrous nanoweb, we applied 15 kV of voltage across 10 cm distance between the collector and the needle tip with a flow rate of the CNT dispersed polymer solution of 1.5 mL/h.

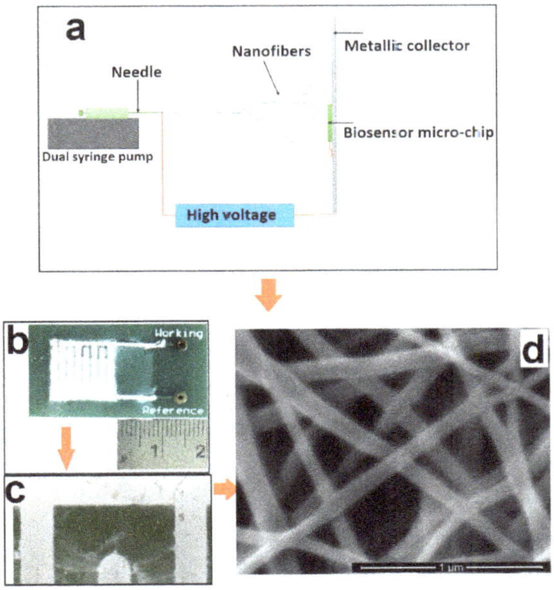

Figure 1. Fabrication of PVA/CNT nanofibers and their deposition on the biosensor: (**a**) Schematic of nanofiber deposition on the biosensor chip using electrospinning; (**b**) Optical image of the actual biosensor with selectively deposited nanofibers; (**c**) Scanning electron microscope (SEM) image of electrospun PVA/CNT nanofibers; (**d**) High resolution SEM of the deposited nanofibers clearly exhibiting the porous nature of the electrospun nanofibers with a diameter of ~250 nm.

The nanoweb biosensor chip was comprised of a base printed circuit platform. The platform is overlaid with the nanoporous nanoweb layer, which is then encapsulated by microfluidic manifolds manufactured using polydimethylsiloxane for fluid encapsulation and confinement. Based on the SEM images, the area of the pores is estimated to be in the range of 100 to 500 square nanometers. CRP is used in phosphate buffered saline as well as 50% human serum for evaluating the sensor performance.

Printed circuit board (PCB) chips comprising gold comb shaped designs of dimensions 10 mm × 10 mm in length; a finger width of 1 mm with spacings of 1 mm were manufactured with a FR4 passivation layer. After cleaning the surface of these chips with 10 mL of isopropyl alcohol

(IPA) and air drying the samples for 10 min, nanowebs constituting 10 mg of CNT in 10% PVA were electrospun (Figure 1b). Figure 1c shows a blown-up image of the individual fingers and the presence of the nanoweb in between.

To ensure that the nanofibers are deposited on the gold microelectrodes and not on the FR4 (glass reinforced epoxy laminate printed circuit board) surface of the microchip itself, both the 'Working' and the 'Reference' electrodes were also grounded. Interestingly, the above mentioned procedure results in selective deposition of nanocomposite fibers only on the conductive electrodes allowing the FR4 surface to be clean without any deposition. The average diameter of the nanocomposite fibers is ~250 nm, as seen from the SEM image in Figure 1d. Based on the flow rate of 1.5 mL/h and the electrospinning time of ~90 min, the thickness of the electrospun nanofibrous substrate was found to be ~0.5 mm. The presence of CNTs in the polymer matrix helps to stretch the nanofibers under the electric field, enabling continuous stretching and thereby forming a uniform fiber diameter without beaded morphology owing to the conductive nature of the CNTs. This ensures uniform fiber distribution, creating a spatially confined environment for better antigen–antibody interactions.

The dimensions of the nanowebs used for the experiments were 13.2 mm × 13.2 mm. The nanowebs were overlaid onto the metallic sensor surface using tweezers. The nanoweb was encapsulated with a polydimethylsiloxane (PDMS) manifold with dimensions of 13 mm × 13 mm. The manifold has a groove of 5 mm depth to enable the localization of the nanoweb onto the sensor surface. All the steps pertaining to the detection assay were performed on the assembled sensor chip.

Detection of the protein binding event was achieved using electrochemical impedance spectroscopy (EIS), and electrical double layer (EDL) capacitance measurement [9,10,12]. Briefly, the EDL comprises two components: the solution resistance (R_s) and the double layer capacitance (C_{dl}). The binding of the protein biomolecules to the covalently anchored antibodies in the nanotextured surface resulted in a modulation to the charge at the electrical double layer formed at the solid/liquid interface [12]. This change in charge produced a change in C_{dl}. The capacitance change was measured as an impedance change using an impedance analyzer (Impedance/Gain Phase Analyzer, Autolabs, Avon, IN, USA).

Oscillating AC fields of 0.05 V were applied, and the frequency was scanned over a range of 40 Hz–10 kHz. Frequencies only of the lower orders (up to 10 kHz) were considered for this study. At these low frequencies C_{dl} undergoes major variations during protein binding at the electrical interface. The sample volume was maintained throughout the experimental trials at 150 μL, sufficient to completely wet the sensor surface. This kept the solution resistance R_s constant within this frequency regime [10]. This EIS technique produced measurements that were purely the result of the changes in C_{dl}, and indicated protein binding.

Two sets of dose response experiments were performed to test the capabilities of the nanoweb sensor platform using electrochemical impedance measurements. The first set was focused on detecting CRP when aliquoted in isotonic buffer solution, 0.15 M phosphate buffer saline (PBS). Aliquots of hs-CRP ranging from 1 fg/mL to 100 ng/mL were prepared on a logarithmic scale. These concentrations were selected because they represent clinically relevant concentrations of CRP in physiological conditions. It is known that dithiobis succinimidyl propionate (DSP), which is a covalent linker, has a thiol (sulphur) group at one end that has high affinity for the gold surface while the other end, which is the NHS group, binds with the biological macromolecule (anti-CRP in this case) [35,36]. Based on the highly porous structure of the nanofibrous substrate, the DSP linker would permeate through these pores and anchor on the underlying gold substrate. This unique substrate now allows the interaction of the analyte (CRP) with the anti-CRP within these pores, increasing the overall sensitivity of the biosensor. Thus, after integrating the nanoweb mat and PDMS manifold onto the chip, 150 μL of 10 mM DSP crosslinker was injected into the manifold and incubated for 30 min at room temperature. The sensor surface was subsequently washed three times with 0.15 M PBS and baseline PBS measurements were taken. Following the crosslinker deposition, 150 μL of 50 μg/mL anti-CRP was incubated onto the sensor surface at 4 °C for 2 h to immobilize the receptors. After immobilizing anti-CRP on the sensing surface, 0.15 M PBS wash was performed three times followed by incubation of super block, a blocking protein that reduces the non-specific binding, on the sensor surface for 30 min at room temperature to minimize non-specific binding, which was again followed by a 0.15 M PBS wash that was performed three times. A zero dose, corresponding to 0.15 M PBS, was injected into the manifold and the measured impedance was considered as the baseline. All the impedance measurements for different dose concentrations of CRP were normalized to this baseline measurement. Starting from the lowest dose within the range, 150 μL of CRP spiked buffer was injected into the manifold, incubated for 15 min and impedance measurement was taken. The change in impedance from baseline measurement was calculated and converted into percentage change of impedance from baseline readings.

The second set of experiments focused on detecting CRP when aliquoted in CRP free human serum (hs). Aliquots of hs-CRP ranging from 1 fg/mL to 100 ng/mL were prepared on a logarithmic scale. After immobilization of anti-CRP onto the sensor surface, followed by blocking and washing, the zero dose, corresponding to CRP-free human serum, was injected into the manifold. EIS measurements were taken after 15 min incubation and considered as the baseline measurement. CRP-spiked human sera of different concentrations were subsequently injected onto the sensor surface and measured impedance was converted to percentage change from baseline impedance.

For finite element analysis, a portion of the microelectrode sensor was taken and electrochemical impedance analysis was carried out to obtain the Nyquist plots. First the area and the volume of the gold surface and the number of gold atoms at the surface of the electrode were calculated by using the concentration of redox couple as 10^{-3} mol/m^3. At the initial stage, only oxidizing agents are present at the electrode surface and, since the electrode is bare, these agents easily get electrons from the gold surface and get reduced. The initial reaction rate, K_0, at this stage was found to be ~2.5×10^{-4} cm/s. Utilizing the transport of diluted species module, DSP was diffused on the gold electrode and the surface concentration of DSP was calculated after 1600 s. From the concentration value obtained, it was found that the DSP monolayer occupied ~30% of the electrode surface with each molecule shielding around 8 gold atoms. According to this insight, ~50% of the gold atoms at the electrode surface are shielded, which affects the forward reaction rate coefficient, K_f, of the redox reaction and in turn reduces the K_0 by another 50%, making it ~1.25×10^{-4} cm/s. In the same manner, the concentrations of CRP and anti-CRP were also calculated. Considering the size of anti-CRP and CRP, it was found that they could shield around 22 and 62 gold atoms respectively. Thus, 60% of the gold atoms would be shielded by anti-CRP and 70% by CRP reducing the K_0 by 40% (1×10^{-4} cm/s of the initial value) after anti-CRP deposition and by 30% (7.5×10^{-5} cm/s) after CRP deposition. In the case of the nanofiber meshed electrode, we can see from the design that some part of the gold surface is occupied by nanofibers and it was estimated to be covering ~55% of the gold atoms. Thus, the initial value for K_0 was found to be ~1.125×10^{-4} cm/s. Taking the same assumption for the number of gold atoms shielded by DSP, anti-CRP and CRP, we get the following values of K_0: 0.5625×10^{-4} cm/s after DSP deposition, 0.45×10^{-4} cm/s and 0.3375×10^{-4} cm/s after anti-CRP and CRP deposition, respectively.

3. Results

3.1. Determination of CRP Using EIS

The binding of the protein biomolecules to the covalently anchored antibodies on the nanotextured surface results in the modulation of charge at the electrical double layer (C_{dl}), as seen in Figure 2. The binding of CRP to its antibody at this interface produces a specific and measurable change of impedance across the electrode. As the binding of the biomolecules occur directly on the sensor surface and is not mediated through a redox probe, the impedance changes are non-inductive in nature.

The charge transfer resistance (R_{ct}) experienced by the leakage charge through EDL at the sensor/analyte interface and the Warburg impedance (Z_w), which is the diffusional impedance experienced by ions in bulk buffer, are not dominant at lower frequencies and mostly treated as constant [33]. The solution resistance (R_s) is also considered constant as the analyte volume is kept constant throughout the experiments. Therefore, impedance measured at the lower frequencies, i.e., below 1 kHz, are representative of the changes to the EDL due to biomolecule binding. Hence, impedance data was represented for 100 Hz, which is a representative frequency of the low frequency range. A higher quantity of target protein in the analyte produces a higher amount of shift in the impedance from baseline and thereby allows for quantitative detection of the protein.

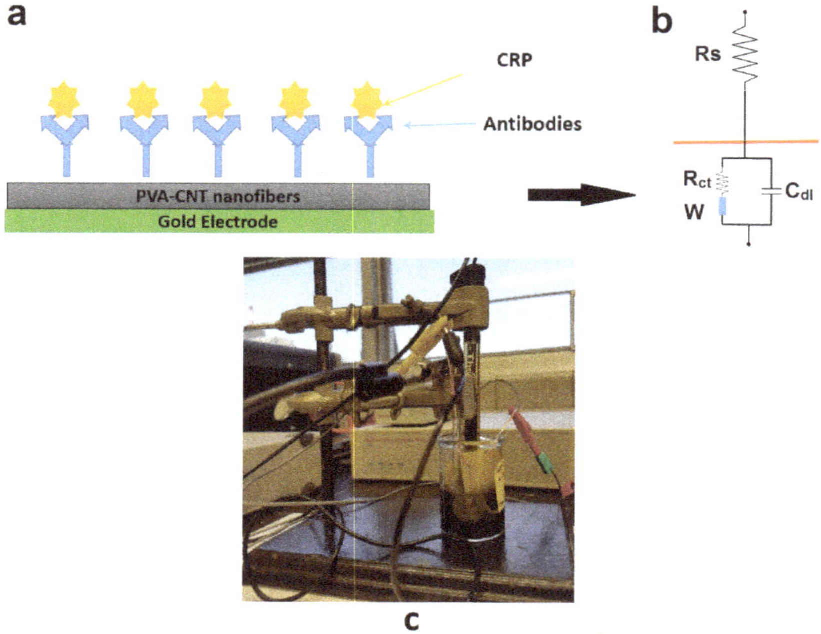

Figure 2. General scheme and electrochemical impedance measurement setup: (**a**) Immobilization of antibody/antigen on the electrospun nanostructured mesh selectively deposited on the interdigitated micro comb gold electrodes; (**b**) Resulting electrical circuit at the electrode/electrolyte interface; (**c**) Experimental setup for the quantification of the electrochemical impedance of the PVA/CNT-coated electrode.

A biosensor dose response analysis was performed based on the two sets of experiments with the CRP in PBS buffer and human serum (hs) to test the capabilities of the nanofibrous sensor platform as shown in Figure 3.

As can be seen from Figure 3a, the synthesized nanofibrous sensor platform was able to detect CRP from 100 ng/mL down to 1 fg/mL, indicating a unique ability of this platform to detect a minimum concentration of CRP. Each concentration was tested on different electrodes to avoid the removal of residual protein and successive regeneration of the sensor surface after every test. Also from Figure 3b, comparing the impedance values across this regime between the PBS and hs, it is found that there is relatively larger change in impedance for hs from ~360 ohms to ~220 ohms compared to PBS (~220 ohms to ~140 ohms), showing that this technique has a lot of potential to detect biological macromolecules in their native environment.

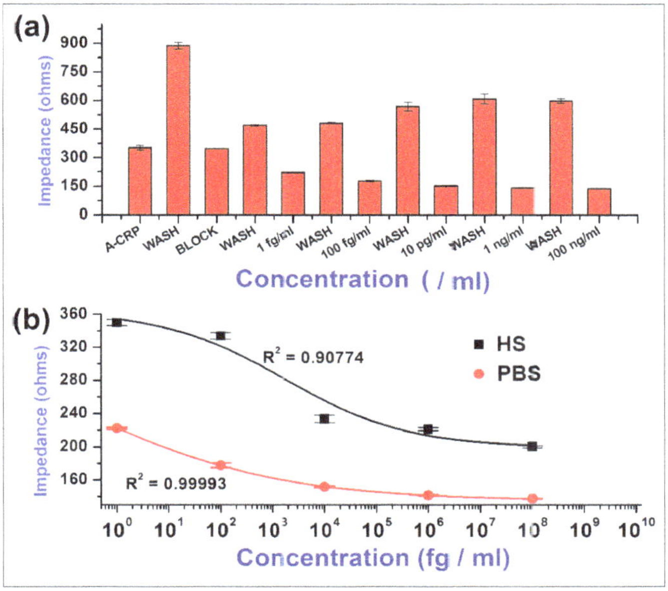

Figure 3. Dose response in PBS and hs: (**a**) Absolute impedance values after each step of antibody immobilization (A-CRP), followed by washing and superblock (Block) deposition to test the sensitivity of the biosensor on individual electrodes; (**b**) Changes in the impedance with respect to the concentration of CRP in both PBS and hs. All the impedance measurements are with respect to the concentration of antigen (CRP) per mL of either buffer (PBS/hs).

3.2. Biosensor Performance

The sensor performance was determined by the impedance change from the antibody saturation baseline expressed as a percentage change (Figure 4a,b). As the dose of CRP in the solution increased, the corresponding measured absolute impedance value from the biosensor decreased. Figure 4a shows the percentage change in the impedance with respect to the various doses of CRP in PBS and hs. The limit of detection (LOD) in this case was estimated to be 1 fg mL^{-1} with a percent change of 38% from the antibody baseline impedance, which is the anti-CRP mounted on the PVA/CNT-coated electrode control. The sensitivity of the sensor was computed by determining the dose of the antigen that gave at least 10% change from the baseline signal, where 10% change correlated to the signal background. Figure 4a also shows the dose response of CRP in 50% human serum. Initially the background signal from the antibody saturated sensor was determined by injecting 50% human serum directly on the sensor surface. The assay on the nanofibers was constructed in a similar manner for the detection of CRP from PBS samples. We observed a 45% change in the case of hs and a 38% change in the case of PBS for the impedance while detecting the lowest dose of 1 fg mL^{-1}. The percentage change in impedance ranged from 45% to 70% (hs) and 38% to 60% (PBS) over a concentration range of 1 fg mL^{-1} to 100 ng mL^{-1}. Saturation in measurement was observed for the CRP dose of 10 pg mL^{-1}, indicating a limit on linearity from the performance standpoint.

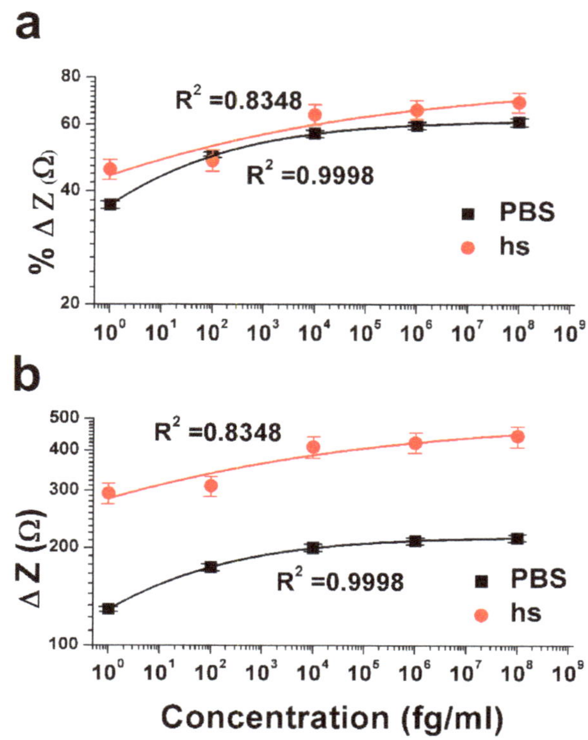

Figure 4. Dose response of CRP in PBS and hs: (**a**) Percentage change in impedance (%ΔZ) corresponding to the concentration of CRP; (**b**) CRP dose dependent change in the impedance (ΔZ).

3.3. Cross-Reactivity Test

Cross-reactivity tests were performed to determine the selectivity of the biosensor using anti-troponin-T as the antibody and CRP as the antigen. The sensor was immobilized with 1 μg mL^{-1} of anti-troponin-T and was prepared in a manner described earlier for the case of anti-CRP. The cross reactivity was studied by immobilizing anti-troponin-T with CRP in PBS and was compared to the performance of chips with anti-CRP and CRP in both hs and PBS (Figure 5). Dose response studies were conducted in a similar manner to those described in the previous subsection. As before, the impedance response for CRP interaction with anti-CRP ranged from ~38% to 60% in PBS and ~45% to 70% in hs. On the other hand, there was a very low response of the CRP interaction with anti-troponin-T, with the percentage change in impedance below 10%. These results indicated the robustness as well as the selectivity of the designed sensor and its response.

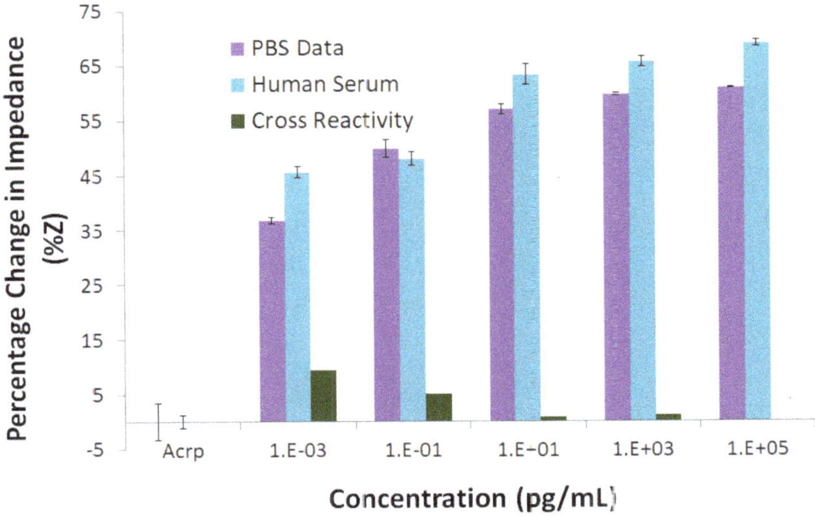

Figure 5. Cross reactivity study. Comparison of CRP binding with anti-CRP in both hs and PBS and cross reactivity of CRP with anti-troponin-T.

3.4. Finite Element Analysis to Determine the Change in Impedance on Detection of CRP

Both the electrodes, bare gold and the one covered with nanofiber mesh, were simulated for the change in impedance when depositing DSP, anti-CRP and CRP and were compared with the controls (absence of these molecules). Figure 6a shows the FEA model that was constructed with the same dimensions as the real electrode. Three layers of randomly oriented nanofibers were modeled on the constructed electrode having the same diameter (~250 nm) as the real nanofibers. As can be seen from Figure 6b,c, the Nyquist plots were obtained for both of these scenarios and the change in impedance was compared. It can be seen that a significant change in impedance with deposition of each layer on the bare gold electrode as well as on the nanofiber meshed gold surface was obtained. In addition, it is found that the impedance due to the CRP deposition on bare gold electrodes is less compared to the one with a nanofiber mesh in agreement with the real-time impedance data that was obtained showing the beneficial effects of the nanofiber mesh owing to its porosity.

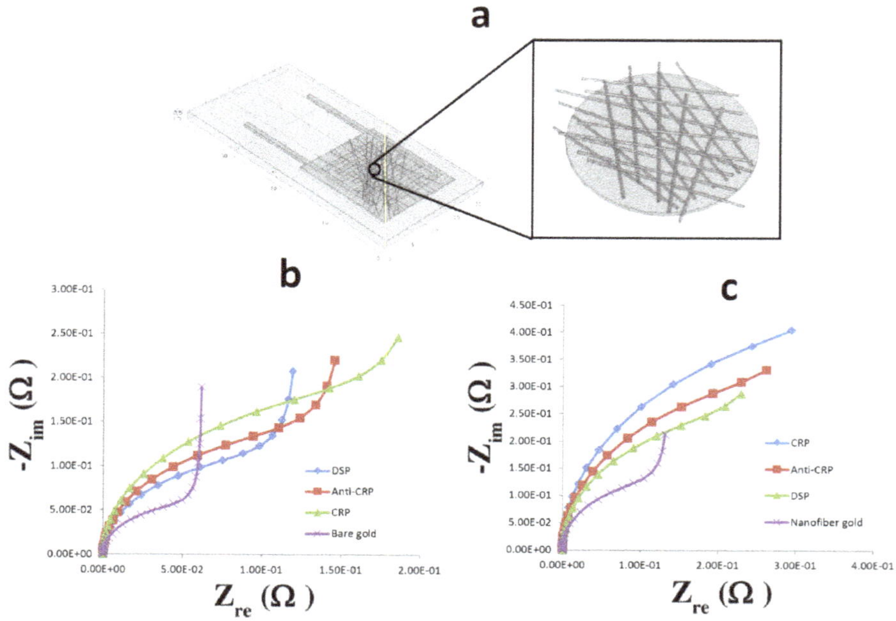

Figure 6. Finite element analysis of the change in impedance due to the presence of CRP, anti-CRP and DSP: (**a**) Modeled gold coated copper electrode on FR4 material; (**b**) Change in impedance due to the presence of DSP, anti-CRP and CRP on bare electrodes; (**c**) Change in impedance due to the presence of DSP, anti-CRP and CRP on nanofiber-coated electrodes.

4. Discussion

The PVA/CNT nanofiber-embedded biosensor device exhibits an amplification of the measured electrochemical impedance signal associated with the protein binding. CRP immobilization is achieved using a standard immunoassay protocol. The transduction is chemo-electro ionic with the protein binding resulting in a modulation of the electrical double layer at the interface of the nanofibers. Charged groups are present both on the hydrophilic surface as well as on the hydrophobic residues that are interior to the protein molecule [37]. Upon the application of voltage, the charged surface facilitates either repulsive or attractive forces on the ions at the electrode interface leading to a change in the dielectric thickness across the electrode [34]. These changes in the dielectric thickness are directly proportional to the thickness of the electrical double layer. When the electrode surface becomes nanotextured as in this particular study, the effect of the modulation on the dielectric thickness is amplified primarily due to the segmentation of the electrical double layer.

The changes in the impedance were comparable in both the ionic buffer as well as in human serum. The minimization of high background in the human serum buffer may be attributed to the sieve-like behavior of the nanofibers that provide size matched confinement of the target proteins. Furthermore, given the nature of the substrate, being nanofibrous and porous, it turns out to be a better microenvironment for the analyte to interact. In addition, the pores within the nanofibrous substrate would provide an environment that may promote the phenomenon of macromolecular crowding that typically exists in confined spaces, thereby altering the behavior of human serum [16,17]. Interestingly, it was observed that the protein association is significantly enhanced at lower concentrations, which in turn enhanced the sensitivity of the biosensor device.

The dimensions of the composite nanofibers are ~100 μm in length and ~250 nm in diameter. The nanofibers make a good biochemical transducer due to the presence of CNT and are highly suitable for achieving electrical signal amplification associated with the detection of the biomolecules. The resulting nanostructure comprises a non-periodic array of nanoscale confined spaces, which are electrically connected through the metallic micro electrode. The nanofibers have been surface functionalized with a protocol similar to the enzyme linked immunosorbent assay.

The most important characteristic of a biosensor is the calibration dependence of the impedance signal (Z) on the concentration of the analyte (C) [38]. The behavior of the calibration curve can be understood using a suitable mathematical equation such as the Hill isotherm model [39], which is used to describe the binding of different species onto homogenous substrates to fit the experimental data. According to this model, the antigen binding ability at one site on the macromolecule may influence different binding sites on the same macromolecule and the adsorption process is a cooperative phenomenon. As seen in Figure 4, the biosensor response in the form of percent change in impedance (%Δ Z) and difference impedance (ΔZ), are closely fitted with the following Hill formula:

$$y = Vmax \tag{1}$$

where y is the biosensor response (signal Z), $Vmax$ is the maximum state of the reaction reached at concentration x, k is the antigen concentration that binds the receptor sites at half concentration and n is the Hill coefficient.

For Figure 4, the values of n (0.2221 ± 0.00374 (PBS) and 0.1050 ± 0.0862 (hs)) are calculated based on Equation (1). This low value of Hill's coefficient at all the concentrations is a measure of the variable free energy of the interactions between biomolecules and surfaces [40–42]. In both PBS and hs, Hill's slope is less than 1, which is an indication of negative cooperative binding, which means that although many binding sites may be present, only specific binding occurs at a given point of time [43]. This observation is also supported by the cross-reactivity study, where at 1 fg mL^{-1} concentration of CRP, the percentage change in the impedance is less than 10%. The values of k are 0.1744 ± 0.0089 fg mL^{-1} (PBS) and 0.0708 ± 0.2357 fg mL^{-1} (hs), respectively.

5. Conclusions

A novel biosensor device is fabricated and tested for the detection of CRP protein. This affinity-based biosensor, working on the principle of electrochemical impedance spectroscopy, demonstrated an excellent signal to noise ratio with a detection limit of 1 fg mL^{-1}. Furthermore, a 45% (hs) and 38% (PBS) change in the electrical impedance was observed while measuring the lowest detectable dose of 1 fg mL^{-1}. It is proposed that within the confinement of the pores of the nanofibrous mesh, the overall specific binding increases, thereby reducing the noise set by non-specific interactions. The signal transduction is chemo-electro ionic with the protein binding resulting in a modulation of the electrical double layer capacitance at the nanofibrous interface. Furthermore, the simulated behavior of the biosensor electrode in the presence and absence of the nanofiber mesh clearly demonstrated a larger change in impedance owing to the nanofiber mesh indicating the role of porosity in the detection of CRP.

Author Contributions: Conceptualization, I.M., A.A. and P.P.; Methodology, A.A., S.P. and P.P.; Software, I.M. and P.B.; Validation, I.M., A.A. and P.B.; Formal analysis, I.M., P.B. and A.A.; Investigation, I.M. and A.A.; Resources, S.P. and P.P.; Data curation, I.M. and A.A.; Writing—original draft preparation, I.M. and A.A.; Writing—review and editing, I.M.; Supervision, S.P. and P.P.; Project administration, I.M., A.A. and P.P. All authors have read and agreed to the published version of the manuscript.

Funding: This research received no external funding.

Conflicts of Interest: The authors declare no conflict of interest.

References

1. Schiess, R.; Wollscheid, B.; Aebersold, R. Targeted proteomic strategy for clinical biomarker discovery. *Mol. Oncol.* **2009**, *3*, 33–44. [CrossRef] [PubMed]
2. Hage, F.G.; McCrory, M.; Szalai, A. C-reactive protein and cardiovascular disease: Lessons learned from studying genetically engineered mice. In *C-Reactive Protein: New Research*; Nagasawa, S., Ed.; Nova Publishers: Hauppauge, NY, USA, 2009; pp. 83–116. ISBN 978-1-60692-237-8.
3. Hage, F.G. C-reactive protein and hypertension. *J. Hum. Hypertens.* **2014**, *28*, 410–415. [CrossRef] [PubMed]
4. Nagai, T.; Anzai, T.; Kaneko, H.; Mano, Y.; Anzai, A.; Maekawa, Y.; Takahashi, T.; Meguro, T.; Yoshikawa, T.; Fukuda, K. C-reactive protein overexpression exacerbates pressure overload–induced cardiac remodeling through enhanced inflammatory response. *Hypertension* **2011**, *57*, 208–215. [CrossRef] [PubMed]
5. Benjamin, E.J.; Blaha, M.J.; Chiuve, S.E.; Cushman, M.; Das, S.R.; Deo, R.; de Ferranti, S.D.; Floyd, J.; Fornage, M.; Gillespie, C.; et al. Heart disease and stroke statistics—2017 update: A report from the american heart association. *Circulation* **2017**, *135*, e146–e603.
6. Peng, G.; Et, A. Detection of lung, breast, colorectal, and prostate cancers from exhaled breath using a single array of nanosensors. *Br. J. Cancer* **2010**, *103*, 524–551. [CrossRef]
7. Koh, I.; Josephson, L. Magnetic nanoparticle sensors. *Sensors* **2009**, *9*, 8130–8145. [CrossRef]
8. Engel, Y.; Elnathan, R.; Pevzner, A.; Davidi, G.; Flaxer, E.; Patolsky, F. Supersensitive detection of explosives by silicon nanowire arrays. *Angew. Chemie Int. Ed.* **2010**, *49*, 6830–6835. [CrossRef]
9. Dan, Y.; Lu, Y.; Kybert, N.J.; Luo, Z.; Johnson, A.T.C. Intrinsic response of graphene vapor sensors. *Nano Lett.* **2009**, *9*, 1472–1475. [CrossRef]
10. Vance, S.A.; Sandros, M.G. Zeptomole detection of C-reactive protein in serum by a nanoparticle amplified surface plasmon resonance imaging aptasensor. *Sci. Rep.* **2014**, *4*, 1–7. [CrossRef]
11. Fakanya, W.; Tothill, I. Detection of the inflammation biomarker c-reactive protein in serum samples: Towards an optimal biosensor formula. *Biosensors* **2014**, *4*, 340–357. [CrossRef] [PubMed]
12. Ridker, P. C-reactive protein a simple test to help predict risk of heart attack and stroke. *Circulation* **2003**, *108*, 81–85. [CrossRef] [PubMed]
13. Ridker, P.M. Clinical application of C-reactive protein for cardiovascular disease detection and prevention. *Circulation* **2003**, *107*, 363–369. [CrossRef] [PubMed]
14. El-Moghazy, A.Y.; Huo, J.; Amaly, N.; Vasylieva, N.; Hammock, B.D.; Sun, G. An innovative nanobody-based electrochemical immunosensor using decorated nylon nanofibers for point-of-care monitoring of human exposure to pyrethroid insecticides. *ACS Appl. Mater. Interfaces* **2020**, *12*, 6159–6168. [CrossRef] [PubMed]
15. El-Moghazy, A.Y.; Zhao, C.; Istamboulie, G.; Amaly, N.; Si, Y.; Noguer, T.; Sun, G. Ultrasensitive label-free electrochemical immunosensor based on PVA-co-PE nanofibrous membrane for the detection of chloramphenicol residues in milk. *Biosens. Bioelectron.* **2018**, *117*, 838–844. [CrossRef]
16. Zimmerman, S.B.; Minton, A.P. Macromolecular Crowding: Biochemical, biophysical, and physiological consequences. *Annu. Rev. Biophys. Biomol. Struct.* **1993**, *22*, 27–65. [CrossRef]
17. Ellis, R.J. Macromolecular crowding: Obvious but underappreciated. *Trends Biochem. Sci.* **2001**, *26*, 597–604. [CrossRef]
18. Dobson, C.M. Chemical space and biology. *Nature* **2004**, *432*, 824–828. [CrossRef]
19. Zimmerman, S.B.; Trach, S.O. Estimation of macromolecule concentrations and excluded volume effects for the cytoplasm of Escherichia coli. *J. Mol. Biol.* **1991**, *222*, 599–620. [CrossRef]
20. Kosaka, P.M.; Tamayo, J.; Ruz, J.J.; Puertas, S.; Polo, E.; Grazu, V.; de la Fuente, J.M.; Calleja, M. Tackling reproducibility in microcantilever biosensors: A statistical approach for sensitive and specific end-point detection of immunoreactions. *Analyst* **2013**, *138*, 863–872. [CrossRef]
21. Nair, P.R.; Alam, M.A. Performance limits of nanobiosensors. *Appl. Phys. Lett.* **2006**, *88*, 233120. [CrossRef]
22. Kosaka, P.M.; Pini, V.; Ruz, J.J.; da Silva, R.A.; González, M.U.; Ramos, D.; Calleja, M.; Tamayo, J. Detection of cancer biomarkers in serum using a hybrid mechanical and optoplasmonic nanosensor. *Nat. Nanotechnol.* **2014**, *9*, 1047–1053. [CrossRef] [PubMed]
23. Froehling, P.E. Dendrimers and dyes—A review. *Dye. Pigment.* **2001**, *48*, 187–195. [CrossRef]
24. Evanko, D. Label-free microscopy. *Nat. Methods* **2010**, *7*, 36. [CrossRef]
25. Sadik, O.A.; Aluoch, A.O.; Zhou, A. Status of biomolecular recognition using electrochemical techniques. *Biosens. Bioelectron.* **2009**, *24*, 2749–2765. [CrossRef] [PubMed]

26. Ma, H.; Wallbank, R.W.R.; Chaji, R.; Li, J.; Suzuki, Y.; Jiggins, C.; Nathan, A. An impedance-based integrated biosensor for suspended DNA characterization. *Sci. Rep.* **2013**, *3*, 2730. [CrossRef]
27. Maruyama, Y.; Terao, S.; Sawada, K. Label free CMOS DNA image sensor based on the charge transfer technique. *Biosens. Bioelectron.* **2009**, *24*, 3108–3112. [CrossRef]
28. Rothberg, J.M.; Hinz, W.; Rearick, T.M.; Schultz, J.; Mileski, W.; Davey, M.; Leamon, J.H.; Johnson, K.; Milgrew, M.J.; Edwards, M.; et al. An integrated semiconductor device enabling non-optical genome sequencing. *Nature* **2011**, *475*, 348–352. [CrossRef]
29. Gonçalves, D.; Prazeres, D.M.F.; Chu, V.; Conde, J.P. Label-free electronic detection of biomolecules using a-Si:H field-effect devices. *J. Non. Cryst. Solids* **2006**, *352*, 2007–2010. [CrossRef]
30. Kim, N.-Y.; Adhikari, K.K.; Dhakal, R.; Chuluunbaatar, Z.; Wang, C.; Kim, E.-S. Rapid, sensitive, and reusable detection of glucose by a robust radiofrequency integrated passive device biosensor chip. *Sci. Rep.* **2015**, *5*, 7807. [CrossRef]
31. Barbaro, M.; Bonfiglio, A.; Raffo, L.; Alessandrini, A.; Facci, P.; BarakBarak, I. A CMOS, fully integrated sensor for electronic detection of DNA hybridization. *IEEE Electron Device Lett.* **2006**, *27*, 595–597. [CrossRef]
32. Stagni, C.; Guiducci, C.; Benini, L.; Ricco, B.; Carrara, S.; Paulus, C.; Schienle, M.; Thewes, R. A Fully Electronic Label-Free DNA Sensor Chip. *IEEE Sens. J.* **2007**, *7*, 577–585. [CrossRef]
33. Sun, T.; Swindle, E.J.; Collins, J.E.; Holloway, J.A.; Davies, D.E.; Morgan, H. On-chip epithelial barrier function assays using electrical impedance spectroscopy. *Lab Chip* **2010**, *10*, 1611–1617. [CrossRef] [PubMed]
34. Daniels, J.S.; Pourmand, N. Label-free impedance biosensors: Opportunities and challenges. *Electroanalysis* **2007**, *19*, 1239–1257. [CrossRef] [PubMed]
35. Balamurugan, S.; Obubuafo, A.; Soper, S.A.; Spivak, D.A. Surface immobilization methods for aptamer diagnostic applications. *Anal. Bioanal. Chem.* **2008**, *390*, 1009–1021. [CrossRef] [PubMed]
36. Riepl, M.; Mirsky, V.M.; Novotny, I.; Tvarozek, V.; Rehacek, V.; Wolfbeis, O.S. Optimization of capacitive affinity sensors: Drift suppression and signal amplification. In *Proceedings of the Analytica Chimica Acta*; Elsevier: Amsterdam, The Netherlands, 1999; Volume 392, pp. 77–84.
37. Gitlin, I.; Carbeck, J.D.; Whitesides, G.M. Why are proteins charged? Networks of charge-charge interactions in proteins measured by charge ladders and capillary electrophoresis. *Angew. Chem. Int. Ed. Engl.* **2006**, *45*, 3022–3060. [CrossRef]
38. Kurganov, B.; Lobanov, A.; Borisov, I.; Reshetilov, A. Criterion for hill equation validity for description of biosensor calibration curves. *Anal. Chim. Acta* **2001**, *427*, 11–19. [CrossRef]
39. Hill, A. The possible effects of the aggregation of the molecules of haemoglobin on its dissociation curves. *J. Physiol.* **1910**, *40*, 4–7.
40. Endrenyi, L.; Fajszi, C.; Kwong, F.H. Evaluation of Hill slopes and Hill coefficients when the saturation binding or velocity is not known. *Eur. J. Biochem.* **1975**, *51*, 317–328. [CrossRef]
41. Wyman, J. Linked functions and reciprocal effects in hemoglobin: A second look. *Adv. Protein Chem.* **1964**, *19*, 223–286.
42. Saroff, H.A.; Minton, A.P. The Hill plot and the energy of interaction in hemoglobin. *Science* **1972**, *175*, 1253–1255. [CrossRef]
43. Kirsch, J.F. Enzyme kinetics and mechanism, by Paul F. Cook and WW Cleland. *Protein Sci.* **2008**, *17*, 380–381. [CrossRef]

biosensors

MDPI

Article

An All-in-One Solid State Thin-Layer Potentiometric Sensor and Biosensor Based on Three-Dimensional Origami Paper Microfluidics

Shiva Pesaran, Elmira Rafatmah and Bahram Hemmateenejad *

Chemistry Department, Shiraz University, Shiraz 71454, Iran; sh.pesaran@shirazu.ac.ir (S.P.); elmirarafatmah@shirazu.ac.ir (E.R.)
* Correspondence: hemmatb@shirazu.ac.ir; Tel.: +98-71-3646-0724; Fax: +98-71-3646-0788

Abstract: An origami three-dimensional design of a paper-based potentiometric sensor is described. In its simplest form, this electrochemical paper-based analytical device (ePAD) is made from three small parts of the paper. Paper layers are folded on each other for the integration of a solid contact ion selective electrode (here a carbon-paste composite electrode) and a solid-state pseudo-reference electrode (here writing pencil 6B on the paper), which are in contact with a hydrophilic channel fabricated on the middle part (third part) of the paper. In this case, the pseudo-reference and working electrodes are connected to the two sides of the hydrophilic channel and hence the distance between them is as low as the width of paper. The unmodified carbon paste electrode (UCPE) and modification with the crown ether benzo15-crown-5 (B15C5) represented a very high sensitivity to Cu (II) and Cd^{2+} ions, respectively. The sensor responded to H_2O_2 using MnO_2-doped carbon paste electrode (CPE). Furthermore, a biosensor was achieved by the addition of glucose oxidase to the MnO_2-doped CPE and hence made it selective to glucose with ultra-sensitivity. In addition to very high sensitivity, our device benefits from consuming a very low volume of sample (10.0 μL) and automatic sampling without need for sampling devices.

Keywords: paper-based origami sensor; three-dimensional microfluidic; potentiometric; biosensor; carbon paste electrode

Citation: Pesaran, S.; Rafatmah, E.; Hemmateenejad, B. An All-in-One Solid State Thin-Layer Potentiometric Sensor and Biosensor Based on Three-Dimensional Origami Paper Microfluidics. *Biosensors* **2021**, *11*, 44. https://doi.org/10.3390/bios11020044

Received: 17 January 2021
Accepted: 7 February 2021
Published: 10 February 2021

1. Introduction

Electroanalytical sensors based on ion-selective potentiometry have gained a lot of interests in analytical chemistry because of their simplicity, high speed, low cost, wide dynamic range and potential for miniaturization. Potentiometry is still the universal approach for pH measurement and is among significant measurement techniques in clinical laboratories. Now-a-day, ion-selective electrodes (ISE) have also become versatile tools for the detection and analysis of organic and inorganic species [1]. In addition, the development of thin-layer potentiometry and solid contact ion-selective electrodes (SCISE) during the last few decades provide potentiometric sensors as an ideal candidate for fabrication of miniaturized sensors with the enhanced sensitivity and simplified construction and operation methods [2,3].

In spite of the drastic properties encountered for the potentiometric ion-selective sensors, further, improvements are required. Fabrication of sensors having a lower cost (for large scale applications and using them as a point of care diagnostic tools), higher sensitivity and diminished memory effects, being easy to prepare and operate and of course being portable are appreciated. In this regard, paper-based analytical devices (PAD) have becoming attractive platforms for the fabrication of novel and affordable sensors ad biosensors [4].

Paper substrate as micro-PAD (μPAD) in the form of microfluidic was first reported in 2007 by Whitesides et al. [5]. Paper as a substrate has recently attracted much attention because it is a flexible, readily available with low cost and is biocompatible [6,7]. The μPADs, which rely on the transportation of solutions by capillary force through a device [8], are fabricated and patterned into one-dimensional (1D), two-dimensional (2D) and more complex three-dimensional (3D) microfluidic devices [9,10]. In the 3D devices, which allow rapid distribution of the sample in the z-direction [11], the origami design can effectively eliminate problems of reagent diffusion by lateral flow in the channels of planar paper devices [12].

Since a few years ago, resourceful μPADs have been developed with a variety of detection methods including but not limited to colorimetry [13–15], fluorescence [16], chemiluminescence [17], electrochemistry [18,19] and electroluminescence [20]. Among the colorimetric and electrochemical methods, which have found more extensive applications, paper-based electrochemical devices (ePAD), which were firstly published by Dungchai [21], have the advantages of insensitivity to color interferences and fluctuation in the environmental light and are being more quantitative with higher sensitivity and wider linear range compared to colorimetric-based PADs.

According to the explanations outlined above, we decided to design an origami paper-based potentiometric sensor applicable in sensing of a wide range of species (both chemical and biochemical). Monitoring of metal ion concentration is very important for prevention of negative impacts that it can have on human health. The performance of the proposed μPAD was first investigated in analysis of inorganic ions such as Cu^{2+}, Cr^{3+}, Ag^{+} and Hg^{2+}. Chromium can cause allergic reactions on the skin and can be carcinogenic, mercury compounds are very strong poisons, copper is necessary for different chemical and bio-chemical processes in the body, but it can be toxic above a certain concentration. After-wards it was decided to assess this origami paper-based potentiometric sensor's application in detecting of glucose as bioanalyte. It was observed that the origami structure of the sensor showed its significance by comparing the obtained detection limits (DLs) with previous similar reported bulk methods.

2. Materials and Methods

2.1. Materials and Equipment

D-Fructose, sucrose, D-maltose, D-galactose, lactose, glycogen, graphite powder, MnO_2, $KMnO_4$, $Na_2C_2O_4$, NH_4Cl, $Cd(NO_3)_2{\cdot}4H_2O$, K_2HPO_4, Nujol oil, benzo15-crown-5, $Fe(NO_3)_3{\cdot}9H_2O$, $Ni(NO_3)_2{\cdot}6H_2O$, KNO_3, sodium acetate, urea, uric acid, NaCl, KCl, $CaCl_2{\cdot}2H_2O$, $MgSO_4{\cdot}7H_2O$, $NaHCO_3$, Na_2SO_4, Na_2HPO_4, $NaH_2PO_4{\cdot}H_2O$, ammonia solution 25%, hydrogen peroxide 30% and murexide were all purchased from Merck. KH_2PO_4, creatinine, $Cu(NO_3)_2{\cdot}3H_2O$, $AgNO_3$, $Hg(NO_3)_2{\cdot}H_2O$, $[Cr(H_2O)_6](NO_3)_3{\cdot}3H_2O$ were purchased from Fluka. L(+)-Cysteine and Glucose oxidase (GOx), from Aspergillus niger type II, were purchased from Riedel-de-Haën and Sigma-Aldrich respectively. $Na_3C_6H_5O_7{\cdot}2H_2O$, D-glucose and glacial acetic acid were purchased from BDH Chemicals while Schleicher & Schuell® (s&s) Grade 2040b qualitative filter paper (with a thickness of is 0.2 mm) was used for device fabrication. Carbon nanotube (CNT) was a gift of Dr. Doroodmand's lab. Koh-i-Noor Hardtmuth pencils 3B, 4B, 6B, 9B, 4H and HB of different commercial brands were collected from local stores. All solutions were prepared using double deionized water.

An HP LaserJet 1320 printer from HP was used to print the devices. A Memmert Incubator Oven INB200 was used for curing the printed μPADs. Electrochemical measurements were made using an AZ-86502 bench top pH meter at room temperature (25.0 °C), Bionime GM110 Blood glucose monitor was used as reference method. A lab-made potentiometer was used for wifi sending of potential data on a mobile phone. The scanning electron micrographs (SEM) were obtained with a TESCAN model VEGA3 instrument.

2.2. Device Fabrication

The designed pattern in AutoCAD software was laser printed on one side of the filter paper. Precision sensor pattern details are presented in Figure 1, where white and black colors represent hydrophilic and hydrophobic area of the sensor, respectively. As shown in Figure 1, in the left and right layers are provided hydrophilic zones for working and reference electrodes, respectively. The hydrophilic zone in the middle layer is used for sampling and plays the role of sample vessel. The device includes two other layers with small hydrophilic area, which are used as channels for connecting electrodes with sample solution. The size of sensor after folding it in the origami shape is around (2 × 2 cm). Hydrophobization of the printed areas was performed by heating them in the oven with a temperature of 200.0 °C for an hour [22].

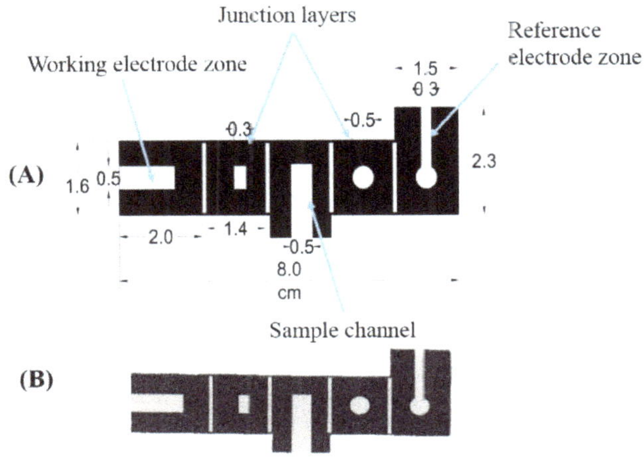

Figure 1. (**A**) Schematic diagram for fabrication of origami potentiometric device pattern. (**B**) The image of fabricated origami potentiometric device. The numbers denote the sizes in cm.

Indicator electrodes were fabricated by pressing the carbon paste mixture through an iron mold attached to the paper with the aim of a magnet. A uniform carbon paste was prepared by mixing an appropriate amount of graphite powder, Nujol oil, and modifier for 20.0 to 25.0 min. Finally, polishing the pressed electrode surface on an oily paper was performed until a compact and smooth electrode was obtained. Pseudo-reference electrodes were simply prepared by drawing a pencil on the filter paper. Six types of pencils (9B, 6B, 4B, 3B, 4H, HB) with various carbon content were used.

2.3. Potentiometric Measurements Using the ePAD

Folding of the sensor (Scheme 1) was performed by first bending the indicator electrode layer counterclockwise. Then the two layers of the indicator electrode and its square hydrophilic junction layer were folded on the sample channel. The last two layers of the pseudo-reference electrode and its circular hydrophilic layer were first folded on each other and then they were bent to the back side of the sample channel.

After folding the prepared sensor, 50.0 µL of sample solution was loaded to the sample channel. This was done by placing the beginning of the sample channel in the solution container. By capillary effect, sample solution moves through the proposed channel in the folded sensor. The sample volume was set to be as much as the channel becomes saturated from the sample. The device was then sandwiched among two glass sheets using clothespins. In order to increase the repeatability, potentiometer connection was stabilized by fixing the potentiometer alligators and the proposed sensor (Figure S1). Alligators were fixed on the transparent Flexi sheet (150.0 × 150.0 mm and thickness of 5.0 mm) with the aim of fast adhesive at a 90-degree angle toward each other. Then the sensor was kept stable in its situation by a small clamp (a wooden clothespin). The potential difference among the indicator and the pseudo-reference electrodes was measured afterward.

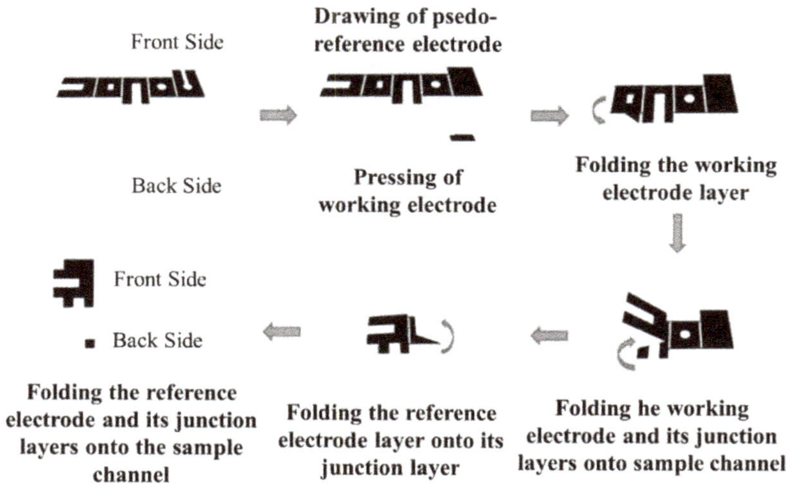

Scheme 1. Schematic representation of the folding processes for a three-dimensional origami paper-based potentiometric device.

2.4. Inorganic Ions Potentiometric Measurement

The basis of the working electrode in this assay is carbon-paste electrode, which has represented selectivity toward potentiometric sensing of some metal ions [23]. So, we firstly applied our device for measurement of metal ions including Cu^{2+}, Cr^{3+}, Ag^{+}, and Hg^{2+} by carbon paste (72 wt % of graphite powder and 28 wt % of Nujol oil) indicator electrodes and benzo15-crown-5 (B15C5) modified electrode (71% graphite, 25% Nujol oil and 4% B15C5) for measurement of Cd^{2+}. The CPE was prepared by hand mixing of pure graphite powder (0.1 g) and Nojul oil (0.04 g), which were placed in a mortar and mixed well for 20–25 min. Moreover, modified CPE with B15C5 were prepared by hand mixing of pure graphite powder (0.1 g), B15C5 (5.6×10^{-3} g) and Nojul oil (0.03 g), placed in a mortar and mixed well for 20–25 min to form a uniform paste. Therefore, indicator electrodes were supposed to be fabricated by pressing the carbon paste mixture through an iron mold attached to the paper with the aim of a magnet, followed by polishing the pressed electrode surface on an oily paper. On the other hand, the reference electrode was fabricated using a 6B pencil to fill the designated area. In addition to electrode preparation, ionic strength and the pH were adjusted by capillary movement of KNO_3 0.1 mol L^{-1} in HAc-NaAc buffer 0.1 mol L^{-1} pH 5.0 in the sample channel. After dryness of the sensor in room temperature for 4.0 min, the sensor was folded in the right manner so it was ready for ion determination. The bottom side of sample channel was placed in the analyte solution so the sample would rise by capillary effect to the top of the channel. By attachment of potentiometer alligators to the sensor electrodes, potential difference related to the ion concentration was read.

2.5. Measurement of H_2O_2 and Glucose

The modified CPE with MnO_2 was prepared by hand mixing well of pure graphite powder (0.1 g), MnO_2 (5.6×10^{-3} g) and Nojul oil (0.03 g) in a mortar for 20–25 min to form a uniform paste. The other procedure was like that explained for sensing of metal ions. However, here, ionic strength and the pH were adjusted by capillary movement of NH_3-NH_4Cl buffer solution 0.1 mol L^{-1} with pH 8.5 in the sample channel.

2.6. Measurement of Glucose

The process of preparation the glucose sensor is similar to the hydrogen peroxide sensor, with the difference that, 2.5 μL of GO_x solution (0.1 g of GO_x in 100.0 μL of

$0.1\ mol \cdot L^{-1}$ phosphate buffer) was dropped cast on the bottom of the indicator electrode before introduction of sample solution.

2.7. Real Sample

Four different blood samples were gathered and diluted by a saline solution with a 1:1000 ratio. Dilution was done to decrease the sample glucose concentration such that it would be in the linear range of the sensor. The related EMF of each sample was measured by directly loading of blood to the ePAD without any pre-separation step.

3. Results and Discussion

3.1. Sensor Design

The main objective of this design to decrease the distance between the indicator and the pseudo-reference electrodes which would improve the detection limit [24]. As a result, a three-dimensional origami device was considered. The device contains 5 paper layers (Figure 1). The first and the last ones are the electrode layers while the middle layer includes the sample channel. The other two layers contain connecting channels. It should be mentioned that the carbon paste indicator electrode had been placed at the backside of the sensor while the pseudo-reference electrode was drawn by pencil on the side which the deigned pattern had been printed.

For optimization of ePAD geometry and investigation of some analytical features of the sensor, unmodified carbon paste electrode (UCPE) was used as an indicator electrode. In bulk solution, UCPE represented selectivity toward metal ions such as Cu^{2+}, Ag^{+}, Cr^{3+} and Hg^{2+} [23,25]. Here, in the preliminary investigations, the response of ePAD to Cu^{2+} ion was followed.

In the preliminary investigation, it was found the width of sampling channel effected significantly the precision. In order to find the optimum dimension, four different sample channel patterns were designed and compared (Figure S2): (A) a sensor without sample channel (all part of the middle layer was used as sampling pad); (B) using a sample channel which is wider than electrode width; (C) channel width equal to the electrodes' width and (D) the sample channel thinner than the electrodes' width. The responses of the ePADs to $1.0 \times 10^{-6}\ mol \cdot L^{-1}$ solution of Cu^{2+} were investigated for 3 replicate measurements, and the relative standard deviation was calculated.

When the width of sampling channel was considered thinner than the width of electrodes, the signal was totally unstable and it was always fluctuating. By increasing the width of the sampling channel to the same size of the electrodes, the most stable results (relative standard deviation of 2.8%) were obtained. However, the signal instability was increased again by increasing the width of sampling channel. For the design, in which no sample channel was fabricated, smaller signals and less stable were achieved. Thus, in the future studies, the width of the sample channel was considered as the same size of electrode width. As shown in Supplementary Materials Figure S3, 50.0 μL of sample solution is required to fully fill the sampling channel.

In the next step, the number of junction layers was examined to see if it had any effect on the results. It has been reported that in the origami designs, increasing the number of layers may increase the quality of measurements by increasing the uniformity of analyte distribution [26]. In this part of study, 5-point calibration curves were plotted for each design and the slope of calibration curve and the goodness of fit (R^2) were used as decision criteria. In the first design (3-layer design), the electrodes are in direct contact with two different sides of the sample pad. Hence, the distance between the electrodes is as the width of the filter paper, which is around 0.23 mm. For this design, a super-Nernstian slope (as is expected for Cu^{2+} ion) [27] with acceptable correlation coefficient was obtained. However, better results were achieved by adding one layer between the electrodes and the sampling pad (5-layer design). The super-Nernstian slope was not affected significantly. However, the quality of fitting, which was the results of better reproducibility was increased from $R^2 = 0.90$ to $R^2 = 0.97$. By adding one more layer to both sides (7-layer design), the quality of fit remained the same, but the slope was decreased significantly. By increasing the distance between the electrodes, the thin-layer condition may break. Moreover, the analyte may not reach to the electrodes in the same concentration as in the sample tube. As we reported previously, silver ions can be coordinated with hydroxy groups of paper cellulose [25] and this more likely to happen for copper ions too.

3.2. Selection of Suitable Pencil for Pseudo-Reference Electrode

Graphite pencils are attractive as an electrode material for paper-based analytical devices [28]. Pencils are cheap, highly conductive and available in many different diameters and lengths [29]. Therefore, the pseudo-reference electrode of the designed three-dimensional ePAD was simply prepared by drawing a graphite pencil on the paper substrate [30]. To find an appropriate pencil for this purpose, various sensors were prepared using 9B, 6B, 4B, 3B, 4H, HB pencils (B-grade have a higher carbon content while H-grade pencil contains more clay) as a pseudo-reference electrode. Obtained calibration curves for Cu^{2+} determination in Figure S6, showed that 6B and 4B pencils would result in a calibration curve with higher sensitivity and better correlation coefficients (R^2). The 9B pencil would cause in a thick graphite network on the paper surface. This graphitic layer is weakly bonded by van der Waals force which its surface condition can be easily affected by the addition of a defined sample solution. On the other hand, the pseudo-reference electrode prepared by H pencils would contain a large amount of clay. This would increase the electrode hardness which also leads to a brittle and fragile electrode [31]. The surface morphology of the electrodes was investigated by SEM (Figure S7). Pseudo-reference electrodes drawn by 4B and 6B pencils were checked as they showed similar calibration curves in the case of Cu^{2+} determination. By increasing the carbon content of the pencil, a more uniform electrode coverage on the paper was seen (6B vs. 4B). For further studies, the 6B pencil was used in the pseudo-reference electrode drawing.

3.3. Inorganic Ions Determination

Determination of four inorganic ions including Cu^{2+}, Cr^{3+}, Ag^{+}, and Hg^{2+} were done by carbon paste indicator electrodes without any modifier while in the case of Cd^{2+} measurement, benzo15-crown-5 (B15C5) was added to the paste.

The experimental conditions of the sensors were the same as reported in the original papers for conventional electrodes, so it would be possible to make a better comparison between ePAD and the previous reported sensors. Measuring the pH value of paper, which is explained in Supplementary Materials, showed that paper was almost neutral ($\approx$6.8). Hence, pH does not play a significant role in measurement and the measurement conditions are similar to those of the bulk system [27,32–35].

3.3.1. Unmodified Carbon Paste Electrode for Measurement of Cu^{2+}, Ag^{+}, Cr^{3+} and Hg^{2+} Ions

Ionic strength and pH were adjusted by capillary movement of 50.0 μL of 0.1 mol·L^{-1} KNO_3 in 0.1 mol·L^{-1} HAc-NaAc buffer with a pH value of 5.0 in the sample channel. After dryness of the sensor at room temperature for 4.0 min, the sensor was folded so it was ready for ion determination. The bottom side of the sample channel was placed in the analyte solution so the sample would rise by the capillary effect to the top of the channel. By attachment of potentiometer alligators to the sensor electrodes, potential difference related to the ion concentration was read.

The effective parameters (type and concentration of buffer, pH value, electrolyte, and carbon paste mixture) were all adjusted the same as the previously reported method for determination of Cu^{2+} ion by carbon paste ISE [32]. For more certainty, a general comparison between other options of electrolyte and buffer was performed in the case of Cu^{2+} determination by ePAD. As it can be seen in the Supplementary Materials, results indicate that the optimum condition for ePAD matches the reported bulk analysis conditions [32] (0.1 mol·L^{-1} acetate buffer pH 5.0 containing 0.1 mol·L^{-1} KNO_3). This optimum condition was used for all studied metal ions in this section. The obtained calibration curves are given in Figure 2.

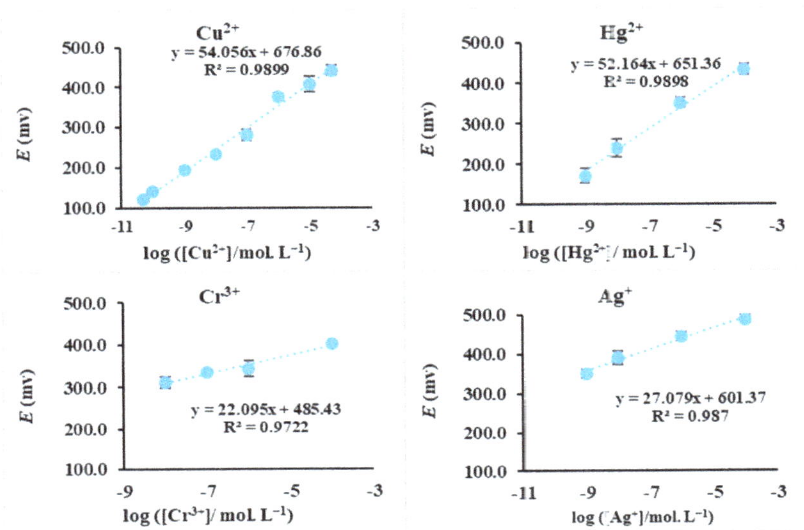

Figure 2. Calibration curves for Cu^{2+}, Hg^{2+}, Cr^{3+} and Ag^{+} ions with UCPE. Experimental conditions: 0.1 mol·L^{-1} KNO_3 in 0.1 mol·L^{-1} HAc-NaAc buffer pH 5.0, room temperature, 6B pencil as a pseudo-reference electrode and CPE (72 wt % of graphite powder and 28 wt % of Nujol oil) as an indicator electrode.

The results obtained from the experiments indicate a super-Nernstian response with slope of 54.1 and 52.2 mV·decade^{-1} for Cu^{2+} and Hg^{2+} respectively. A Nernstian behavior with a slope of 22.1 mV·decade^{-1} for Cr^{3+} and a sub-Nernstian response (slope 27.1 mV·decade^{-1}) for Ag^{+} is seen. The slopes of Cu^{2+}, Hg^{2+} and $Cr^{3+}$are almost the same as those reported for these metal ions in the bulk solution analysis [27,32–34]. However, unlike previous reports, a sub-Nernstain behavior was observed for Ag^{+} determination. This can be attributed to the adsorption of Ag^{+} ions on the paper surface (Figure S8) [27]. So, the concentration of silver ion at the end of sampling channel, where the solution is in contact with electrodes, might be lower than that in the sample solution. In the other words, the electrode senses lower concentrations of Ag^{+} than in the bulk solution. Fortunately, as shown in Figure 2, this a reproducible behavior, an in the presence of adsorption, a well-defined calibration curve was obtained for Ag^{+}.

In Figure 3 are shown the response's dynamic of the sensor at different concentrations of Cu^{2+}. Beginning of potential reading was just after putting the end of sampling pad into sample solution. As seen the response time depends on the concentration of Cu^{2+} such that the response time was decreased by increasing in concentration. This plot shows that changes in the response of the sensor (even for trace concentrations of Cu^{2+}) are significant comparing to the blank solution.

Comparison between the results of the reported bulk analysis [27] with the results for Cu^{2+} using ePAD, suggests that a wider linear range is accessible utilizing the origami paper-based ISE sensor (5.0×10^{-11}–5.0×10^{-5} mol·L^{-1} by ePAD vs. 1.0×10^{-6}–1.0×10^{-3} mol·L^{-1} in bulk analysis). Moreover, DL for Cu^{2+} is 4.0×10^{-11} by ePAD (Figure S9).

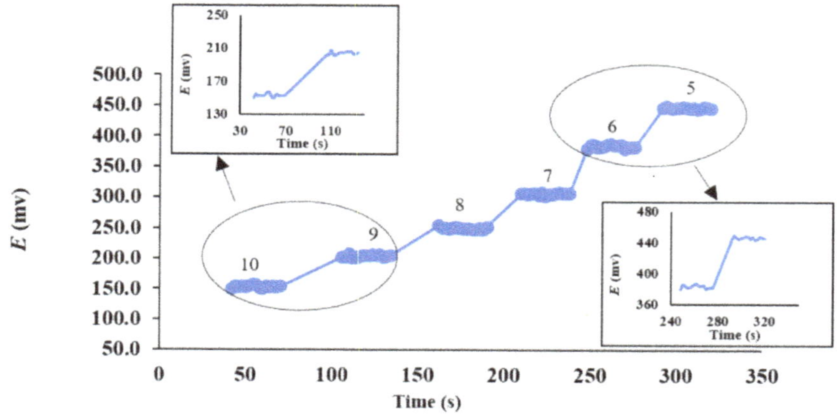

Figure 3. Dynamic response time of copper sensor at different concentration of Cu^{2+}: The numbers denote the concentration ($mol{\cdot}L^{-1}$) of Cu^{2+} in pM ($-\log[Cu^{2+}]$) format. Experimental conditions: KNO_3 0.1 $mol{\cdot}L^{-1}$ in HAc-NaAc buffer 0.1 $mol{\cdot}L^{-1}$ pH 5.0, room temperature, 6B pencil as a reference electrode and CPE (72 wt % of graphite powder and 28 wt % of Nujol oil) as an indicator electrode. For each concentration, a separate sensor was used.

3.3.2. Benzo15-Crown-5 Modified Carbon Paste Electrode for Measurement of Cd^{2+} Ion

The ability of crown ethers to complex with cations is well known [36]. The cavity size of the B15C5 (1.10 Å) suits well for the uptake of Cd^{2+} (radius 0.97 Å). Therefore, this crown ether had been used previously as a modifier in carbon paste electrode for sensing of Cd^{2+} [35].

For Cd^{2+} sensing by the ePAD, the indicator electrode was drawn using Benzo15-crown-5 modified carbon paste (B15C5-MCPE). For preparation of B15C5-MCPE, the same composition as the original work was used [35]. B15C5-MCPE includes 71% graphite, 25% Nujol oil and 4% B15C5. The hydrophilic area of pseudo-reference electrode was filled by 6B pencil. Ionic strength and the pH were adjusted by capillary movement of 50.0 µL of 0.1 $mol{\cdot}L^{-1}$ KNO_3 in 0.1 $mol{\cdot}L^{-1}$ HAc-NaAc buffer with a pH value of 5.0 in the sample channel. After the dryness and folding of the sensor, the bottom side of sample channel was placed in the analyte solution.

First, the response of the ePAD to different transition metal ions was investigated by obtaining the calibration curves (Figure 4). The potentiometric selectivity coefficient of the cadmium sensor was also determined by the separate solution method (SSM) (Table S3). It was observed that among the studied metal ions, the ePAD respond selectively toward Cd^{2+} ions. The obtained calibration curve for Cd^{2+} is shown in Figure 4, suggesting linearity in the concentration range from 1.0×10^{-11}–1.0×10^{-6} $mol{\cdot}L^{-1}$. The slope of the calibration curve is 34.7 mV $decade^{-1}$, showing a Nernstian behavior. The response time of this sensor was measured to be around 50 s, which is comparable with that of the bulk system (which was around 30 s).

Overall, this ePAD represented much higher sensitivity (DL of 6.3×10^{-12}) compared to the previously conventional potentiometric cell (DL of 3.2×10^{-5}) [35].

3.4. *Bio-Sensing by ePAD*

In this section, the usability of the ePAD in H_2O_2 and glucose determination is investigated and discussed.

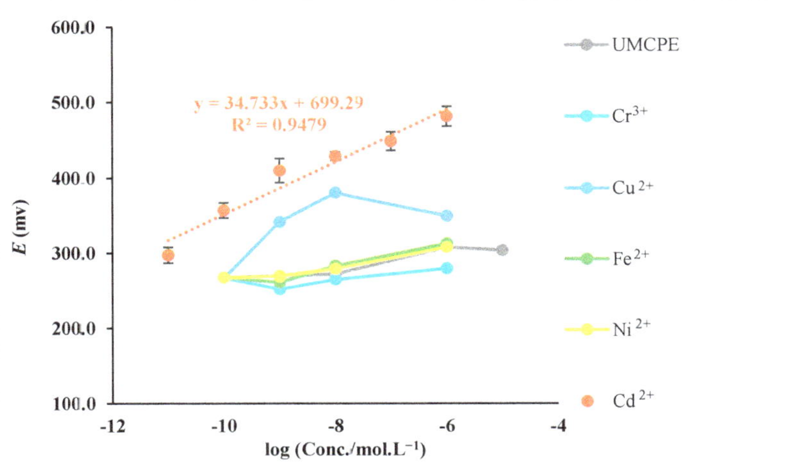

Figure 4. Investigation of B15C5-MCPE selectivity toward Cd^{2+} ion and Calibration curve for Cd^{2+} ion determination. Experimental conditions: channel containing 0.1 mol·L^{-1} KNO_3 in 0.1 mol·L^{-1} HAcNaAc buffer pH 5.0; 6B pencil as a pseudo-reference electrode and B15C5-MCPE as an indicator electrode.

3.4.1. Measurement of H_2O_2 with MnO_2 Modified Carbon Paste Electrode (MnO_2-MCPE)

The main steps for measurement of H_2O_2 concentration were the same as those explained for inorganic ions determinations. However, the modifier in the indicator electrode was changed to the manganese dioxide and the pH adjustment was performed by 0.1 mol·L^{-1} NH_3-NH_4Cl buffer solution with pH 8.5 [37]. MnO_2-MCPE prepared as same as previously reported [37], which included 71% graphite, 25% Nujol oil and 4% MnO_2. The SEM image of MnO_2-MCPE is shown in Figure S10B. A mixture of needle-shaped and flake-shaped of MnO_2 with extended surface area is obvious compared to the unmodified carbon paste electrode.

The potentiometric response of the ePAD to the H_2O_2 standard solutions of different concentration was evaluated. The calibration curve was obtained by plotting potential responses against logarithm of H_2O_2 concentration. As can be seen from Figure 5A, a linear regression with the slope of 26.98 mV decade^{-1} exists in the H_2O_2 concentration range of 1.0×10^{-4}–1.0×10^{-10} mol·L^{-1}. The unmodified CPE was not sensitive to the changes in H_2O_2 concentration. The analytical appraisals of ePAD are compared with those of bulk potentiometry in Table S4. Similar to the metal ion sensing, much higher sensitivity has been achieved by ePAD with DL of 4.0×10^{-11} mol·L^{-1}. The response time of ePAD for H_2O_2 sensing is 12 s, which is comparable to bulk potentiometry.

The possible response mechanism of MnO_2-MCPE toward H_2O_2 has been explored previously by Zhen and Gao [37]. Briefly, H_2O_2 oxidize MnO_2 to produce MnO_4^{2-} then the organic materials existed in the paraffin oil reduces the MnO_4^{2-} to MnO_2. As a result, the potentiometric response of this sensor is probably a mixture of redox potential between the MnO_4^{2-} and MnO_2.

3.4.2. Using ePAD As a Biosensor for Glucose

In order to prepare a biosensor for glucose, glucose oxidase enzyme (GOx) was fixed on the MnO_2-MCPE indicator electrode. To do so, a 2.5 µL of GOx solution, which was prepared in 0.1 mol·L^{-1} phosphate buffer pH = 7.4, was dropped caste on the surface of the indicator electrode. The enzyme concentration was optimized by fixing GOx solutions with various concentrations on the electrode and observing the responses of the sensors to the glucose sample with 1.0 µmol L^{-1} concentration. Four various GOx concentrations of 0.24,

0.74, 1.24 and 1.72 U µL^{-1} were tested. The results (Figure S11) showed an improvement in the response by increasing the Gox concentration up to 1.24 U µL^{-1}. Further additions in GOx concentration showed no effect on the recorded potential response. As a result, 1.24 U µL^{-1} had been chosen as an optimum GOx concentration.

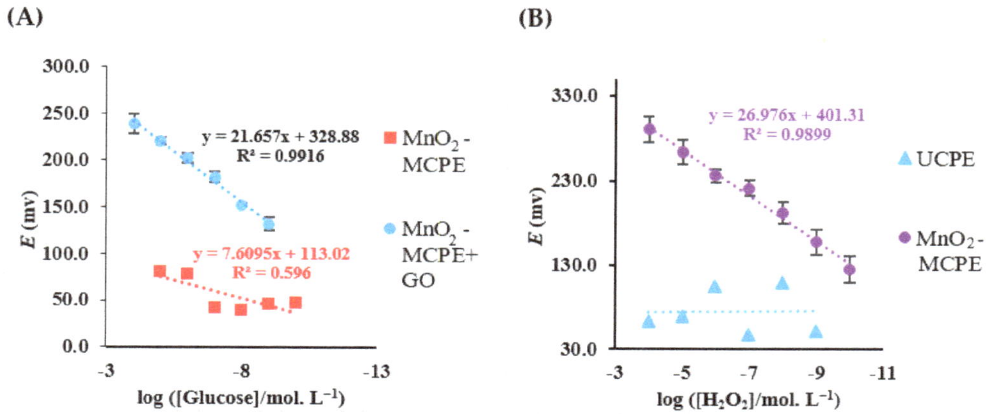

Figure 5. (**A**) Calibration curves for H_2O_2 using different indicator electrode composition; UCPE (blue triangles) and modified CPE (71 wt % of graphite powder, 25 wt % of Nujol oil and 4 wt % of manganese dioxide) (purple circles). (**B**) Calibration curve for glucos represented in the presence (blue circle) and absence (red square) of GOx. Experimental conditions: NH_3-NH_4Cl buffer solution 0.10 mol·L^{-1} with pH = 8.5, 6B pencil as a pseudo-reference electrode and modified CPE (71 wt % of graphite powder, 25 wt % of Nujol oil and 4 wt % of manganese dioxide) as an indicator electrode.

The changes in potential difference of ePAD as a function of glucose concentration were monitored. For comparison, the response of ePADs without the addition of GOx was also measured. The results are given in Figure 5B, representing a well-defined linear relationship for glucose concentration in the range of 1.0×10^{-9}–1.0×10^{-4} mol·L^{-1}. However, in the absence of GOx, the ePAD did not respond to the glucose variations. Similar to the other studied analytes, the ePAD represented a super sensitivity to glucose with DL of 4.6×10^{-10}. The response time was 15–25 s, which is comparable to the previously reported glucose biosensor [38–40].

To reduce sample consumption in the glucose determination, the sensor dimension was decreased to the one-third of the original size (Figure S12). As a result, the device preserved its performance while the sample volume needed for the analysis was reduced from 50.0 to 10.0 µL and the response time was decreased e.g., from 20 s to 15 s for 1.0×10^{-5} mol·L^{-1} of glucose.

The sensitivity of the sensor (slope in calibration curves) during this interval, was chosen as a factor representing the sensor stability. As it can be seen from Table S6, the sensor is stable for 17 days after preparation. In longer times, the sensor sensitivity is reduced and also the sensor needs more time to provide the consistent potential difference as a sensor response.

3.4.3. Real Sample Analysis

Recently, attentions have been directed toward the development of non-invasive methods for diabetes monitoring. For non-invasive measurements, the glucose level in body fluids other than blood (such as urine, tear, sweat, and saliva) should be measured. However, the glucose level in these fluids is very low (<8.0×10^{-4} mol·L^{-1}) [41] and most of the developing biosensors for glucose suffer from low sensitivity. The super-sensitivity of our ePAD allows non-invasive measurement of glucose in biological fluids other than blood. To do so, glucose testing was performed on various body fluid samples such as a tear, blood, and urine.

The results of glucose concentration values in Table 1 showed an excellent agreement with the reference method (Bionime GM110 Blood glucose monitor).

Table 1. Concentration of glucose predicted by ePAD biosensor and commercial blood glucometer.

Commercial Glucometer (mmol·L^{-1})	ePAD Biosensor (mmol·L^{-1})	%Relative Error
7.5	6.7	−10.7
5.8	5.4	−6.9
7.4	7.5	+1.4
5.0	4.8	−4.0

A commercial Tearlose tear solution and artificial human urine were prepared [41]. The procedure for preparation of artificial urine is explained in the Supplementary Materials. The defined concentration of glucose was spiked to these solutions and then the recovery values were obtained by comparing the spiked and the sensor determined amounts. Recoveries from 86.0% to 110.0% expresses appropriate performance of the designed ePAD (Table 2).

Table 2. Result of recovery studied of glucose in tear solution and an artificial human urine.

Sample	Added (µmol·L^{-1})	Found (µmol·L^{-1})	%Recovery	%Relative Error
Artificial urine	0.0010	0.0011	110	10.0
	1.00	1.07	107	7.0
	10.00	8.64	86.4	−13.6
Artificial tear	0.0010	0.00098	97.7	−2.0
	0.010	0.0099	98.9	−1.0
	10.00	11.20	112	12

3.5. Comparison with Other Microfluidics Devices

A comparison between our ePAD and some of recently published paper-based microfluidic glucose sensors is given in Table 3. It can be observed that the detection limit and response time of our device has been improved compared to other paper-based sensors. Moreover, sample volume used by this ePAD, has been reduced compared to most of the assays, which is a great advantage in biological studies.

Table 3. A comparison between this work and several electrochemical ePADs for measurement of glucose.

Sensors	Linear Range (mol·L^{-1})	DL (mol·L^{-1})	Response Time (s)	Ref.
Nafion/GOx/platinized filter paper	1.0×10^{-4}–3.2×10^{-3}	3.2×10^{-5}	50	[42]
platinized filter paper/Nafion/GOx	3×10^{-4}–3×10^{-3}	1×10^{-4}	60	[37]
platinised paper/polyvinyl alcohol/chitosan/GOx	3.0×10^{-5} to 1.0×10^{-3}	2.0×10^{-5}	20–30	[39]
Au dendrites/Whatman filter papers	1.0×10^{-5}–1.5×10^{-2}	6×10^{-7}	240	[43]
MnO_2-doped CPE/GOx	1.0×10^{-9}–1.0×10^{-4}	4.6×10^{-10}	15–25	In this work

4. Conclusions

A 3D origami paper-based potentiometric device had been introduced. The origami design of the sensor made it possible to have lower limits of detection for inorganic ions and also biological analytes. The selectivity of the sensor response to some analytes was improved by the addition of modifiers to the indicator electrode. A glucose biosensor with super sensitivity was obtained by using MnO_2–modified carbon paste electrode, which measured H_2O_2, and addition of glucose oxidase, which convert glucose to H_2O_2. This ultra-sensitivity allowed measurement of glucose in biological fluids other than blood (urine and tear). In addition, reproducibility and response time of the sensor is improved compared to the bulk analysis. The accurate results obtained for measurement of a very low level of glucose in these fluids, suggests that the ePAD biosensor has the potential to be used for non-invasive monitoring of diabetes.

Supplementary Materials: The following are available online at https://www.mdpi.com/2079-6374/11/2/44/s1, Figure S1: Photograph of a potentiometric paper-based ion selective sensor. The reference and indicator electrodes were positioned vertical and horizontal, respectively. (The reference electrode is located behind the paper and under clothespins); Figure S2. Pictures of the sensors with different width of sample channel: (A) Without sample channel; (B) The sample channel which is wider than electrode width; (C) Channel width equal to the electrode's and (D) The sample channel thinner than the electrode width; Figure S3. Different volumes of murexide solution was loaded and then R values (in RGB space) of the first, end and the middle sections of the channel; Figure S4. Image of sensor with (a) 20.0 μL (b) 30.0 μL and (c) 50.0 μL of murexide solution; Figure S5. Pictures of sensors in different designs A, B, C the number of layers is odd and D, E, F, G the number of layers is even; Figure S6. Effect of the type of pencil used as reference electrode on the performance of the potentiometric ePAD: (A), (B), (C), (D), (E) and (F) related to 6B, 4B, 3B, 9B, HB and 4H pencil leads, respectively. Experimental conditions: KNO_3 0.1 mol·L^{-1} in HAc-NaAc buffer 0.1 mol L^{-1} pH 5.0, room temperature, 6B pencil as a reference electrode and CPE (72 wt % of graphite powder and 28 wt % of Nujol oil) as an indicator electrode; Figure S7. SEM images of surface of electrode. Reference electrode with (A) 6B and (B) 4B pencil and indicator electrode with UCPE; Figure S8. Possible conformation of the compound formed from cellulose binding to silver; Figure S9. Detection limit for Cu^{2+} ion. Experimental conditions: KNO_3 0.1 mol·L^{-1} in HAc-NaAc buffer 0.1 mol·L^{-1} pH 5.0, room temperature, 6B pencil as a reference electrode and CPE (72 wt % of graphite powder and 28 wt % of Nujol oil) as an indicator electrode; Figure S10. The SEM image of surface of working electrode with A) UCPE and B) MnO_2 modified CPE; Figure S11. Optimization of GOx concentration in phosphate buffer (0.1 mol·L^{-1}) by fixing glucose concentration at 1.0 μM. Experimental conditions: NH_3-NH_4Cl buffer solution 0.10 mol·L^{-1} with pH = 8.5, 6B pencil as a reference electrode and modified CPE (72 wt % of graphite powder, 25 wt % of Nujol oil and 4 wt % of manganese dioxide) as an indicator electrode; Figure S11. The image of sensor in different sizes; Figure S13. Comparison between different conditions of electrolyte and identical buffer in Cu^{2+} determination by ePAD. (A) NaCl 1.0 mol L^{-1} in HAc-NaAc buffer 0.1 mol L^{-1} (B) KNO_3 0.01 mol L−1 in HAc-NaAc buffer 0.1 mol L^{-1} (C) KNO_3 0.1 mol L−1 in HAc-NaAc buffer 0.1 mol L^{-1}. 6B pencil as a reference electrode and CPE (72 wt % of graphite powder and 28 wt % of Nujol oil) as an indicator electrode. Table S1. Effect of the size of sample channel width on the precision of potentiometric measurements (1.0×10^{-6} mol·L^{-1} of Cu^{2+} and three times repeat for each design). Experimental conditions: KNO_3 0.1 mol·L^{-1} in HAc-NaAc buffer 0.1 mol·L^{-1} pH 5.0 and room temperature; Table S2. Analytical characterization of the sensor with different junction layer design. Experimental conditions: KNO_3 0.1 mol·L^{-1} in HAc-NaAc buffer 0.1 mol·L^{-1} pH 5.0, room temperature and difference concentration of copper (1.0×10^{-5}–1.0×10^{-9} mol·L^{-1}); Table S3. Potentiometric selectivity coefficient of the Cd^{2+} ion using SSM. Experimental conditions: KNO_3 0.1 mol·L^{-1} in HAc-NaAc buffer 0.1 mol·L^{-1} pH 5.0, room temperature. 6B pencil as a reference electrode and modified CPE (mixing 71 wt % of graphite powder, 25 wt % of Nujol oil and 4% B15C5) as an indicator electrode; Table S4. A comparison between the ePAD in this work and various bulk potentiometric method for measurment of H_2O_2; Table S5. The effect of the hydrophobicity of the back of the working electrode on the precision of potentiometric measurements (1.0×10^{-6} mol·L^{-1} of Cu^{2+} and repeated three times for each design). Experimental conditions: KNO_3 0.1 mol·L^{-1} in HAc-NaAc buffer 0.1 mol·L^{-1} pH 5.0 and room temperature; Table S6. Stability of glucose sensor (The response of sensor was recorded in the presence of three concentration of glucose). Experimental conditions: NH_3-NH_4Cl buffer solution 0.10 mol L^{-1} with pH = 8.5. Sensor was kept in the dark condition before use.

Author Contributions: S.P.: Methodology, investigation, writing—original draft preparation; E.R.: Methodology, validation; B H.: Conceptualization, supervision, writing—review and editing. All authors have read and agreed to the published version of the manuscript.

Funding: This research was funded by Shiraz University Research Council.

Institutional Review Board Statement: Not applicable.

Informed Consent Statement: Not applicable.

Data Availability Statement: Not applicable.

Conflicts of Interest: The authors declare no conflict of interest.

References

1. Thapliyal, N.; Chiwunze, T.E.; Karpoormath, R.; Goyal, R.N.; Patel, H.; Cherukupalli, S. Research progress in electroanalytical techniques for determination of antimalarial drugs in pharmaceutical and biological samples. *RSC Adv.* **2016**, *6*, 57580–57602. [CrossRef]
2. Umezawa, Y.; Sofue, S.; Takamoto, Y. Thin-layer ion-selective electrode detection of anticardiolipin antibodies in syphilis serology. *Talanta* **1984**, *31*, 375–378. [CrossRef]
3. Zou, X.U.; Cheong, J.H.; Taitt, B.J.; Bühlmann, P. Solid contact ion-selective electrodes with a well-controlled Co (II)/Co (III) redox buffer layer. *Anal. Chem.* **2013**, *85*, 9350–9355. [CrossRef]
4. Santhiago, M.; Nery, E.W.; Santos, G.P.; Kubota, L.T. Microfluidic paper-based devices for bioanalytical applications. *Bioanalysis* **2014**, *6*, 89–106. [CrossRef]
5. Cate, D.M.; Adkins, J.A ; Mettakoonpitak, J.; Henry, C.S. Recent developments in paper-based microfluidic devices. *Anal. Chem.* **2014**, *87*, 19–41. [CrossRef]
6. Nge, P.N.; Rogers, C.I.; Woolley, A.T. Advances in microfluidic materials, functions, integration, and applications. *Chem. Rev.* **2013**, *113*, 2550–2583. [CrossRef]
7. Siegel, A.C.; Phillips, S.T.; Dickey, M.D.; Lu, N.; Suo, Z.; Whitesides, G.M. Foldable printed circuit boards on paper substrates. *Adv. Funct. Mater.* **2010**, *20*, 28–35. [CrossRef]
8. Feng, L.; Li, H.; Niu, L.Y.; Guan, Y.S.; Duan, C.F.; Guan, Y.F.; Tung, C.-H.; Yang, Q.-Z. A fluorometric paper-based sensor array for the discrimination of heavy-metal ions. *Talanta* **2013**, *108*, 103–108. [CrossRef]
9. Gharaghani, F.M.; Akhond, M.; Hemmateenejad, B. A three-dimensional origami microfluidic device for paper chromatography: Application to quantification of Tartrazine and Indigo carmine in food samples. *J Chromatogr. A* **2020**, *1621*, 461049–461052. [CrossRef] [PubMed]
10. Sharifi, H.; Tashkhourian, J.; Hemmateenejad, B. A 3D origami paper-based analytical device combined with PVC membrane for colorimetric assay of heavy metal ions: Application to determination of Cu (II) in water samples. *Anal. Chim. Acta* **2020**, *1126*, 114–123. [CrossRef] [PubMed]
11. Martinez, A.W.; Phillips, S.T.; Whitesides, G.M. Three-dimensional microfluidic devices fabricated in layered paper and tape. *Proc. Natl. Acad. Sci. USA* **2008**, *105*, 19606–19611. [CrossRef]
12. Ding, J.; Li, B.; Chen, L.; Qin, W. A three-dimensional Origami paper-based device for potentiometric biosensing. *Angew. Chem. Int. Ed.* **2016**, *55*, 13033–13037. [CrossRef] [PubMed]
13. Shariati, S.; Khayatian, G. The colorimetric and microfluidic paper-based detection of cysteine and homocysteine using 1, 5-diphenylcarbazide-capped silver nanoparticles. *RSC Adv.* **2021**, *11*, 3295–3303. [CrossRef]
14. Hidayat, M.A.; Maharani, D.A.; Purwanto, D.A.; Kuswandi, B.; Yuwono, M. Simple and sensitive paper-based colorimetric biosensor for determining total polyphenol content of the green tea beverages. *Biotechnol. Bioprocess Eng.* **2020**, *25*, 255–263. [CrossRef]
15. Taghizadeh-Behbahani, M.; Hemmateenejad, B.; Shamsipur, M. Colorimetric determination of acidity constant using a paper-based microfluidic analytical device. *Chem. Pap.* **2018**, *72*, 1239–1247. [CrossRef]
16. Chen, H.; Hu, O.; Fan, Y.; Xu, L.; Zhang, L.; Lan, W. Fluorescence paper-based sensor for visual detection of carbamate pesticides in food based on CdTe quantum dot and nano ZnTPyP. *Food Chem.* **2020**, *327*, 127075. [CrossRef]
17. Mesgari, F.; Beigi, S.M.; Fakhri, N.; Hosseini, M.; Aghazadeh, M.; Ganjali, M.R. based chemiluminescence and colorimetric detection of cytochrome c by cobalt hydroxide decorated mesoporous carbon. *Microchem. J.* **2020**, *157*, 104991. [CrossRef]
18. Cao, L.; Han, C.; Xiao, H.; Chen, Z.; Fang, C. A novel 3D paper-based microfluidic electrochemical glucose biosensor based on rGO-TEPA/PB sensitive film. *Anal. Chim. Acta* **2020**, *1096*, 34–43. [CrossRef] [PubMed]
19. Novell, M.; Guinovart. T.; Blondeau, P.; Rius, F.X.; Andrade, F.J. A paper-based potentiometric cell for decentralized monitoring of Li levels in whole blood. *Lab Chip* **2014**, *14*, 1308–1314. [CrossRef] [PubMed]
20. Yakoh, A.; Siangproh, W.; Chailapakul, O.; Ngamrojanavanich, N. Optical bioelectronic device based on a screen-printed electroluminescent transducer. *ACS Appl. Mater. Interfaces* **2020**, *12*, 22543–22551. [CrossRef] [PubMed]

21. Dungchai, W.; Chailapakul, O.; Henry, C.S. Electrochemical detection for paper-based microfluidics. *Anal. Chem.* **2009**, *81*, 5821–5826. [CrossRef] [PubMed]
22. Dey, R.; Kar, S.; Joshi, S.; Maiti, T.K.; Chakraborty, S. Ultra-low-cost 'paper-and-pencil'device for electrically controlled micromixing of analytes. *Microfluid. Nanofluid.* **2015**, *19*, 375–383. [CrossRef]
23. Shamsipur, M.; Tashkhourian, J.; Hemmateenejad, B.; Sharghi, H. Application of artificial neural network to simultaneous potentiometric determination of silver (I), mercury (II) and copper (II) ions by an unmodified carbon paste electrode. *Talanta* **2004**, *64*, 590–596. [CrossRef]
24. Stanic, Z.; Girousi, S. Carbon paste electrodes in potentiometry: The state of the art and applications in modern electroa-nalysis (A review). *Sens. Electroanal.* **2011**, *6*, 89–128.
25. Hu, X.; Leng, Z. Highly sensitive potentiometry for determination of chromium (VI) with carbon paste electrode. *Anal. Proc. Incl. Anal. Commun. RSC* **1995**, *32*, 521–522. [CrossRef]
26. Liana, D.D.; Raguse, B.; Gooding, J.J.; Chow, E. Recent advances in paper-based sensors. *Sensors* **2012**, *12*, 11505–11526. [CrossRef]
27. Pei, J.; Yin, Q.; Zhong, J. Potentiometric determination of trace silver based on the use of a carbon paste electrode. *Talanta* **1991**, *38*, 1185–1189. [CrossRef]
28. Hu, X.; Leng, Z. Highly selective and super-Nernstian potentiometry for determination of Cu^{2+} using carbon paste electrode. *Anal. Lett.* **1995**, *28*, 979–989. [CrossRef]
29. Kim, J.; Kwon, S.; Ostler, E. Antimicrobial effect of silver-impregnated cellulose: Potential for antimicrobial therapy. *J. Biol. Eng.* **2009**, *3*, 3–20. [CrossRef]
30. Santhiago, M.; Henry, C.S.; Kubota, L.T. Low cost, simple three dimensional electrochemical paper-based analytical device for determination of p-nitrophenol. *Electrochim. Acta* **2014**, *130*, 771–777. [CrossRef]
31. Kurra, N.; Kulkarni, G.U. Pencil-on-paper: Electronic devices. *Lab Chip* **2013**, *13*, 2866–2873. [CrossRef] [PubMed]
32. Radu, A.; Fayose, T.; Mendecki, L.; Ullah, S. Single strip solid contact ion selective electrodes on pencil-drawn electrode substrate. *Anal. Methods* **2017**, *9*, 1213–1220.
33. Kawahara, R.; Sahatiya, P.; Badhulika, S.; Uno, S. Based potentiometric pH sensor using carbon electrode drawn by pencil. *Jpn. J. Appl. Phys.* **2018**, *57*, 04FM08. [CrossRef]
34. Srivastava, S.K.; Gupta, V.K.; Jain, S. A PVC-based benzo-15-crown-5 membrane sensor for cadmium. *Electroanalysis* **1996**, *8*, 938–940. [CrossRef]
35. Weber, E. *New Developments in Crown Ether Chemistry: Lariat, Spherand and Second-Sphere Complexes, Crown Ethers and Analogs*; John Wiley & Sons Ltd.: Hoboken, NJ, USA, 1989; pp. 305–357.
36. Zheng, X.; Guo, Z. Potentiometric determination of hydrogen peroxide at MnO_2-doped carbon paste electrode. *Talanta* **2000**, *50*, 1157–1162. [CrossRef]
37. Cánovas, R.; Parrilla, M.; Blondeau, P.; Andrade, F.J. A novel wireless paper-based potentiometric platform for monitoring glucose in blood. *Lab Chip* **2017**, *17*, 2500–2507. [CrossRef]
38. Ali, S.M.U.; Nur, O.; Willander, M.; Danielsson, B. Glucose detection with a commercial MOSFET using a ZnO nanowires extended gate. *IEEE Trans. Nanotechnol.* **2009**, *8*, 678–683.
39. Guadarrama-Fernández, L.; Novell, M.; Blondeau, P.; Andrade, F.J. A disposable, simple, fast and low-cost paper-based biosensor and its application to the determination of glucose in commercial orange juices. *Food Chem.* **2018**, *265*, 64–69. [CrossRef]
40. Liu, Q.; Liu, Y.; Wu, F.; Cao, X.; Li, Z.; Alharbi, M.; Abbas, A.N.; Amer, M.R.; Zhou, C. Highly sensitive and wearable In_2O_3 nanoribbon transistor biosensors with integrated on-chip gate for glucose monitoring in body fluids. *ACS Nano* **2018**, *12*, 1170–1178. [CrossRef] [PubMed]
41. Chutipongtanate, S.; Thongboonkerd, V. Systematic comparisons of artificial urine formulas for in vitro cellular study. *Anal. Biochem.* **2010**, *402*, 110–112. [CrossRef]
42. Parrilla, M.; Cánovas, R.; Andrade, F.J. Based enzymatic electrode with enhanced potentiometric response for monitoring glucose in biological fluids. *Biosens. Bioelectron.* **2017**, *90*, 110–116. [CrossRef] [PubMed]
43. Rafatmah, E.; Hemmateenejad, B. Dendrite gold nanostructures electrodeposited on paper fibers: Application to electrochemical non-enzymatic determination of glucose. *Sens. Actuators B Chem.* **2020**, *304*, 127335. [CrossRef]

biosensors

MDPI

Communication

Smartphone and Paper-Based Fluorescence Reader: A Do It Yourself Approach

Laura Alejandra Ireta-Muñoz and Eden Morales-Narváez *

Biophotonic Nanosensors Laboratory, Centro de Investigaciones en Óptica A. C., León 37150, Mexico; laura.iret@gmail.com

* Correspondence: eden@cio.mx

Received: 18 May 2020; Accepted: 29 May 2020; Published: 2 June 2020

Abstract: Given their photoluminescent character, portable quantum dot readers are often sophisticated and relatively expensive. In response, we engineered a "do it yourself" fluorescence reader employing paper materials and a mid-range smartphone camera. Black paperboard facilitated a versatile, lightweight and foldable case; whereas cellophane paper was observed to behave as a simple, yet effective, optical bandpass filter leading to an advantageous device for the quantitative interrogation of quantum dot nanocrystals concentrations (from 2.5 to 20 nM), which are suitable for optical point-of-care biosensing. The streptavidin-coated nanocrystals employed are commercially available and the developed reader was benchmarked with a standard portable quantum dot reader, thereby demonstrating advantages in terms of cost and linear analytical range.

Keywords: biosensing; point-of-care; photoluminescence; portable devices; miniaturization

1. Introduction

Paper is a versatile material to work with; it is relatively cheap and easy to handle given its lightweight and flexible character. In this regard, paper-based analytical devices are amenable to simple and effective on-site testing in different applications, including diagnostics, environmental monitoring and food analysis [1–5]. Moreover, paper-based analytical devices are also amenable to the integration with portable technologies such as smartphones and drones [4,6].

Highly sensitive point-of-care biosensing can be critical to enable timely healthcare decisions [7–10]. Quantum dot (QD) nanocrystals have been proven advantageous in these approaches due to their highly efficient fluorescence, size-tunable Gaussian emission spectrum, excellent stability against photobleaching, large Stokes shift and low background signal [11–13]. However, given their photoluminescent character, portable QD readers are often sophisticated and relatively expensive; for example, they may require filters and/or lenses [14,15], or involve high-range mobile phones obviating the need for external optical filters [16]. Hence, generally, QD readers are not particularly accessible in relatively low-resource settings. In response, employing paper materials and a mid-range smartphone camera, we engineered a "do it yourself" QD reader.

2. Materials and Methods

Black matboard, cellophane paper, and electronic components were purchased in the local market (León, Guanajuato, Mexico). The laminated cards, nitrocellulose membrane, sample and absorbent pads for the production of the lateral flow strips (LFS) were purchased from Millipore (Billerica, MA, USA, https://www.merckmillipore.com). Streptavidin–quantum dot 655 (CdSe@ZnS) was from Life Technologies (Carlsbad, CA, USA, https://www.thermofisher.com). LFS were spotted with 2 μL of QDs at different concentrations, from 2.5 nM to 20 nM. After the spotting process, the LFS were dried at room temperature overnight to be then analyzed. A USB4000 UV–Vis (ultraviolet–visible) spectrometer (Ocean Optics, Inc., Largo, FL, USA, https://www.oceaninsight.com) was utilized to acquire the light-emitting diode (LED) emission spectrum. A Cytation 5 multimodal spectrometer (BioTek Instruments, Inc., Winooski, VT, USA, https://www.biotek.com) was employed to record the emission spectra of the QDs and nitrocellulose autofluorescence, as well as the absorbance spectra of the paper-based filters. Limit of detection (LOD) was estimated by interpolating the average of the intensity value of the blank sample plus 3 times its standard deviation within the respective calibration plot. Limit of quantification (LOQ) was estimated by interpolating the average of the intensity value of the blank sample plus 10 times its standard deviation within the respective calibration plot.

Safety. In order to avoid damage of the reader in long-term use due to possible LED overheating, it is recommended to use the reader less than 20 min per analysis. If the analysis requires more time, the device is recommended to be turned off for 5 min and the reader can be then utilized again. No heating effect was spotted on the analyzed sample during fluorescence interrogation. Moreover, all the internal edges of the case should be cautiously sealed (for instance, using a black tape) in order to avoid possible UV radiation exposure.

3. Results and Discussion

3.1. Design of the Paper-Based Quantum Dot (QD) Reader

Firstly, considering the dimensions, camera and universal serial bus (USB) specifications of the employed smartphone [Moto G5, see Table S1 in the Supplementary Materials, (SM)], we designed a foldable paper model of the reader case. Figures S1 and S2 (detailed in SI) detail the characteristics of such a papercraft, which was implemented using black matboard. The color of this type of paper was chosen to exhibit a low background when exposed to the excitation source, that is, a violet LED; thus potentially minimizing undesired noise during the imaging process. The papercraft includes paper based filters, an external filter holder, a tray to introduce the sample into the reader, an illumination angle control, as well as a USB connector to take advantage of the smartphone battery to power the excitation source via the on-the-go configuration, see Figure 1 and Figure S2 (detailed in SM).

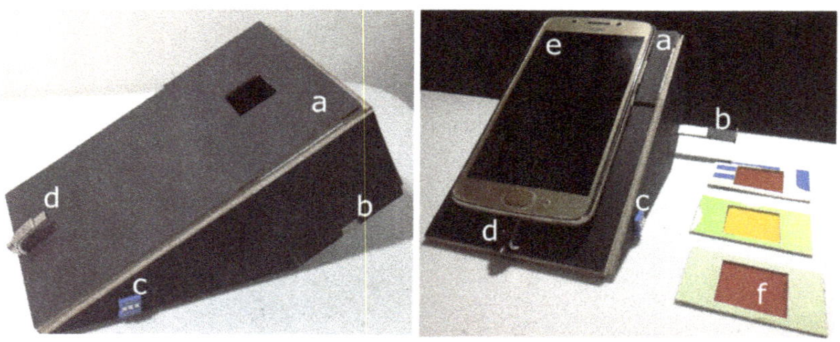

Figure 1. Pictures of the paper-based quantum dot (QD) reader. (**a**) External filter holder. (**b**) Tray to introduce the sample into the reader. (**c**) Illumination angle control. (**d**) Universal serial bus (USB) connector to power the excitation source. (**e**) Moto G5. (**f**) Paper-based filters.

With this hardware in hand, we proceeded to study the performance of the proposed QD reader. The optical path and the involved components are illustrated in Figure 2A. Two 5 mm round LEDs with emission wavelength centered at 400 nm, full width-half maximum (FWHM) around 14 nm, were employed to excite the QDs emitting around 655 nm. Figure S3A (depicted in SM) details the electronic circuit employed to power these light sources. The employed streptavidin-coated QDs are commercially available (Life Technologies, Carlsbad, CA, USA) and exhibit a rice-like shape with an average size around 14 nm [17], Figure 2B displays the respective emission spectrum. As model samples,

LFS were manufactured following previous procedures [18]. To endow the imaging acquisition process with a relatively even illumination, we studied 3 illumination angles (180, 45 and 90°) and measured the coefficient of variation (CV) of the pixel intensities centered in the detection pads of the LFS, see Figure S3B,C (detailed in SM). Figure S3C (included in SM) demonstrates that the illumination angle at 45° resulted to show the lower CV, which accounted for 11.55%. Consequently, this illumination angle was chosen as optimal among the available illumination angles.

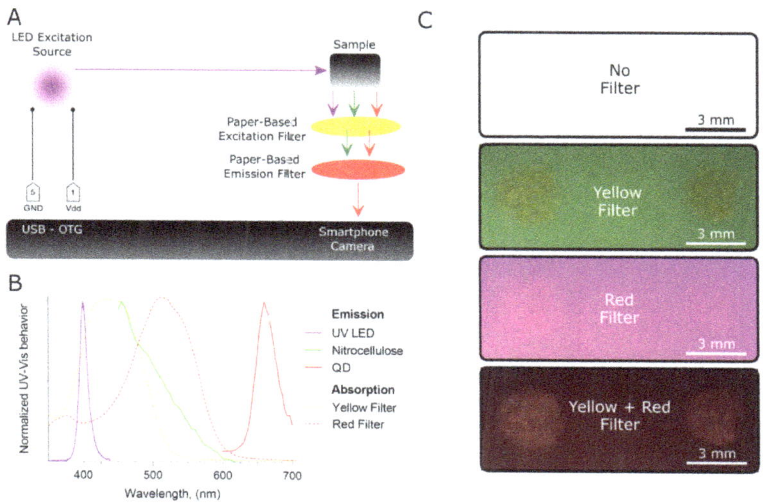

Figure 2. Optical components of the paper-based QD reader and their characterization via ultraviolet–visible (UV–Vis) spectroscopy. (**A**) Schematic representation of the optical path. (**B**) UV-Vis behavior of the optical components. (**C**) Images recorded with different filter configurations. The images were acquired through the smartphone camera.

3.2. Ultraviolet–Visible (UV–Vis) Characterization of the Paper-Based Filters

We also characterized the absorbance of the proposed paper-based filters using UV–Vis spectroscopy. Conveniently, the studied yellow cellophane exhibited an optical bandpass filter-like behavior with a central wavelength at c.a. 435 nm and a FWHM around 103 nm [19], see Figure 2B. Hence, this material was proposed as a paper-based excitation filter. The detection pad of the LFS is made of nitrocellulose. Using the proposed yellow filter during the imaging process, these detection pads were observed to display a strong green autofluorescence when excited with the employed light source emitting at 400 nm; see the corresponding emission spectrum in Figure 2B and a picture recorded under these conditions in Figure 2C. Hence, following color theory [19], we envisaged that a paper-based emission filter might be convenient to remove such a green noise. In this context, we explored the UV–Vis absorbance of a piece of red cellophane. This material also exhibited an optical bandpass filter-like behavior with a central wavelength at c.a. 510 nm and a FWHM around 110 nm, see Figure 2B. Eventually, using the proposed filters, we managed to acquire an image of the respective red emission of QDs spotted onto nitrocellulose at a relatively low concentration (2.5 nM), see Figure 2C.

3.3. Analytical Behavior of the Resulting QD Reader

Upon the aforementioned optical characterization, LFS were drop-casted with 2 µL of several QD concentrations (2.5, 5, 7.5, 10 and 20 nM) and we recorded the respective images using the paper-based QD reader under different filter configurations (red filter, yellow filter, yellow + red filter), see Figure 3.

It is worth mentioning that hydrophobic walls created within paper via wax printing can enhance the variability of spots when drop-casted within wax-printed wells [20]. Although the pixel intensity of the acquired images can be directly analyzed in the smartphone using IJ_Mobile [21], we preferred to extract and handle these data by using MATLAB aiming at performing a controlled statistical analysis. Briefly, to extract pixel intensities, an image binarization process was performed using Otsu's method [22], 0 values were considered background and 1 values were considered the QDs' signal to build a binary mask. This binary mask was used to define regions of interests and extract the studied pixel intensities, Figure S4 (included in SM) shows an example of this process. By means of the resulting pixel intensities, we performed the respective calibration plots, see Figure 3. The value of the inverse of the slope of the resulting curves sheds light on the sensitivity of the respective configuration, whereas the resulting Y-intercept value offers information on the baseline. Generally, images recorded with the red filer configuration show a strong violet background, triggering a baseline accounting for c.a. 186 pixel intensity units at the blank signal. The sensitivity of this configuration accounts for 0.5 nM of QDs per pixel intensity units. As mentioned before, images captured with the yellow filter configuration show a green background. However, the corresponding baseline accounts for c.a. 79 pixel intensity units and the respective sensitivity is around 0.19 nM of QDs per pixel intensity units. As depicted in Figure 3C, the yellow + red filter configuration showed a relatively cleaner imaging process. The resultant baseline was around 69 pixel intensity units and the corresponding sensitivity accounted for 0.17 nM of QDs per pixel intensity units. Table 1 summarizes these analytical details.

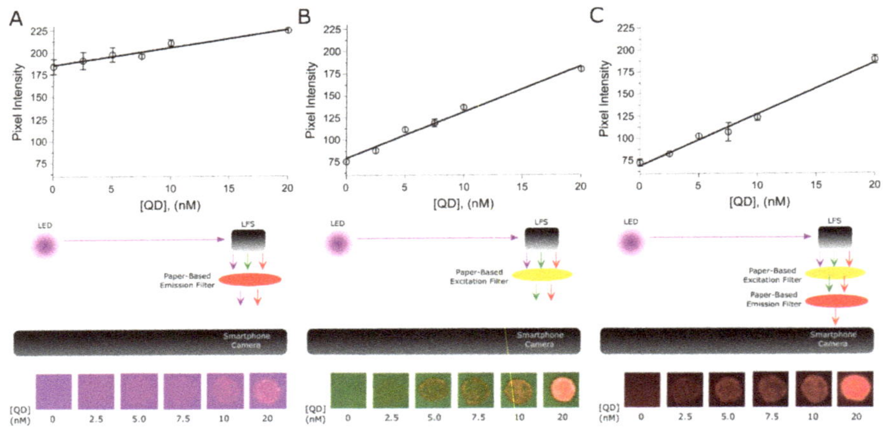

Figure 3. The resulting calibration plots under different paper-based filter configurations and the corresponding images. (**A**) Red filter. (**B**) Yellow filter. (**C**) Red + yellow filter. The error bars represent the standard deviation of two QD spots drop-casted onto lateral flow strips (LFS). The size of the spots is c.a. 3 mm.

Prompted by these results, we performed a comparative study by analyzing the same LFS using a commercially available equipment specially designed to measure QDs emitting at 655 nm onto LFS (ESEQuant LR3, QIAGEN, Hilden, Germany). Figure S5 (included in SM) displays the profiles resulting from the photoluminescent intensity of the QD spots onto the LFS analyzed by ESEQuant LR3. The average value of the profile corresponding to the analyzed QD spot was chosen as the analytical parameter to build the corresponding calibration plot (arbitrary units). However, the analyzed concentration range (2.5–20 nM) did not fit a linear response with an acceptable coefficient of determination (R^2), which accounted for c.a. 0.9127, see Figure S6A (detailed in SM). Eventually, we sought a linear response within a QD concentration range from 2.5 to 10 nM, which offered an improved R^2 (0.9772) and a sensitivity of 0.06 nM of QD per arbitrary unit, see Figure S6B (displayed in SI) and Table 1. Hence, in terms of the 1/slope value, ESEQuant LR3 was observed to be 2.75 times more sensitive than the paper-based reader incorporating the red + yellow filter configuration. In contrast, the paper-based reader offered a broader linear range in the explored concentrations as detailed in Table 1. LODs and LOQs were also estimated. As observed in Table 1, in terms of the LOD, ESEQuant LR3 was observed to be 3.02 times more sensitive than the paper-based reader incorporating the yellow filter configuration. ESEQuant LR3 was also observed to offer a LOQ 15 times lower than that of the paper-based reader incorporating the yellow filter configuration. Table 1 also highlights that the reader resulted to be more sensitive with the red + yellow filter configuration in terms of the 1/slope value, whereas the lowest LOD offered by the paper-based reader was achieved with the yellow filter configuration. Hence, the resulting paper-based bandpass filters were proven to be technically sound by evaluating these analytical parameters.

Table 1. Analytical performance of the studied QD readers.

Configuration	Linear Model	R^2	1/slope	LOD [1] (nM)	LOQ [2] (nM)	Linear Range (nM)
Red filter	Y = 1.994*X + 185.6	0.9455	0.5014	11.856	42.201	2.5–20
Yellow filter	Y = 5.206*X + 79.12	0.9834	0.1921	0.918	4.744	2.5–20
Red + yellow filter	Y = 5.81*X + 69.06	0.9892	0.1721	2.773	7.778	2.5–20
ESEQuant	Y = 15.98*X − 3.454	0.9772	0.06256	0.303	0.314	2.5–10

[1] Limit of detection. [2] Limit of quantification.

Importantly, the proposed paper-based QD reader can be considered a low-cost device, it is approximately 1877-fold cheaper than ESEQuant LR3 and does not require a high-range smartphone camera. Furthermore, this paper-based device obviates the usage of expensive filters or lenses. Table 2 highlights these competitive advantages.

Table 2. Photoluminescence readers for point-of-care applications.

Reported Price (USD)	Filters/Lens	Smartphone	Reference
8450	Not specified	–	ESEQuant LR3
10 [1]	No	High-range (iPhone SE or Nexus 5)	[16]
5 [2]	Yes/Yes	High-range (iPhone 5s)	[23]
4.5 [3]	Yes/No	Mid-range (Moto G5)	This work

[1] Smartphone not included. [2] Filters, lens and smartphone not included. [3] Paper-based filters included, smartphone not included.

4. Conclusions

A cost-effective paper-based photoluminescent QD reader is reported. Black paperboard facilitated a lightweight and foldable case. Given the versatility offered by this material, the case can be redesigned easily to be compatible with other mobile phones. Moreover, cellophane paper was observed to behave as a simple optical bandpass filter leading to an advantageous device for quantitative interrogation of QD concentrations that are suitable for optical point-of-care biosensing [18]. Although the fabrication of this device may require previous skills in electronics and engineering, other types of readings such as optical density or chemiluminescence can be performed by properly adapting the design of the proposed reader. This reader is also amenable to the analysis of different disposable sensors based on fluorescence, including LFS, vertical flow, dip-stick and microfluidic paper-based analytical devices.

5. Patents

Patent application under preparation.

Supplementary Materials: The following are available online at http://www.mdpi.com/2079-6374/10/6/60/s1: Figure S1. Foldable papercraft of the proposed QDs reader; Figure S2. A. Lateral view of the reader case. B. Top view of the reader case; Figure S3. A. Scheme of the employed electronic circuit; Figure S4. Image processing and pixel intensity estimation; Figure S5. Experimental evidence. Lateral flow strips with different QDs concentrations analyzed using ESEQuant LR3 (QIAGEN, Hilden, Germany); Figure S6. Calibration plots resulting from the calculation of the area below the curves representing the QDs intensities measured by ESEQuant LR3 and Table S1. Smartphone (Moto G5) specifications.

Author Contributions: Conceptualization, E.M.-N. and L.A.I.-M.; methodology, E.M.-N.; software, L.A.I.-M.; validation, E.M.-N. and L.A.I.-M.; formal analysis, E.M.-N. and L.A.I.-M.; investigation, L.A.I.-M.; resources, E.M.-N.; writing—original draft preparation, L.A.I.-M.; writing—review and editing, E.M.-N.; visualization, E.M.-N. and L.A.I.-M.; supervision, E.M.-N.; funding acquisition, E.M.-N. All authors have read and agreed to the published version of the manuscript.

Funding: The authors acknowledge the financial support from CONACYT (Mexico, Grant No. 293523, Apoyo al Fortalecimiento de la Infraestructura Científica y Tecnológica 2018; Grant No. 299058, Laboratorio Nacional de Micro y Nanofluídica; Grant No. 297497, FORDECYT, "Generación de Plataformas tecnológicas basadas en Microdispositivos para el sector Industrial de los estados de Aguascalientes, Guanajuato, Puebla, Querétaro y San Luis Potosí").

Acknowledgments: The authors acknowledge the technical support provided by Alfredo Benítez-Lara and Ricardo Valdivia-Hernandez.

Conflicts of Interest: Patent application under preparation. The authors declare no conflict of interest.

References

1. Fu, L.-M.; Wang, Y.-N. Detection methods and applications of microfluidic paper-based analytical devices. *TrAC Trends Anal. Chem.* **2018**, *107*, 196–211. [CrossRef]
2. Marquez, S.; Morales-Narváez, E. Nanoplasmonics in paper-based analytical devices. *Front. Bioeng. Biotechnol.* **2019**, *7*, 69. [CrossRef] [PubMed]
3. Liang, J.; Jiang, C.; Wu, W. Toward fiber-, paper-, and foam-based flexible solid-state supercapacitors: Electrode materials and device designs. *Nanoscale* **2019**, *11*, 7041–7061. [CrossRef] [PubMed]
4. Marquez, S.; Liu, J.; Morales-Narváez, E. Paper-based analytical devices in environmental applications and their integration with portable technologies. *Curr. Opin. Environ. Sci. Health* **2019**, *10*, 1–8. [CrossRef]
5. Chang, J.; Li, H.; Hou, T.; Li, F. Paper-based fluorescent sensor for rapid naked-eye detection of acetylcholinesterase activity and organophosphorus pesticides with high sensitivity and selectivity. *Biosens. Bioelectron.* **2016**, *86*, 971–977. [CrossRef]
6. Sun, H.; Jia, Y.; Dong, H.; Fan, L.; Zheng, J. Multiplex quantification of metals in airborne particulate matter via smartphone and paper-based microfluidics. *Anal. Chim. Acta* **2018**, *1044*, 110–118. [CrossRef]
7. Dincer, C.; Bruch, R.; Costa-Rama, E.; Fernández-Abedul, M.T.; Merkoçi, A.; Manz, A.; Urban, G.A.; Güder, F. Disposable sensors in diagnostics, food, and environmental monitoring. *Adv. Mater.* **2019**, *31*. [CrossRef]
8. Christodouleas, D.C.; Kaur, B.; Chorti, P. From point-of-care testing to eHealth diagnostic devices (eDiagnostics). *ACS Cent. Sci.* **2018**, *4*, 1600–1616. [CrossRef]
9. Gao, X.; Li, X.; Sun, X.; Zhang, J.; Zhao, Y.; Liu, X.; Li, F. DNA tetrahedra-cross-linked hydrogel functionalized paper for onsite analysis of dna methyltransferase activity using a personal glucose meter. *Anal. Chem.* **2020**, *92*, 4592–4599. [CrossRef]
10. Liu, X.; Li, X.; Gao, X.; Ge, L.; Sun, X.; Li, F. A universal paper-based electrochemical sensor for zero-background assay of diverse biomarkers. *ACS Appl. Mater. Interfaces* **2019**, *11*, 15381–15388. [CrossRef]
11. Liu, L.; Yang, D.; Liu, G. Signal amplification strategies for paper-based analytical devices. *Biosens. Bioelectron.* **2019**, *136*, 60–75. [CrossRef] [PubMed]
12. Howes, P.D.; Chandrawati, R.; Stevens, M.M. Colloidal nanoparticles as advanced biological sensors. *Science* **2014**, *346*. [CrossRef] [PubMed]
13. Resch-Genger, U.; Grabolle, M.; Cavaliere-Jaricot, S.; Nitschke, R.; Nann, T. Quantum dots versus organic dyes as fluorescent labels. *Nat. Methods* **2008**, *5*, 763–775. [CrossRef] [PubMed]
14. Ming, K.; Kim, J.; Biondi, M.J.; Syed, A.; Chen, K.; Lam, A.; Ostrowski, M.; Rebbapragada, A.; Feld, J.J.; Chan, W.C.W. Integrated quantum dot barcode smartphone optical device for wireless multiplexed diagnosis of infected patients. *ACS Nano* **2015**, *9*, 3060–3074. [CrossRef]
15. Rajendran, V.K.; Bakthavathsalam, P.; Bergquist, P.L.; Sunna, A. Smartphone detection of antibiotic resistance using convective PCR and a lateral flow assay. *Sens. Actuators B Chem.* **2019**, *298*. [CrossRef]

16. Shah, K.G.; Singh, V.; Kauffman, P.C.; Abe, K.; Yager, P. Mobile phone ratiometric imaging enables highly sensitive fluorescence lateral flow immunoassays without external optical filters. *Anal. Chem.* **2018**, *90*, 6967–6974. [CrossRef]
17. Morales-Narváez, E.; Montón, H.; Fomicheva, A.; Merkoçi, A. Signal enhancement in antibody microarrays using quantum dots nanocrystals: Application to potential alzheimer's disease biomarker screening. *Anal. Chem.* **2012**, *84*, 6821–6827. [CrossRef]
18. Hassan, A.H.A.; Bergua, J.F.; Morales-Narváez, E.; Mekoçi, A. Validity of a single antibody-based lateral flow immunoassay depending on graphene oxide for highly sensitive determination of *E. coli* O157:H7 in minced beef and river water. *Food Chem.* **2019**, *297*. [CrossRef]
19. Reinhard, E.; Khan, E.A.; Akyz, A.O.; Johnson, G.M. *Color Imaging Fundamentals and Applications*; A K Peters/CRC Press: Natick, MA, USA, 2008; ISBN 1-56881-344-9.
20. Carrilho, E.; Martinez, A.W.; Whitesides, G.M. Understanding wax printing: A simple micropatterning process for paper-based microfluidics. *Anal. Chem.* **2009**, *81*, 7091–7095. [CrossRef]
21. IJ_Mobile for Android—APK Download. Available online: https://apkpure.com/ij-mobile/com.ij_mobile (accessed on 11 December 2019).
22. Otsu, N. A threshold selection method from gray-level histograms. *IEEE Trans. Syst. Man Cybern.* **1979**, *9*, 62–66. [CrossRef]
23. Paterson, A.S.; Raja, B.; Mandadi, V.; Townsend, B.; Lee, M.; Buell, A.; Vu, B.; Brgoch, J.; Willson, R.C. A low-cost smartphone-based platform for highly sensitive point-of-care testing with persistent luminescent phosphors. *Lab Chip* **2017**, *17*, 1051–1059. [CrossRef] [PubMed]

biosensors

MDPI

Article

On-Site Detection of Carcinoembryonic Antigen in Human Serum

Tohid Mahmoudi [1], Mohammad Pourhassan-Moghaddam [2,3], Behnaz Shirdel [1], Behzad Baradaran [1], Eden Morales-Narváez [4,*] and Hamed Golmohammadi [5,*]

1 Immunology Research Center, Tabriz University of Medical Sciences, Tabriz 5166-15731, Iran; mahmoodit@tbzmed.ac.ir (T.M.); behnaz.shirdel201467@gmail.com (B.S.); baradaranb@tbzmed.ac.ir (B.B.)
2 School of Life Sciences, Faculty of Science, University of Technology Sydney, Sydney, NSW 2007, Australia; m.pourhassan_moghaddam@unsw.edu.au
3 ARC Research Hub for Integrated Device for End-User Analysis at Low-Levels (IDEAL Research Hub), Faculty of Science, University of Technology Sydney, Sydney, NSW 2007, Australia
4 Biophotonic Nanosensors Laboratory, Centro de Investigaciones en Óptica, A. C. Loma del Bosque 115, Lomas del Campestre, León 37150, Guanajuato, Mexico
5 Nanosensor Bioplatforms Laboratory, Chemistry and Chemical Engineering Research Center of Iran, Tehran 1496-813151, Iran
* Correspondence: eden@cio.mx (E.M.-N.); golmohammadi@ccerci.ac.ir (H.G.)

Abstract: Real-time connectivity and employment of sustainable materials empowers point-of-care diagnostics with the capability to send clinically relevant data to health care providers even in low-resource settings. In this study, we developed an advantageous kit for the on-site detection of carcinoembryonic antigen (CEA) in human serum. CEA sensing was performed using cellulose-based lateral flow strips, and colorimetric signals were read, processed, and measured using a smartphone-based system. The corresponding immunoreaction was reported by polydopamine-modified gold nanoparticles in order to boost the signal intensity and improve the surface blocking and signal-to-noise relationship, thereby enhancing detection sensitivity when compared with bare gold nanoparticles (up to 20-fold in terms of visual limit of detection). Such lateral flow strips showed a linear range from 0.05 to 50 ng/mL, with a visual limit of detection of 0.05 ng/mL and an assay time of 15 min. Twenty-six clinical samples were also tested using the proposed kit and compared with the gold standard of immunoassays (enzyme linked immunosorbent assay), demonstrating an excellent correlation (R = 0.99). This approach can potentially be utilized for the monitoring of cancer treatment, particularly at locations far from centralized laboratory facilities.

Keywords: lateral flow immunoassay; carcinoembryonic antigen; cancer diagnosis; smartphone-based sensors; point-of-care testing

Citation: Mahmoudi, T.; Pourhassan-Moghaddam, M.; Shirdel, B.; Baradaran, B.; Morales-Narváez, E.; Golmohammadi, H. On-Site Detection of Carcinoembryonic Antigen in Human Serum. *Biosensors* **2021**, *11*, 392. https://doi.org/10.3390/bios11100392

Received: 15 September 2021
Accepted: 12 October 2021
Published: 14 October 2021

Publisher's Note: MDPI stays neutral with regard to jurisdictional claims in published maps and institutional affiliations.

1. Introduction

Cancer is the main cause of morbidity and mortality around the world, with an approximated 18.1 million recent patients and 9.6 million deaths in 2018 [1]. Despite the significant progress in cancer treatment in recent years, the current methodologies have still failed to reach completely satisfactory results, mainly due to late detection. Hence, the early diagnosis of cancer via the quantification of some biomarkers can be considered the golden step for its timely treatment, since it increases the successful treatment rate and, consequently, reduces the related costs and health burden. Moreover, the monitoring of cancer biomarkers is necessary during the treatment process, which further intensifies the importance of the development of on-site diagnostic devices [2].

The carcinoembryonic antigen (CEA) is a collection of glycoproteins that are usually produced for the duration of fetal development, but its production ends prior to birth. Its increase is mostly utilized as a tumor marker to monitor the treatment of colorectal carcinoma or other carcinomas, to recognize recurrences, and for the staging of tumors [3]. Enzyme-linked immunosorbent assay (ELISA) [4], radioimmunoassay [5], chemiluminescence immunoassay [6], and chemiluminescent enzyme immunoassay [7] are the conventional methods for CEA quantification. However, these methodologies are mainly based on expensive devices and time-consuming procedures, and require skilled personnel, hindering their widespread application for patient monitoring—especially at the point-of-care, as well as in resource-limited settings. Therefore, the development of easy-to-use, fast, affordable, but efficient cancer diagnostic methods obviating the need for sophisticated, expensive, bulky equipment and skilled technicians is still in high demand [8,9].

The ASSURED criteria, proposed by the World Health Organization (WHO) in 2003 [10], are a collection of conditions for the ideal tests that can be employed at all levels of health care systems. These criteria emphasize that the ideal diagnostic devices should be affordable, sensitive, specific, user-friendly, rapid and robust, equipment-free, and deliverable to the end-users. However, subsequently, Peeling et al. added two further criteria—real-time connectivity (R), and ease of specimen collection (E)—to the aforementioned conditions, so that these REASSURED criteria enable real-time monitoring of patients by diagnostic systems and enhance the efficiency of health care systems [11]. In this context, lateral flow immunoassays (LFIAs) are simple-to-use immunochromatographic test strip devices with broad applications in clinical analysis, food safety control, environmental monitoring, and drug-abuse assessment [12], which satisfy the WHO's ASSURED criteria to some extent. Interestingly, the integration of smartphone technology with LFIAs can effectively promote their further potential applications through meeting the real-time connectivity (R) criterion. Benefiting from simplicity and portability, a plethora of smartphone-based (bio)sensing platforms have been reported for the detection of various analytes [13]; however, little attention has been paid to the detection of cancer biomarkers [8]. Although some LFIAs have been developed recently for CEA quantification using quantum dot beads [14,15] and magnetic nanoparticles [16,17] as tags, those studies do not satisfy the REASSURED criteria completely. Meanwhile, despite the broad applications of bare gold nanoparticles (GNPs) as the most straightforward tags in LFIAs, these tags suffer from two traits: (1) the low efficiency of formed immunocomplexes due to the random orientation of antibodies (Abs) prepared by passive adsorption, and (2) their bright-red color, which impedes their visual detection and the interpretation of results—especially in target concentrations near to the limit of quantification—even for analysis via strip readers [18]. To remedy this issue, the traditional strategy involves the application of thiol-containing linkers that bind spontaneously to the surface of GNPs and remain an active carboxyl or amine group for subsequent covalent coupling to Abs [19]. The other approach is the utilization of functional polymers for the encapsulation of GNPs, which not only stabilize nanoparticles, but also provide an active layer for subsequent immobilization of Abs. In this regard, Xu et al. [20] modified the surface of GNPs with a polydopamine layer (GNP@PDA) and used it in a competitive-type LFIA for the sensitive detection of zearalenone in maize, where there was an indirect relation between the visual signal and the target concentration. To the best of our knowledge, the efficiency and behavior of the GNP@PDA tag has not been investigated in a sandwich-type LFIA, and its advantageous quantification via smartphone-based module has not yet been reported.

Herein, a highly sensitive sandwich-type GNP@PDA-based LFIA was developed for the quantification of CEA in sera. The GNPs were first synthesized via the Turkevich method, and then coated with a nanolayer of polydopamine, via the self-polymerization of dopamine in alkaline media, to act as an antibody immobilization layer. By employing GNP@PDA as a tag, in the presence of CEA, sandwich immunocomplexes were formed at the test zone, providing a slightly dark-red color proportional to the concentration of the target. In addition, the fabricated plasmonic LFIA was further coupled with a 3D-printed smartphone-based colorimetric imaging device to capture digital images of the strips' test zones and, subsequently, to quantify the target concentration via a detection algorithm with a self-developed smartphone app. The developed platform provides a cost-effective, easy-to-use, portable smartphone-based LFIA kit for the quantification of CEA in serum samples down to 0.05 nM, and satisfies the REASSURED criteria, for which it needs only to capture an image of the test zone to show the respective concentration of the target in a sample.

2. Materials and Methods

2.1. Reagents and Instruments

Hydrogen tetrachloroaurate (III) hydrate ($HAuCl_4 \cdot 3H_2O$), trisodium citrate ($Na_3C_6H_5O_7 \cdot 2H_2O$), Tween-20, bovine serum albumin (BSA), and dopamine hydrochloride (DA·HCl) were all purchased from Sigma-Aldrich. The MF1 membrane was obtained from Whatman International Ltd. (Maidstone, UK). The mouse monoclonal antibody (MAB1393) and goat polyclonal antibody (PAB7939)—both against human CEA—were obtained from Abnova (Taipei City, Taiwan). Based on the manufacturer's information, these Abs are specific to human CEA, and have no cross-reactivity with human nonspecific cross-reacting antigen (NCA, NCA2) or biliary glycoprotein-1 (BGP1). The stock solutions of CEA involved in the CEA ELISA kit (MONOKIT)—solely or by dilution—were used as targets. For the preparation of all solutions, Milli-Q-grade water was used. The components of the strips—including pads (SP08, SCL0020215), nitrocellulose membrane (LFNC-C-SS03-15 μm, JCN476015), and backing card (type-L)—were all acquired from Nupore Filtration Systems Pvt. Ltd. (Ghaziabad, India).

The UV–Vis spectra of the solutions were recorded using Cytation 5 (BioTek, Winooski, VT, USA) with a quartz microplate. The structure of synthesized GNP@PDA was observed via transmission electron microscopy (TEM) with a 100 kV running voltage (Zeiss-EM10C-Germany, Jena, Germany). The ζ-potentials were acquired using a Zetasizer Nano ZS (Malvern Instruments, Malvern, UK). A smartphone containing a 13 MP rear camera (Samsung galaxy C8) was utilized for image capturing and analysis. The Fourier-transform infrared–attenuated total reflectance (FTIR–ATR) spectra were measured using an AVATAR (Thermo Scientific, Minneapolis, MN, USA).

2.2. Synthesis of GNP@PDA

The GNPs were firstly synthesized via the conventional Turkevich method [21]. In brief, 50 mL of 0.01% $HAuCl_4$ solution was heated until boiling, and then 5 mL of 0.04 mol/L sodium citrate was quickly added to it under stirring. After heating for 10 min, the mixture was cooled down to room temperature with constant stirring. The obtained colloidal gold was filtered with a 0.22 μm syringe filter and then kept in a dark bottle at 4 °C.

The GNP@PDA was then synthesized as stated by an earlier work [20], with some modifications. In brief, the pH of the prepared GNPs (1 mL) was set to 7.5 by adding 0.1 mol/L K_2CO_3 solution, followed by the addition of 5 μL of 3% H_2O_2. Following vigorous stirring for 5 min, 5 μL of freshly prepared DA·HCl solution (10 mg/mL) was added to the solution and stirred for 1 h. The addition of dopamine and stirring were repeated in order to control the graft polymerization of dopamine on the GNPs surfaces. After centrifugation at 9000 rpm for 20 min and removal of the supernatant, the prepared GNP@PDA was washed twice and re-suspended in borate buffer (0.005 mol/L, pH 7.5).

2.3. Preparation of GNP–mAb and GNP@PDA–mAb Conjugates

The mouse monoclonal antibody against human CEA (mAb) was conjugated to the GNPs (OD = 1) or corresponding GNP@PDA via simple mixing [20]. To this end, different values of mAb were added dropwise to 1 mL of GNPs or GNP@PDA solution, and after gentle shaking at 4 °C overnight, the mixture was blocked with 2% BSA for 1 h. Subsequently, via centrifugation at 9000 rpm for 20 min, the conjugates were re-suspended in 50 μL of 0.005 mol/L borate buffer containing 5% sucrose and 1% BSA. The conjugates were then kept in microtubes at 4 °C for further use.

2.4. Preparation of the Lateral Flow Test Strips

Before assembling the strip components, the sample pad and conjugate pad were pre-treated by dipping them in blocking solutions (0.005 mol/L borate buffer pH 7.5, 0.05% Tween-20 for the former, and the same solution with 3% sucrose for the latter). The pads were then dried out in an oven at 50 °C for 2 h. Thereafter, considering an overlap of 2 mm between the pads and the NC membrane, they were subsequently laminated onto the backing card. Finally, the assembly was cut into strips with a 3.5 mm width. The lengths of the sample pad, conjugate pad, NC membrane, and adsorption pad were 30, 5, 20, and 27 mm, respectively. The desired value of the conjugate was pipetted onto the conjugate pad in each test. For the test zone (TZ), 0.2 μL of polyclonal antibody (pAb) (0.25, 0.5, 1 mg/mL), and for the control zone (CZ), 0.2 μL of 100 μg/mL Protein G (PG), were pipetted. The ready-to-test strips were then dried at room temperature for 1 h and stored for up to one week.

2.5. The Fabrication and Setup of the Smartphone-Based Colorimetric Imaging Device

The smartphone-based colorimetric imaging device was composed of two major constituents: a strip cartridge, and an optical imaging box containing a cartridge-placing section, USB cable, and smartphone-reading section (Figure S1). The strip cartridge was designed to hold the strips within the imaging box, with reproducible sections of the TZ being imaged with the camera. The optical imaging box was also composed of an internal light source (high-power white LED, TOP-1BD1, from Epileds, Taiwan, with λ = 7000 nm, light intensity 100–140 Lm, and forward voltage 3–3.4 V) and an electric circuit fixed on a stand powered by municipal electricity along with an adaptor, or by the smartphone itself; it was made with a 3D printer using acrylonitrile butadiene styrene polymer. The 3D CAD file of the platform was first prepared using SolidWorks software. More details of the design and dimensions are given in the Supplementary Materials (Figure S2). The total cost of this platform was USD ~2.5 (Table S1).

2.6. The App Development

The smartphone app—TBZMed Sensor (Figure S3)—was planned for the Android platform (version 7.1.1) in the Android studio environment and installed on a Samsung Galaxy C8 smartphone. The app benefits from a user-friendly interface (Figure S3a,b) for selecting and cropping the images of the test zone. For imaging, the strip was placed in the strip cartridge and then inserted in the cartridge-placing section within the imaging box. The light source was then turned on and an image was captured after setting the camera to manual mode, autofocus, and zoom to 4×, and clicking on the strip on the screen. After capturing the image and cropping the test zone with the same area for all samples, the TBZMed Sensor extracted the RGB and grayscale values from the JPEG digital image. Therefore, a calibration curve was first established based on the relationship between the concentration of CEA in standard serum samples and the grayscale value of the TZ. To this end, the area of interest within a TZ could be selected by a circle frame (Figure S3c), and the color information of the selected area was then recorded. The mean values of the R, G, B, and grayscale channels were calculated automatically. The calibration graphs for standard serum samples were investigated and calculated in Microsoft Excel based on different R, G, B, and grayscale data. Then, the best calibration equation was obtained, and was entered into the TBZMed Sensor environment. The channel type was also saved in the app environment (Figure S3d). Finally, the concentration of CEA in an unknown sample could be calculated easily in ng/mL using the saved calibration curve in the app, by simply capturing an image from the TZ and selecting the equation from the app (Figure S3e). In addition, an image of the test result could be easily shared by clinking on the sharing icon (<) for possible real-time connectivity (Figure S3f).

2.7. Lateral Flow Immunoassay Procedure

The standard CEA solutions, or dilutions thereof, were used as targets by mixing them with buffer or standard serum samples. Firstly, the assay procedure consisted of dispensing 5 µL of sample solution onto the sample pad and then adding 60 µL of running buffer (Tris buffer 0.05 mol/L, 0.05% Tween-20). After complete running of the test, images were captured by the smartphone-based platform and then analyzed via the developed app. The data were represented as mean ± SD for three replicates. The calibration graph obtained for the serum samples was used for insertion into the app environment and reading the concentrations of clinical samples. For serum and clinical samples, the sample pad's performance was improved by putting the MF1 membrane on it. According to the manufacturer's information, this pad is a bound glass-fiber filter that can be used for lateral flow assays, and is typically used for whole-blood analysis.

2.8. Clinical Samples Analysis

This study was approved by the Medical Ethics Committee of Tabriz University of Medical Sciences, Iran, and all methods were performed following the relevant guidelines and regulations. The clinical serum samples were obtained from Shahid Ghazi Tabatabaei Hospital of Tabriz, Iran, and included 12 positive samples and 14 negative samples. Real and false negative samples were determined according to a report from the hospital during a one-year follow-up of the patients' conditions. The assay procedure for clinical sample analysis was as described in Section 2.7. The values of CEA in clinical samples were measured using a CEA ELISA kit (MONOKIT) in the hospital laboratory.

3. Results and Discussion

Scheme 1 displays the schematic diagram of CEA detection by the GNP@PDA-based lateral flow immunosensor and its components (Scheme 1a). In the absence of CEA (Scheme 1b), only a slightly dark red at the CZ was formed for the conjugate bound to PG; however, in the presence of CEA (Scheme 1c), a GNP@PDA–mAb–CEA–pAb immunocomplex was formed at the TZ, and GNP@PDA–mAb–PG was formed at the CZ; hence, a slightly dark-red color appeared on them. The color intensity at the TZ was recorded with the developed platform. A higher concentration of CEA in the sample resulted in the stronger color intensity of the TZ up to the hook effect region, which then turned to low signal intensities. Eventually, the results can be easily quantified via the smartphone readout and subsequently shared via the Internet with a physician, family, and or even emergency services (Scheme 1d).

3.1. Characterization of GNP–mAb and GNP@PDA–mAb Conjugates

The physicochemical properties of nanoparticles and nanobioconjugates have a considerable effect on the performance of the test strip. The full characterization of these nanostructures is given in the Supplementary Materials, Section S2 and Figure S4, showing successful functionalization of GNPs by a nanometer layer of PDA. The average ζ-potential for GNPs is −15.6 ± 1.4 mV, which is reduced by the formation of the PDA layer around them, reaching as low as −29.2 ± 2.3 mV. This is mainly attributed to the abundance of hydroxyl groups in the PDA layer [20]. After the immobilization of Abs on the nanoparticles' surface, due to the positive charge of Abs at pH 7.5, the ζ-potential is increased slightly (−28.0 ± 1.6 mV). Using Tukey's multiple comparisons test clearly shows significant differences between the zeta potential values obtained for all materials except for GNP@PDA and GNP@PDA–Ab, which showed no significant statistical differences (p-value = 0.7095) (See Section S2, Table S2 in the Supplementary Materials), possibly due to the small values of Ab compared to the PDA layer. The TEM image of GNP@PDA (Figure S4f) shows a thin layer of PDA with lower contrast (~2 nm) around the GNPs, indicating that the structure of the GNP@PDA is core–shell with an average diameter of ~27.49 ± 7.6 nm.

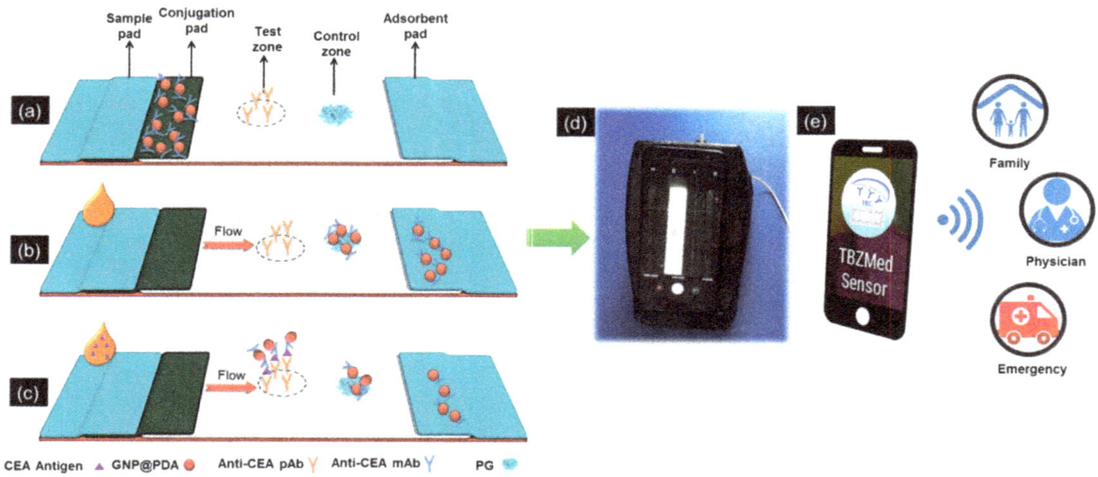

Scheme 1. Schematics of (**a**) ready-to-use test strip and its components, with GNP@PDA–mAb conjugates deposited on the conjugate pad, pAb on the test zone, and PG on the control zone; (**b**) the formation of GNP@PDA–mAb–PG immuno-sandwich structures in the control zone in the absence of CEA; (**c**) the formation of a GNP@PDA–mAb–CEA–pAb immunocomplex at the test zone and GNP@PDA–mAb–PG at control zone in the presence of CEA; (**d**) smartphone-based quantification and (**e**) real-time connectivity using our developed app.

3.2. Optimization of Effective Factors on the Performance of the Developed Smartphone-Based LFIA Kit

The blocking of sample and conjugation pads, the type of running buffer, the amount of mAb immobilized on the GNP@PDA, the amount of GNP@PDA–mAb dropped on the conjugation pad, the amount of pAb at the TZ, and the reading time are some of the most important factors that impact on the performance of our developed LFIA kit. Hence, the effects of the mentioned parameters were investigated and, subsequently, the optimal levels were used for the next experiments. The optimization results and further descriptions are given in Section S3 of the Supplementary Materials, and in Figures S5 and S6. The results show that blocking of the sample pad with 0.005 mol/L borate buffer pH 7.5 containing 0.05% Tween-20, and the conjugate pad with 0.005 mol/L borate buffer pH 7.5 containing 0.05% Tween-20 and 3% sucrose, are necessary for the successful functioning of the strips. Among the three buffers—including borate (0.005 mol/L, pH 7.5), phosphate (1X, pH 7.4), and Tris (0.005 mol/L, pH 7.5), all with 0.1% Tween-20—the third one showed a good flow rate and the least nonspecific adsorption of conjugates on the TZ. The amount of mAb immobilized on GNP@PDA affects the immunosensor performance in two ways: Firstly, sufficient mAb on the nanoparticles is necessary in order to be able to form immuno-sandwich complexes on the TZ. Secondly, the high amounts of mAb cause steric hindrance, or even capture more target molecules and prevent their immunocomplex formation within the TZ [22]. The values of 3.2 and 6.41 μg per mL of GNPs/PDA showed favorable behavior in the analysis of buffer and serum samples, respectively. The volume of the conjugate also played a critical role in the assay—by its increase, the signal at the TZ was increased, but the background signal was also boosted. Hence, 2.5 μL of the probe was selected as the optimal level. The amount of pAb at the TZ is essential for the efficient capturing of GNP@PDA–mAb–CEA immunocomplexes; its performance was increased until the advent of steric hindrance at 0.5 mg/mL, which reduced the immuno-sandwich formation. Therefore, 0.5 mg/mL was selected as the optimal value for the amount of pAb at the TZ. Since the contrast of the TZ was decreased over time because of the drying of the membrane, the images should be captured within 2 min after completion of the test. For better comparison, the performance of GNPs as tags was examined under the same conditions obtained for GNP@PDA.

3.3. Analytical Performance of the Developed Smartphone-Based LFIA Kit

The citrate-capped GNPs show bright-red color; in contrast, the color of GNP@PDA is slightly darker due to the black color of PDA self-polymerized by dopamine on GNPs in an alkaline medium. Moreover, the PDA layer enables more efficient immobilization of Ab on the tags' surface. These advantageous features can boost the signal intensity and detection sensitivity. To check this, a comparison was first made between the utility of GNPs and GNP@PDA tags in the visual analysis of CEA via the strips in the buffer. Under the optimized conditions, the same volumes of CEA at various concentrations were added to the strips, and the visual signals from the TZ were assessed. As seen in Figure 1a,b, the dots formed at the TZ for strips prepared using GNP@PDA as a tag have a slightly darker red color compared to strips prepared using GNPs alone and, interestingly, the visual limits of detection for GNPs and GNP@PDA as tags are 1 and 0.05 ng/mL, respectively. This observation proves the better efficiency of GNP@PDA as tags in a sandwich-type lateral flow test strip compared to GNPs, with 20 times reduction in the visual limit of detection. Another advantage of GNP@PDA compared to GNPs is its better surface blocking and, hence, lower background signal, which enhances the sensitivity and reduces the limit of detection. Such improved blocking behavior may be a result of the interaction of BSA with PDA chains anchored on the surface of GNPs. The hook effect was observed for concentrations higher than 125 ng/mL for GNPs, compared to 50 ng/mL for GNP@PDA. This is a common phenomenon in sandwich-type LFIAs because of the occupation of fragment antigen-binding sites for both capture and reporting of Abs in the presence of high concentrations of the target, which reduces the immunocomplex formation and, hence, deceases the color of the TZ [23]. This is also an important issue in the use of commercial CEA strips. As for commercial strips, herein, in the case of suspicious samples, the user could dilute the sample and re-check it by strip. If the signal increased after dilution of the sample, this showed the high concentration of CEA and the occurrence of the hook effect. Therefore, the user should consider the result obtained for the diluted sample to be a correct answer.

Visual detection of LFIA results suffers from some limitations: firstly, the user's eyesight impacts on the reading of the TZ, and can become challenging, especially at low concentrations of analyte; secondly, different environmental illumination conditions affect the interpretation of colors and results [24,25]. However, the imaging box developed herein provides constant lightning conditions, and the smartphone quantification platform provides quantitative results in a facile way, preventing the user's eyesight error. Since the matrix of real samples is more complex than the buffer, the same quantitative studies performed with the buffer were conducted in standard serum samples containing the desired values of CEA, with GNP@PDA as a tag, to obtain the desired calibration curve. The images of strips and the calibration curve are given in Figure 1c. After elucidating various calibration curves for the color intensities of the R, G, B, and Gray channels vs. the concentration of the target, as seen in Figure 1c, a logarithmic relationship was obtained between the concentration of CEA in serum and the (1/Gray value)*10,000 in the range of 0.05–50 ng/mL with R = 0.99. The relative standard deviation values were in the range of 2.5% to 12.5%, confirming the good reproducibility of the results. For concentrations higher than 50 ng/mL, the hook effect was observed. Trials for concentrations lower than 0.05 ng/mL failed to show considerable changes compared to the buffer, clearly showing the effect of corona formation in serum samples, as stated elsewhere [26].

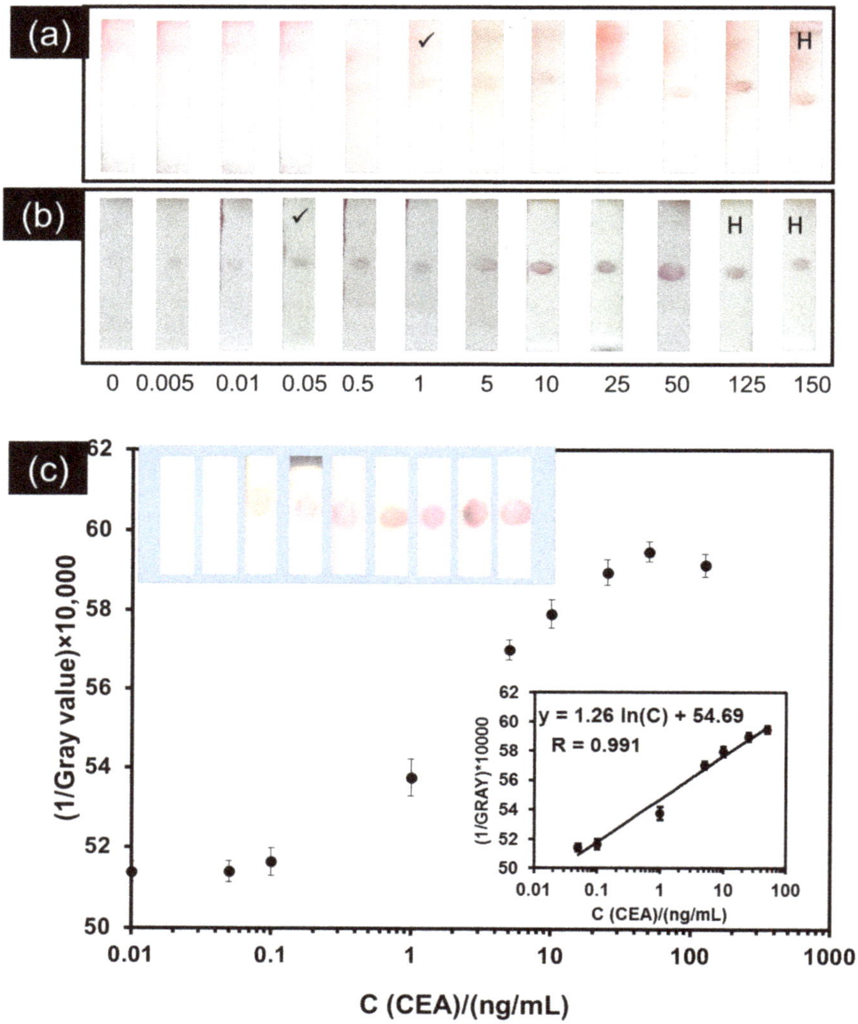

Figure 1. Images of strips taken without the imaging platform: (**a**) GNPs and (**b**) GNP@PDA as tags for different values of CEA in buffer solution (in ng/mL), (√ shows the visual limit of detection, H represents the occurrence of the hook effect), and (**c**) (1/Gray value)* × 10,000 vs. concentration of CEA (ng/mL) in standard serum samples. Insets: calibration curve for analysis of CEA, and images of strips taken with the imaging platform.

Table 1 indicates the intra-assay and inter-assay recoveries for samples spiked with CEA at different concentrations using GNP@PDA as a tag. The experiments were carried out in triplicate for each spiked concentration. The intra-assay studies were performed within 1 day, and the inter-assay studies were conducted for 3 days. The average recoveries and coefficient of variation (CV) values for the intra-assay tests were within 95–120% and 6.3–10%, respectively. The inter-assay recoveries ranged from 90 to 120%, with a CV ranging from 8.33 to 11.66%. These results prove that this immunosensor benefits from high accuracy and precision [27]. Since long-term stability studies were not the purpose of our work, we did not check it, but taking into account the several weeks of stability of commercially available pregnancy tests, we can expect similar performance [28].

Table 1. The obtained recovery and coefficient of variation (CV) results for CEA-spiked serum samples using the developed kit.

Spiked CEA (ng/mL)	Intra-Assay			Inter-Assay		
	Mean ± SD	Recovery (%)	CV (%)	Mean ± SD	Recovery (%)	CV (%)
0.5	0.6 ± 0.06	120	10	0.6 ± 0.07	120	11.66
2	2.3 ± 0.15	115	6.5	2.4 ± 0.25	120	10.41
20	19 ± 1.2	95	6.3	18 ± 1.5	90	8.33
35	34 ± 2.6	97	7.6	35 ± 3.4	100	9.70
45	46 ± 3.4	102	7.4	46 ± 5.2	102	11.30

3.4. Application of the Developed Smartphone-Based LFIA Kit for the Detection of CEA in Clinical Serum Samples

After quantifying CEA in real serum samples, the extracted calibration curve was inserted into the developed "TBZMed Sensor" app and used to analyze the clinical serum samples. Among these samples, 12 samples had CEA values higher than 5 ng/mL (positive samples), while 14 samples had CEA values lower than 5 ng/mL (negative samples). A comparison of the results obtained by our developed smartphone-based LFIA kit and ELISA is given in Figure 2a. As seen, an excellent correlation (R = 0.99) between these two approaches was attained, confirming a good agreement between them. In addition, the sensitivity and specificity of the developed smartphone-based LFIA kit were further calculated. The obtained results (sensitivity of 91% and specificity of 93%) in Table 2 and Figure 2a illustrate that the developed smartphone-based LFIA kit can potentially be utilized for accurate and rapid monitoring of cancer treatment at sites far from centralized laboratory facilities. Using an unpaired *t*-test for the analysis of CEA (+) and CEA (−) clinical samples revealed a considerable ability of the developed platform to discriminate between these two groups (p-value < 0.0001) (Figure 2b). Receiver operating characteristic (ROC) analysis is a valuable means of assessing the performance of diagnostic tests, acting as a simple graphical tool for displaying the accuracy of a medical diagnostic test. ROC analysis was also carried out on clinical serum samples. The TP, TN, FP, and FN values were obtained according to a report from the hospital during a one-year follow-up of the patients' conditions [29]. The area under the ROC curve is a summary measure that essentially averages diagnostic accuracy across the spectrum of test values. As seen in Figure 2c, with an area of 0.99, the developed platform benefits from high diagnostic accuracy [30].

A comparison was also made between our developed cost-effective smartphone-based LFIA kit and currently reported lateral-flow-based portable detection systems for the quantification of CEA, in terms of analytical performance and meeting the REASSURED criteria (Table 3). Since the commercial magnetic and fluorescent strip readers do not have ability to connect to the Internet, such assays suffer in terms of real-time connectivity (lack of the R criterion), as described for [14,17] in Table 3. The approaches developed in [16,17] may suffer in terms of the user-friendly (U), equipment-free (E), and deliverable to the end-user (D) criteria due to the complexity of the quantification procedure. In addition, because of the use of external image analysis software, [16] could not satisfy the necessity of easy quantification and sharing via smartphone (lack of the R criterion). The complexity of the quantification process is also another drawback of some previous works, such as [15,16,31] (lack of the E and U criteria). As seen in Tables 2 and 3, with a comparable linear range of 0.05–50 ng/mL, considerable sensitivity and specificity of 91% and 93%, respectively, the capability of analyzing serum samples and potential application to blood samples, easy access to the Internet for transferring results via the developed app, low cost of the assembly (USD ~2.5, see Table S1), and its user friendly characteristics—i.e., quantification of CEA in clinical samples by image capturing alone—this device satisfies the REASSURED criteria [32–34]. Although our developed platform benefits from the REASSURED criteria, it may suffer from camera-dependent output results, which is an intrinsic drawback of smartphone-based quantification systems. To remedy this issue, a simple approach is to attach the smartphone or a camera to the black box system, providing all users with the same image capturing system.

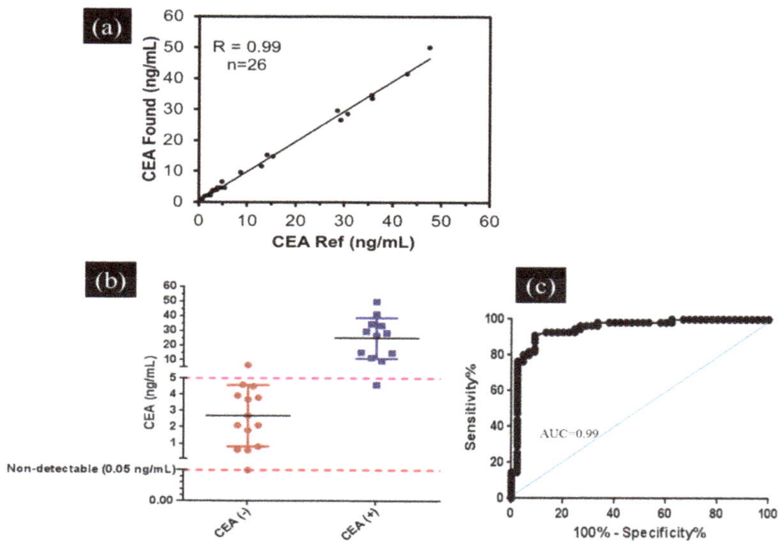

Figure 2. (**a**) Comparison of the results obtained with the developed smartphone-based LFIA kit (CEA found) and CEA reference (commercial ELISA kit) in the measurement of CEA in clinical samples. (**b**) The estimated CEA levels corresponding to each sample, resulting from the analysis facilitated by the smartphone-based platform. The clinical serum samples were classified in two groups: CEA-positive (+, n = 12 samples), and CEA-negative (−, n = 14 samples). (**c**) ROC analysis, with area under the curve (AUC) of 0.99.

Table 2. Characteristics of the developed smartphone-based LFIA kit in the analysis of clinical serum samples.

Samples	Number of Positive Results	Number of Negative Results	Characteristic Parameter	
12 (+)	11	1	Sensitivity	91%
14 (−)	1	13	Specificity	93%

Sensitivity = (TP/TP + FN)∗100; specificity = (TN/TN + FP)∗100; TP: true positive; FN: false negative; TN: true negative; FP: false positive.

Table 3. Comparison of some characteristics of our developed smartphone-based LFIA kit with recently reported LFIA-based portable systems, in terms of the quantification of CEA.

Detection Strategy	Used Tag	LR * (ng/mL)	LOD ** (ng/mL)	Detection Time (min)	REASSURED * Criteria	Ref.
Commercial magnetic strip reader	Magnetic particles	1–100	0.045	30	-EASS-R- -	[17]
Smartphone-based colorimetric image analysis	Magnetic NPs containing Ab and biotinylated DNA	0.25–100	0.0375	15	-EASS-R- -	[16]
Fluorescent handing system	Quantum dots	1–100	5	20	REASS-R-D	[15]
Fluorescent handing system	Quantum dot nanobeads	1–50	0.049	15	REASS-R-D	[32]
Commercial fluorescent reader	Quantum-dot-doped polystyrene nanoparticles	2.8–680	0.35	15	-EASSURED	[14]
Smartphone-based colorimetric image analysis	GNP@PDA	0.05–50	0.05	15	REASSURED	this work

* LR: linear range; ** LOD: limit of detection; * REASSURED criteria: real-time connectivity (R), ease of specimen collection (E), affordable (A), sensitive (S), specific (S), user-friendly (U), rapid and robust (R), equipment-free (E), deliverable to the end-user (D).

4. Conclusions

In conclusion, by combining the desired characteristics of GNP@PDA as an efficient tag in a sandwich-type LFIA, and by the development of a portable smartphone-based platform, a simple, cost-effective (USD ~2.5, see Table S1), and easy-to-use immunosensing device was developed for the smartphone-based detection and quantification of CEA in human serum. With a linear range of 0.05–50 ng/mL and a low LOD of 0.05 ng/mL, sensitivity of 91%, and specificity of 93%, the developed smartphone-based LFIA kit provides the desired characteristics for point-of-care evaluation of cancer biomarkers. The linear range of the strip satisfies the requirements for clinical analysis. The employed MF1 membrane in the sample pad provides the capability to perform whole-blood-sample analysis; hence, our developed sensing platform can potentially be employed for blood sample tests without any pretreatment or enrichment, addressing the REASSURED criteria. In addition to the desired characteristics of the platform, the automation of image capturing via our developed app may further improve utility and user-friendliness. Building upon the satisfactory results of our developed smartphone-based LFIA platform in the detection of CEA in human serum samples—especially in comparison with the results of reference methods—and its other advantageous features as an assay kit that meets the World Health Organization's REASSURED criteria, we believe it could potentially be widely exploited for patient monitoring, particularly at sites far from centralized laboratory facilities, for point-of-care applications, and in resource-limited settings. Although our developed platform shows some benefits in terms of the REASSURED criteria, it should be noted that this system may suffer from camera-dependent output results, which is an intrinsic drawback of smartphone-based quantification systems. To remedy this issue, there are two solutions: (1) some calibration samples can be required within the final kit to calibrate the quantification module before analyzing a real sample, due to differences in the optics of the camera and the qualities of the employed CCDs; (2) a camera can be attached to the black box system, which provides all users with the same image capturing system. Our group is also working in this area to offer solutions to this drawback.

Supplementary Materials: The following are available online at https://www.mdpi.com/article/10.3390/bios11100392/s1: Section S1: Figure S1: Photos of smartphone-based colorimetric imaging device: (a) strip cartridge containing a test strip, the location of the cartridge-placing section within the imaging box, and the USB cable; (b) the location of the smartphone holder and imaging aperture; (c) the whole assembly of the platform, showing the placed strip within the box and the electric circuit connected to the smartphone, and imaging of the strip; Figure S2: (a) photo; (b) design and dimensions of 3D-printed LED holder—i: hole for placing LED; ii cartridge-placing section. Illustration and dimensions of (c) strip cartridge, and (d,e) imaging box; Table S1: Estimated cost of fabrication of each imaging platform; Figure S3: The main interface of TBZMed Sensor: (a) icon; (b) the main interface for selecting an image from the gallery or taking an image via the camera; (c) selecting the test zone and cropping the image; (d) selecting the calibration equation; (e) reading the concentration of CEA in ng/mL; and (f) real-time sharing of the test result; Section S2: Characterization of GNP-mAb and GNP@PDA-mAb conjugates; Figure S4: Characterization of nanostructures, UV–Vis spectra of different nanoparticles and respective bioconjugates: (a) initial; (b) in the presence of 1% NaCl; (c) numerical values of λmax; (d) FTIR–ATR spectra of GNPs and GNP@PDA; (e) ζ-potential values; (f) TEM image of GNP@PDA; Table S2: Results of Tukey's multiple comparisons test on ζ-potential values; Section S3: Optimization of effective factors on the performance of the developed smartphone-based LFIA kit; Figure S5: The flow behavior of strips containing (a) unblocked conjugate pad, (b) blocked conjugate pad (in lateral flow format), (c) both the sample and conjugate pads blocked, and (d) unblocked sample and conjugate pads (in dipstick format); Figure S6: (a) Effect of running buffer on strip performance—T depicts Tris (0.005 mol/L, pH 7.5), B depicts borate (0.005 mol/L, pH 7.5), and P depicts phosphate (1X, pH 7.4), all with %0.1 Tween-20; 1 and 2 indicate that the concentration of CEA is 0 and 5 ng/mL, respectively. (b) Effect of the amount of mAb immobilized on GNP@PDA, including 1.6, 3.2, and 6.4 μg for buffer sample testing (b1 to b3), and 3.2, 6.4, and 9.6 μg for serum sample testing (s1 to s3) on strip performance. (c) Effect of the volume of GNP@PDA–mAb dropped on the conjugation pad—where V1 to V3 depict 1.5, 2.5, and 5 μL, respectively (in each case the left image is for blank and the right image is for the sample containing CEA 5 ng/mL)—on strip performance. (d) Effect of different amounts of polyclonal antibody immobilized on the test zone on the performance of the strip. C1 to C4 represent concentrations of 0, 0.25, 0.5, and 1 mg/mL, respectively.

Author Contributions: Conceptualization, T.M., B.B., and H.G.; methodology, T.M.; formal analysis, T.M., H.G., and B.S.; investigation, T.M., H.G., and B.S.; resources, H.G., E.M.-N., and B.B.; writing—original draft preparation, T.M.; writing—review and editing, T.M., E.M.-N., B.S., M.P.-M., and H.G. All authors have read and agreed to the published version of the manuscript.

Funding: This research was funded by the Iran National Science Foundation (INSF) (grant number of 97001910) and Tabriz University of Medical Sciences (grant numbers 60836 and 63253).

Institutional Review Board Statement: This study was conducted according to the guidelines of the ethical principles and the national norms and standards for conducting Medical Research in Iran, and approved by the Institutional Review Board (or Ethics Committee) of Tabriz University of Medical Sciences (protocol code IR.TBZMED.VCR.REC.1397.161 and date of approval 20 August 2018).

Informed Consent Statement: Informed consent was obtained from all subjects involved in the study. Written informed consent was obtained from the patient(s) to publish this paper.

Data Availability Statement: Not applicable.

Acknowledgments: We acknowledge support from the "Iran National Science Foundation (INSF)", "Immunology Research Center, Tabriz University of Medical Sciences, Tabriz, Iran, and the "Chemistry and Chemical Engineering Research Center of Iran (CCERCI)", and E.M.-N. acknowledges financial support by CONACYT (Mexico, Grant No. 312271, and Grant No. 376135).

Conflicts of Interest: The authors declare no conflict of interest.

References

1. Bray, F.; Ferlay, J.; Soerjomataram, I.; Siegel, R.L.; Torre, L.A.; Jemal, A. Global cancer statistics 2018: GLOBOCAN estimates of incidence and mortality worldwide for 36 cancers in 185 countries. *CA Cancer J. Clin.* **2018**, *68*, 394–424. [CrossRef]
2. Wender, R.C.; Brawley, O.W.; Fedewa, S.A.; Gansler, T.; Smith, R.A. A blueprint for cancer screening and early detection: Advancing screening's contribution to cancer control. *CA Cancer J. Clin.* **2019**, *69*, 50–79. [CrossRef]
3. Duffy, M.J. Carcinoembryonic antigen as a marker for colorectal cancer: Is it clinically useful? *Clin. Chem.* **2001**, *47*, 624–630. [CrossRef]
4. Zhao, L.; Xu, S.; Fjaertoft, G.; Pauksen, K.; Håkansson, L.; Venge, P. An enzyme-linked immunosorbent assay for human carcinoembryonic antigen-related cell adhesion molecule 8, a biological marker of granulocyte activities in vivo. *J. Immunol. Methods* **2004**, *293*, 207–214. [CrossRef]
5. Kuroki, M.; Yamaguchi, A.; Koga, Y.; Matsuoka, Y. Antigenic reactivities of purified preparations of carcinoembryonic antigen (CEA) and related normal antigens using four different radioimmunoassay systems for CEA. *J. Immunol. Methods* **1983**, *60*, 221–233. [CrossRef]
6. Qu, S.; Liu, J.; Luo, J.; Huang, Y.; Shi, W.; Wang, B.; Cai, X. A rapid and highly sensitive portable chemiluminescent immunosensor of carcinoembryonic antigen based on immunomagnetic separation in human serum. *Anal. Chim. Acta* **2013**, *766*, 94–99. [CrossRef] [PubMed]
7. Falzarano, R.; Viggiani, V.; Michienzi, S.; Longo, F.; Tudini, S.; Frati, L.; Anastasi, E. Evaluation of a CLEIA automated assay system for the detection of a panel of tumor markers. *Tumor Biol.* **2013**, *34*, 3093–3100. [CrossRef] [PubMed]
8. Mahmoudi, T.; de la Guardia, M.; Baradaran, B. Lateral Flow Assays towards Point-of-Care Cancer Detection: A Review of Current Progress and Future Trends. *TrAC Trends Anal. Chem.* **2020**, *125*, 115842. [CrossRef]
9. Golmohammadi, H.; Hamzei, Z.; Hosseinifard, M.; Ahmadi, S.H. Smart Fully Integrated Lab: A Smartphone-Based Compact Miniaturized Analytical/Diagnostic Device. *Adv. Mater. Technol.* **2020**, *5*, 2000742. [CrossRef]
10. Mabey, D.; Peeling, R.W.; Ustianowski, A.; Perkins, M.D. Diagnostics for the developing world. *Nat. Rev. Microbiol.* **2004**, *2*, 231–240. [CrossRef] [PubMed]
11. Land, K.J.; Boeras, D.I.; Chen, X.-S.; Ramsay, A.R.; Peeling, R.W. REASSURED diagnostics to inform disease control strategies, strengthen health systems and improve patient outcomes. *Nat. Microbiol.* **2019**, *4*, 46–54. [CrossRef]
12. Mahmoudi, T.; de la Guardia, M.; Shirdel, B.; Mokhtarzadeh, A.; Baradarn, B. Recent Advancements in Structural Improvements of Lateral Flow Assays towards Point-of-Care Testing. *TrAC Trends Anal. Chem.* **2019**, *116*, 13–30. [CrossRef]
13. Younis, M.R.; Wang, C.; Younis, M.A.; Xia, X.H. Smartphone-Based Biosensors. In *Nanobiosensors: From Design to Applications*; Wiley: Hoboken, NJ, USA, 2020; pp. 357–387. [CrossRef]

14. Chen, Z.; Liang, R.; Guo, X.; Liang, J.; Deng, Q.; Li, M.; An, T.; Liu, T.; Wu, Y. Simultaneous quantitation of cytokeratin-19 fragment and carcinoembryonic antigen in human serum via quantum dot-doped nanoparticles. *Biosens. Bioelectron.* **2017**, *91*, 60–65. [CrossRef] [PubMed]
15. Qin, W.; Wang, K.; Xiao, K.; Hou, Y.; Lu, W.; Xu, H.; Wo, Y.; Feng, S.; Cui, D. Carcinoembryonic antigen detection with "Handing"-controlled fluorescence spectroscopy using a color matrix for point-of-care applications. *Biosens. Bioelectron.* **2017**, *90*, 508–515. [CrossRef] [PubMed]
16. Liu, F.; Zhang, H.; Wu, Z.; Dong, H.; Zhou, L.; Yang, D.; Ge, Y.; Jia, C.; Liu, H.; Jin, Q. Highly sensitive and selective lateral flow immunoassay based on magnetic nanoparticles for quantitative detection of carcinoembryonic antigen. *Talanta* **2016**, *161*, 205–210. [CrossRef] [PubMed]
17. Lu, W.; Wang, K.; Xiao, K.; Qin, W.; Hou, Y.; Xu, H.; Yan, X.; Chen, Y.; Cui, D.; He, J. Dual immunomagnetic nanobeads-based lateral flow test strip for simultaneous quantitative detection of carcinoembryonic antigen and neuron specific enolase. *Sci. Rep.* **2017**, *7*, 42414. [CrossRef]
18. Zhou, Y.; Ding, L.; Wu, Y.; Huang, X.; Lai, W.; Xiong, Y. Emerging Strategies to Develop Sensitive AuNP-based ICTS Nanosensors. *TrAC Trends Anal. Chem.* **2019**, *112*, 147–160. [CrossRef]
19. Khashayar, P.; Amoabediny, G.; Larijani, B.; Hosseini, M.; Vanfleteren, J. Fabrication and verification of conjugated aunp-antibody nanoprobe for sensitivity improvement in electrochemical biosensors. *Sci. Rep.* **2017**, *7*, 1–8. [CrossRef]
20. Xu, S.; Zhang, G.; Fang, B.; Xiong, Q.; Duan, H.; Lai, W. Lateral Flow Immunoassay Based on Polydopamine-Coated Gold Nanoparticles for the Sensitive Detection of Zearalenone in Maize. *ACS Appl. Mater Interfaces* **2019**, *11*, 31283–31290. [CrossRef]
21. Mahmoudi, T.; Shirdel, B.; Mansoori, B.; Baradaran, B. Dual sensitivity enhancement in gold nanoparticle-based lateral flow immunoassay for visual detection of carcinoembryonic antigen. *Anal. Sci. Adv.* **2020**, *1*, 161–172. [CrossRef]
22. Byzova, N.A.; Safenkova, I.V.; Slutskaya, E.S.; Zherdev, A.V.; Dzantiev, B.B. Less is more: A comparison of antibody–gold nanoparticle conjugates of different ratios. *Bioconjug. Chem.* **2017**, *28*, 2737–2746. [CrossRef] [PubMed]
23. Schiettecatte, J.; Anckaert, E.; Smitz, J. Interferences in immunoassays. *Adv. Immunoassay Technol.* **2012**, *3*, 45–62.
24. Mahmoudi, T.; Tazehkand, A.P.; Pourhassan-Moghaddam, M.; Alizadeh-Ghodsi, M.; Ding, L.; Baradaran, B.; Bazaz, S.R.; Jin, D.; Warkiani, M.E. PCR-free paper-based nanobiosensing platform for visual detection of telomerase activity via gold enhancement. *Microchem. J.* **2020**, *154*, 104594. [CrossRef]
25. Jung, Y.; Heo, Y.; Lee, J.J.; Deering, A.; Bae, E. Smartphone-based lateral flow imaging system for detection of food-borne bacteria E. coli O157: H7. *J. Microbiol. Methods* **2020**, *168*, 105800. [CrossRef] [PubMed]
26. De Puig, H.; Bosch, I.; Gehrke, L.; Hamad-Schifferli, K. Challenges of the nano–bio interface in lateral flow and dipstick immunoassays. *Trends Biotechnol.* **2017**, *35*, 1169–1180. [CrossRef] [PubMed]
27. Qie, Z.; Yan, W.; Gao, Z.; Meng, W.; Xiao, R.; Wang, S. An anti-BSA antibody-based immunochromatographic assay for chloramphenicol and aflatoxin M 1 by using carboxy-modified CdSe/ZnS core–shell nanoparticles as label. *Microchim. Acta* **2020**, *187*, 10. [CrossRef]
28. Parolo, C.; de la Escosura-Muñiz, A.; Merkoçi, A. Enhanced lateral flow immunoassay using gold nanoparticles loaded with enzymes. *Biosens. Bioelectron.* **2013**, *40*, 412–416. [CrossRef]
29. Litvak, A.; Cercek, A.; Segal, N.; Reidy-Lagunes, D.; Stadler, Z.K.; Yaeger, R.D.; Kemeny, N.E.; Weiser, M.R.; Pessin, M.S.; Saltz, L. False-positive elevations of carcinoembryonic antigen in patients with a history of resected colorectal cancer. *J. Natl. Compr. Cancer Netw.* **2014**, *12*, 907–913. [CrossRef]
30. Zou, K.H.; O'Malley, A.J.; Mauri, L. Receiver-operating characteristic analysis for evaluating diagnostic tests and predictive models. *Circulation* **2007**, *115*, 654–657. [CrossRef]
31. Saisin, L.; Amarit, R.; Somboonkaew, A.; Gajanandana, O.; Himananto, O.; Sutapun, B. Significant sensitivity improvement for camera-based lateral flow immunoassay readers. *Sensors* **2018**, *18*, 4026. [CrossRef] [PubMed]
32. Xiao, K.; Wang, K.; Qin, W.; Hou, Y.; Lu, W.; Xu, H.; Wo, Y.; Cui, D. Use of quantum dot beads-labeled monoclonal antibody to improve the sensitivity of a quantitative and simultaneous immunochromatographic assay for neuron specific enolase and carcinoembryonic antigen. *Talanta* **2017**, *164*, 463–469. [CrossRef]
33. Lee, S.; Kim, G.; Moon, J. Development of a Smartphone-based reading system for lateral flow immunoassay. *J. Nanosci. Nanotechnol.* **2014**, *14*, 8453–8457. [CrossRef]
34. Ruppert, C.; Phogat, N.; Laufer, S.; Kohl, M.; Deigner, H.-P. A smartphone readout system for gold nanoparticle-based lateral flow assays: Application to monitoring of digoxigenin. *Microchim. Acta* **2019**, *186*, 119. [CrossRef] [PubMed]

MDPI
St. Alban-Anlage 66
4052 Basel
Switzerland
Tel. +41 61 683 77 34
Fax +41 61 302 89 18
www.mdpi.com

Biosensors Editorial Office
E-mail: biosensors@mdpi.com
www.mdpi.com/journal/biosensors

www.ingramcontent.com/pod-product-compliance
Lightning Source LLC
Chambersburg PA
CBHW040749250926
51850CB00007BB/1447
9783036524009